Ecological Considerations in Wetlands Treatment of Municipal Wastewaters

ECOLOGICAL CONSIDERATIONS IN WETLANDS TREATMENT OF MUNICIPAL WASTEWATERS

Edited by

Paul J. Godfrey, *Principal Investigator*

Edward R. Kaynor, *Project Coordinator*

Sheila Pelczarski, *Staff Assistant*

Water Resources Research Center
University of Massachusetts
Amherst, Massachusetts

and

Jay Benforado, *Project Officer*

Eastern Energy and Land Use Team
National Water Resources Analysis Group
Kearneysville, West Virginia

VAN NOSTRAND REINHOLD COMPANY
——————— *New York* ———————

Contract No. RFP FWS-9-82-001
U.S. Environmental Protection Agency, cosponsoring agency
Performed for the Eastern Energy and Land Use Team, Office of Biological
    Services, Fish and Wildlife Service, U.S. Department of the Interior,
    Washington, D.C. 20240

Disclaimer

The opinions and recommendations expressed in these proceedings are
those of the authors and do not necessarily reflect the views of the U.S. Fish
and Wildlife Service or the U.S. Environmental Protection Agency, nor
does the mention of trade names constitute endorsement or recommen-
dation for use by the federal government.

Manufactured in the United States of America.

Published by Van Nostrand Reinhold Company Inc.
135 West 50th Street, New York, New York 10020

Van Nostrand Reinhold Company Limited
Molly Millars Lane, Wokingham, Berkshire RG11 2PY, England

Van Nostrand Reinhold
480 Latrobe Street, Melbourne, Victoria 3000, Australia

Macmillan of Canada
Division of Gage Publishing Limited, 164 Commander Boulevard,
Agincourt, Ontario MIS 3C7, Canada

15   14   13   12   11   10   9   8   7   6   5   4   3   2   1

Library of Congress Cataloging in Publication Data
Main entry under title:
Ecological considerations in wetlands treatment of municipal wastewaters.
    Includes index.
    1. Sewage—Environmental aspects—Congresses.   2. Sewage disposal
in rivers, lakes, etc.—Congresses.   3. Wetland ecology—Congresses.   I.
Godfrey, Paul J.
QH545.S49E3   1985   628.3'9   84-25833
ISBN 0-442-23009-5

CONTENTS

PREFACE

This report contains a series of professional papers prepared to address the general question, What are the possible or probable environmental consequences of applying municipal wastewater to wetlands?

Because most of these papers were presented at a workshop held at the University of Massachusetts in Amherst on June 23–25, 1982, they are presented here in the form of a proceedings, with discussion sections following most of the papers. After an opening address by Edward T. LaRoe, deputy chief, Office of Biological Services in the Fish and Wildlife Service, 31 papers are presented within the format of seven topical "sessions" and an eighth session that attempts to pull together and synthesize the diverse findings and conclusions of the conference.

Session I deals with state-of-the-art engineering applications of waste-water to wetlands.

Session II deals with the multiplicity of environmental values and perspectives that must be considered.

Session III shifts the emphasis to specifics of ecosystem dynamics, particularly the interaction between hydrology and nutrients.

Session IV addresses the general topic of possible wetland community changes resulting from wastewater applications.

Session V focuses on environmental health problems for humans and wildlife that could result from accumulations of pathogens or toxic materials.

Session VI considers the question of long-term effects by examining several case histories of wetlands that have received wastewater or simulated wastewater and have been studied for many years.

Session VII returns in part to engineering considerations by addressing the topic of wetland management potential.
Session VIII attempts to synthesize the varying points of view and
scientific documentation contained in the papers of the preceding sessions.

Because these papers approach the topic of wastewater treatment
in wetlands from such diverse perspectives, it is not possible to
represent a single view. However, if majority opinion were to serve as
our criterion for generalization, the following four broad conclusions
might summarize these 31 papers:

1. Both natural and (especially) artificially created wetlands are efficient and usually cost-effective processors of both the nutrients
   and toxins in municipal wastewater.
2. For relatively small communities with natural wetlands near at
   hand or with good potential for creation of artificial wetlands, the
   economic stakes are high, promising savings nationally in the
   billions of dollars compared with wastewater treatment by conventional means.
3. While there is evidence that the negative impacts on wetland
   values from applications of wastewater are not severe, at least in
   the short term, breadth of knowledge about ecosystem effects lags
   behind sophistication of engineering design. This situation creates
   a motivational impetus that could lead to widespread use of the
   technique in advance of reasonably complete knowledge about
   environmental consequences.
4. Intensive further investigation of critical unknowns should be
   undertaken at once by the scientific community. Careful monitoring
   of existing wastewater applications should be the cornerstone for
   such further investigations.

ACKNOWLEDGMENTS

The editors gratefully acknowledge the assistance of the conference moderators: Joseph Larson, James Male, Paul Shuldiner, Cleve
Willis, Bernard Berger, William Niering, and Charlene D'Avanzo for
their assistance in organizing and conducting the conference and in
the preliminary editing of the manuscripts. We wish to thank Robert

Bastian of the U.S. Environmental Protection Agency for his contributions to the organization of the conference and for stimulating a lively exchange between the participants.

Thanks also to Ron Reid of the UMASS Conference Center, Peter Saunders and Hisashi Ogawa for help with the logistical details of the conference, and John R. Cole for preparation of the index. Our very special appreciation goes to the other Water Resources Research Center staff members for their diligent efforts throughout the project.

We wish to thank all of the authors of the papers presented in these proceedings for being both very helpful and tolerant of our editorial efforts and deadline constraints. And, finally, we thank all the participants for making the conference and its proceedings a true exchange of ideas and results crossing the barriers of academic disciplines and agency missions.

PAUL J. GODFREY
EDWARD R. KAYNOR
SHEILA PELCZARSKI
JAY BENFORADO

PARTICIPANTS

Chris Ansell
Office of Technology Assessment
United States Congress
Washington, D.C. 20510

Robert K. Bastian
EPA Office of Marine Discharge
WH546 M Street, SW
Washington, D.C. 20460

Suzanne E. Bayley
Freshwater Institute
University of Winnipeg
Winnipeg, Manitoba R3T 2N6
Canada

Barbara Bedford
Ecosystems Research Center
468 Hollister Hall
Cornell University
Ithaca, NY 14853

Jay Benforado
EPA Office of Research and
   Development
401 M Street, SW
Washington, D.C. 20460

Bernard B. Berger
Department of Civil Engineering
Marston Hall
University of Massachusetts
Amherst, MA 01003

Greg Bourne
Claude Terry Associates
1955 Cliff Valley Way, NE
Atlanta, GA 30029

Kathleen M. Brennan
Lake Michigan Federation
53 West Jackson Boulevard
Chicago, IL 60604

Mark M. Brinson
Department of Biology
East Carolina University
Greenville, NC 27834

Charlene D'Avanzo
School of Natural Sciences
Cole Science Center
Hampshire College
Amherst, MA 01002

John W. Day, Jr.
Coastal Ecology Laboratory
Center for Wetland Resources
Louisiana State University
Baton Rouge, LA 70803

Katherine Carter Ewel
Institute of Food and
   Agricultural Sciences, Forestry
   Resources and Conservation
University of Florida
Gainesville, FL 32611

John T. Finn
Department of Forestry and
  Wildlife Management
Holdsworth Hall
University of Massachusetts
Amherst, MA 01003

Milton Friend
Director
National Wildlife Health
  Laboratory
U.S. Fish and Wildlife Service
6006 Schroeder Road
Madison, WI 53711

Catherine Garra
EPA Region 5
230 South Dearborn Street
Chicago, IL 60604

Anne E. Giblin
Marine Biological Laboratory
Woods Hole, MA 02543

Paul Joseph Godfrey
Director
Water Resources Research
  Center
Blaisdell House
University of Massachusetts
Amherst, MA 01003

James G. Gosselink
Department of Marine Sciences
Center for Wetlands Resources
Louisiana State University
Baton Rouge, LA 70803

D. Jay Grimes
Department of Microbiology
University of Maryland
College Park, MD 20742

Herb Grover
Ecosystems Research Center
Dale Corson Building
Cornell University
Ithaca, NY 14853

Glenn R. Guntenspergen
Department of Botany
  Box 414
University of Wisconsin
Milwaukee, WI 53201

Norman N. Hantzsche
Ramlit Associates, Inc.
2437 Durant Avenue
Berkeley, CA 94705

Mark A. Harwell
Ecosystems Research Center
Dale Corson Building
Cornell University
Ithaca, NY 14853

Harry Hemond
Department of Civil Engi-
  neering
Building 48—Room 429
Massachusetts Institute of
  Technology
Cambridge, MA 02139

Robert E. Hodson
Department of Microbiology
University of Georgia
Athens, GA 30602

William B. Jackson
Bowling Green State University
Bolling Green, OH 43402

Victor W. Kaczynski
CH 2M Hill, Inc.
200 Southwest Fourth Avenue
Portland, OR 97201

John A. Kadlec
Department of Wildlife Sciences
Utah State University
Logan, UT 84321

Robert H. Kadlec
Department of Chemical
  Engineering
University of Michigan
Ann Arbor, MI 48109

William Kappel
United States Geological Survey
521 West Seneca Street
Ithaca, NY 14850

Edward R. Kaynor
Water Resources Research
  Center
Blaisdell House
University of Massachusetts
Amherst, MA 01003

John R. Kelly
Ecosystems Research Center
Dale Corson Building
Cornell University
Ithaca, NY 14853

Timothy J. Kubiac
Socio-Ecology Building
 —Room 480
University of Wisconsin
Green Bay, WI 54302

Edward T. LaRoe
Office of Biological Services
U. S. Fish and Wildlife
  Service
Riddell Building
1730 K Street, NW
Washington, D.C. 20240

Joseph S. Larson, Chairman
Department of Forestry and
  Wildlife Management
Holdsworth Hall
University of Massachusetts
Amherst, MA 01003

Bobbie Lively-DeBold
EPA Region 5
230 South Dearborn St.
Chicago, IL 60604

James W. Male
Department of Civil Engineering
Marston Hall
University of Massachusetts
Amherst, MA 01003

Ronald Mikulak
EPA Region 4
345 Courtland Street
Atlanta, GA 30308

Gordon Miller
Ministry of Environment
Box 213 Resdale
Ontario, M9W 5L1
Canada

William A. Niering
Department of Botany
Connecticut College
New London, CT 06320

Howard T. Odum
Department of Environmental
  Engineering Science
University of Florida
Gainesville, FL 32611

J. Ross Pilling
EPA—WAPORA, Inc.
Suite 490, 35 East Wacker Drive
Chicago, IL 60601

Donald M. Reed
Southeast Wisconsin Regional
  Planning Commission
916 North East Avenue
Box 769
Waukesha, WI 53187

Sherwood C. Reed
United States Army Corps of
  Engineers
CRREL Box 282
Hanover, NH 03755

Curtis J. Richardson
Department of Forestry and
  Environmental Studies
Duke University
Durham, NC 27706

Michael P. Shiaris
Department of Biology
Harbor Campus
University of Massachusetts
Boston, MA 02125

Paul W. Shuldiner
Department of Civil Engineering
Marston Hall
University of Massachusetts
Amherst, MA 01003

Forest Stearns
Department of Botany
  —Box 413
University of Wisconsin
Milwaukee, WI 53201

Jeffrey C. Sutherland
Williams and Works, Inc.
611 Cascade West Parkway, SE
Grand Rapids, MI 49506

Charles Terrell
U. S. Soil Conservation
  Service
Box 2890
Washington, D.C. 20013

Ivan Valiela
BUMPS
Marine Biological Laboratory
Woods Hole, MA 02543

Dennis F. Whigham
Smithsonian Environmental
  Research Center
P. O. Box 28
Edgewater, MD 21037

Ivy Wile
Policy and Planning Branch
Research and Coordination
  Office
Ministry of Environment
135 St. Clair Avenue
Toronto, Ontario, M4V 1P5
Canada

Cleve E. Willis
Department of Agricultural
  and Resource Economics
Draper Hal
University of Massachusetts
Amherst, MA 01003

Joy Zedler
Department of Biology
San Diego State University
San Diego, CA 92182

CONTRIBUTORS

Sarah Allen
Marine Program—MBL
Boston University
Woods Hole, MA   02543

Stephen D. Bach
WAPORA
Suite F
5980 Unity Drive
Norcross, GA   30071

Robert K. Bastian
EPA Office of Marine Discharge
WH 546 M Street, SW
Washington, D.C.   20460

Suzanne E. Bayley
Freshwater Institute
University of Winnipeg
Winnipeg, Manitoba   R3T 2N6
Canada

Ronald Benner
Department of Microbiology
University of Georgia
Athens, GA   30602

Steven Black
1 St. Claire Avenue, West
Toronto, Ontario   M4V 1P5
Canada

Kathleen M. Brennan
Lake Michigan Federation
53 West Jackson Boulevard
Chicago, IL   60604

Mark M. Brinson
Department of Biology
East Carolina University
Greenville, NC   27834

Charlotte Cogswell
Department of Biological
    Sciences
University of Connecticut
Storrs, CT   06268

John W. Day, Jr.
Coastal Ecology Laboratory
Center for Wetland Resources
Louisiana State University
Baton Rouge, LA   70803

Katherine Carter Ewel
Institute of Food and
    Agricultural Sciences
Forestry Resources and
    Conservation
University of Florida
Gainesville, FL   32611

John T. Finn
Department of Forestry and
  Wildlife Management
Holdsworth Hall
University of Massachusetts
Amherst, MA   01003

Milton Friend
Director
National Wildlife Health
  Laboratory
U.S. Fish and Wildlife Service
6006 Schroeder Road
Madison, WI   53711

Anne E. Giblin
Marine Biological Laboratory
Woods Hole, MA   02543

Dale Goehringer
Woods Hole Oceanographic
  Institute
Woods Hole, MA   02543

James G. Gosselink
Department of Marine Sciences
Center for Wetlands Resources
Louisiana State University
Baton Rouge, LA   70803

Leila Gosselink
Department of Marine Sciences
Center for Wetlands Resources
Louisiana State University
Baton Rouge, LA   70803

D. Jay Grimes
Department of Microbiology
University of Maryland
College Park, MD   20742

Glenn R. Guntenspergen
Department of Botany
Box 414
University of Wisconsin
Milwaukee, WI   53201

Norman N. Hantzsche
Ramlit Associates, Inc.
2437 Durant Avenue
Berkeley, CA   94705

Jean Hartman
Langley Research Center
NASA
Hampton, VA   23665

Mark A. Harwell
Ecosystems Research Center
Dale Corson Building
Cornell University
Ithaca, NY   14853

Harry Hemmond
Department of Civil
  Engineering
Building 48—Room 429
Massachusetts Institute of
  Technology
Cambridge, MA   02139

Robert E. Hodson
Department of Microbiology
University of Georgia
Athens, GA   30602

William B. Jackson
Department of Biology
Bowling Green State University
Bowling Green, OH   43402

Victor W. Kaczynski
CH 2M Hill, Inc.
200 Southwest Fourth Avenue
Portland, OR   97201

John A. Kadlec
Department of Wildlife Sciences
Utah State University
Logan, UT   84321

Robert H. Kadlec
Department of Chemical
   Engineering
University of Michigan
Ann Arbor, MI   48109

John R. Kelly
Ecosystems Research Center
Dale Corson Building
Cornell University
Ithaca, NY   14853

G. Paul Kemp
Coastal Ecology Laboratory
Center for Wetlands Resources
Louisiana State University
Baton Rouge, LA   70803

Timothy J. Kubiac
Socio-Ecology Building—Room
   480
University of Wisconsin
Green Bay, WI   54302

Edward T. LaRoe
Office of Biological Services
U.S. Fish and Wildlife Service
Riddell Building
1730 K Street, NW
Washington, D.C.   20240

Alexander E. Maccubbin
North Apartment
527 Parkhurst Boulevard
Buffalo, NY   14223

Gordon Miller
Ministry of Environment
Box 213
Resdale, Ontario   M9W 5L1
Canada

Robert E. Murray
Department of Microbiology
University of Georgia
Athens, GA   30602

Dale S. Nichols
Research Soil Scientist
U.S. Department of Agriculture
   —Forest Service
North Central Forest
   Experiment Station
1831 Highway 169 East
Grand Rapids, MN   55744

William K. Nuttle
Department of Civil
   Engineering
Building 48—Room 429
Massachusetts Institute of
   Technology
Cambridge, MA   02139

Howard T. Odum
Department of Environmental
   Engineering Science
University of Florida
Gainesville, FL   32611

Donald M. Reed
Southeast Wisconsin Regional
   Planning Commission
916 North East Avenue
Box 769
Waukesha, WI   53187

Sherwood C. Reed
U.S. Army Corps of Engineers
CRREL
Box 282
Hanover, NH   03755

Curtis J. Richardson
Department of Forestry and
   Environmental Studies
Duke University
Durham, NC   27706

Frank Rusincovitch
U.S. Environmental Protection
   Agency (A-104)
401 M Street
Washington, D.C.   20460

Gregory L. Seegert
Ecological Analysts, Inc.
Suite 306
1535 Lake Cook Road
Northbrook, IL   60062

Michael P. Shiaris
Department of Biology
Harbor Campus
University of Massachusetts
Boston, MA   02125

Forest Stearns
Department of Botany
Box 413
University of Wisconsin
Milwaukee, WI   53201

Rich Stowell
Dewante and Stowell
1910 S Street
Sacramento, CA   95814

Jeffrey C. Sutherland
Williams and Works, Inc.
611 Cascade West Parkway, SE
Grand Rapids, MI   49506

George Tchobanoglous
Department of Civil
   Engineering
University of California
Davis, CA   95616

John M. Teal
Woods Hole Oceanographic
   Institute
Woods Hole, MA   02543

Kenneth R. Townzen
Vector Biology Control Branch
Department of Health Services
744 P Street
Sacramento, CA   95814

Ivan Valiela
BUMPS
Marine Biological Laboratory
Woods Hole, MA   02543

Richard Van Etten
Woods Hole Oceanographic
   Institute
Woods Hole, MA   02543

Scott Weber
Civil Engineering Department
State University of New York
Buffalo, NY   14214

Dennis F. Whigham
Smithsonian Environmental
   Research Center
P.O. Box 28
Edgewater, MD   21037

Ivy Wile
Policy and Planning Branch
Research and Coordination
  Office
Ministry of Environment
135 St. Claire Avenue
Toronto, Ontario   M4V 1P5
Canada

Barbara A. Wilson
Vector Biology Control Branch
Department of Health Services
744 P Street
Sacramento, CA   95814

Joy Zedler
Department of Biology
San Diego State University
San Diego, CA   92182

Ecological Considerations in Wetlands Treatment of Municipal Wastewaters

Opening Remarks

Edward T. Laroe

Benforado: This morning I'd like to introduce Dr. Ted LaRoe, who is
the deputy chief of the Office of Biological Services in the U.` S.
Fish and Wildlife Service. He formerly has worked with NOAA in its
Office of Coastal Zone Management and also in the Coastal Zone
Management Agencies of Florida and Oregon.

I asked him to give the welcome today because he has a significant background in wetlands. He was, for example, a coauthor of the
wetlands classification system used now by the Fish and Wildlife
Service, and he has done other work in coastal wetlands in Florida
and on the West Coast.

LaRoe: It's with great pleasure that I welcome you here to this
workshop on ecological considerations in wetlands treatment of
municipal wastewater on behalf of both the Fish and Wildlife Service
and the Environmental Protection Agency. I'd like to thank you all
for arranging your schedules in order to participate in this
conference.

My pleasure in addressing you and welcoming you is both professional and personal. This is a subject in which I have long had a
personal interest, and I see some old familiar faces out in the
audience. I also see some new ones, and participants I do not know
except by reputation.

For those of you who do not know me, for the past decade I have
watched with great interest much of the work that has gone on in
wetlands in general, and in the attributes of wetlands for wastewater
disposal as one item of particular interest. I must admit that at
many times I have been frustrated as my career has moved me more and
more toward administration and management, and further and further
away from science and fieldwork and many of you. Given my gaps in
contact with most of you, I'm looking forward to hearing of the
progress you have made and the problems that remain in the use of
wetlands for wastewater disposal. I'm looking forward to hearing the
concerns you have and the recommendations you have to offer as to how
we should proceed with the concept--how we, as resource managers,
should proceed.

The Fish and Wildlife Service is involved in this workshop
because of its concern with the conservation of wetlands and wetland

1

resources. We have seen an unprecedented commitment to wetlands during the past decade--a commitment by scientists, resource agencies, regulatory agencies, and resource managers. We have made tremendous strides in the last 10 or 15 years in being able to understand wetland functions and ecological processes and the benefits and values that wetlands contribute to humankind. One of the values we've recognized--indeed, one of the values we frequently use in our arguments to provide protection for wetlands--is that wetland systems help maintain water quality. We've talked about how they filter out sediments and solids, remove nutrients, and how they absorb toxic substances; and we have argued--often very successfully before regulatory boards and administrative boards--that the removal of wetlands for development not only adds a new source of pollution-- runoff, at a minimum--but destroys the natural cleansing abilities of wetlands and their capacities for waters. This argument has been tightened by people like Gene Odum who have tried to quantify that capacity in dollar terms and we've used those dollar terms in a lot of the arguments to promote the dollar value of wetland areas.

Having argued that wetlands are natural water cleansers, the engineers soon began to suggest that we use that capacity more effec- tively, that we manipulate and "manage" wetland areas to treat waste- water. In some ways it's a very attractive notion, and in some ways it's a very frightening notion. It's attractive for several reasons: first, the cost. We have committed an enormous sum in the past decade to the construction of wastewater treatment facilities. One year's worth of EPA construction grant funds for the construction of new wastewater treatment facilities would fund the entire Fish and Wildlife Service--all aspects of it--for several years. And yet, we are already at the point--notwithstanding that the new construction program is incomplete--where EPA has municipalities coming to it to ask for new construction grants to replace the old, outworn, and outmoded plants that they built just 10 years ago. We haven't even completed the first generation of treatment facilities and we are already having to look at how we are going to bear the cost of the second generation.

A second reason for the attractiveness of using wetlands is that regardless of cost, it may indeed be one of the best ways or even the only way to deal with some forms of pollution, particularly urban runoff and other non-point sources. We've often recommended leaving wetland buffers around development projects to provide some cleansing ability for adjacent waters.

And, a third attractive part of the concept, one that appeals particularly to me, is that if it works--if we can use wetlands successfully while still retaining the range of benefits that the wetlands provide--we will gain an additional incentive--that is, a dollar incentive--to protect those wetlands. And that, again, is the goal of the service.

The key to all this is "if it works": if we can load wetlands with wastewater without destroying their resource value and benefits, without destroying their ecological processes and functions. Given the great importance of wetlands, we will not deliberately set out to alter the system so as to lose the variety of benefits that wetlands provide. I want to emphasize that this is not to say that we shouldn't proceed with the concept of wastewater disposal in wet- lands, just that we should be careful and cautious and know what we're doing.

There will be many, including some in this room, who will be skeptical of the idea, of our ability to manipulate wetlands for this purpose. There will be many who will be skeptical of our understanding of wetlands--of wetland ecosystems, and of the long-term impacts of the use and management of wetlands. It is important that you in this workshop review the issue carefully and completely; that you air your concerns and whatever solutions you can identify; that you identify issues and problems that must be addressed before we proceed with the new technology; that you perhaps design a series of tests--field tests--that can be used to study unresolved problems, develop appropriate procedures, and ultimately make a decision whether or not to proceed with the entire program for potential use of wetlands.

As you go through your discussions in the next few days, I would like to ask you to focus your attention on the system, not just water quality, not just wetland plants, but on the wetland ecosystem, and on the full array of functions and benefits that system provides, for that is what we must balance.

The hosts of the conference here have put together a great diversity of outstanding talent. I think that is the strength of this workshop. We have people ranging from engineers to ecologists together at the beginning to discuss a potential new technology. We can identify and avoid problems from the outset rather than confront each other in an adversarial role _after_ the problems crop up and it's really too late. I urge you to use the opportunity you have here constructively.

I'd like to make one last observation in closing. I know that many members of the scientific community feel frustrated by the apparent inability of the "bureaucracy" to comprehend--or at least to give appropriate attention to--science and scientific concepts. Your contributions--the contributions of the group in this workshop--have been the cause, more than anything else, of our commitment to wetlands today. I've been very pleased by the way in which scientists have been able to convey an understanding and appreciation of wetlands to the decisionmakers and resource managers. It may seem to you that it has taken a long time, and some of us may feel it has taken too long, but the appreciation and understanding is there. It may be temporarily set back on occasion, but it's there. We look forward to your continuing contribution in this field, and particularly as applied to this subject.

I hope you all have a rewarding and productive workshop. I look forward to your results. And I thank you very much.

SESSION I

Engineering

1

Wetland Systems for Wastewater Treatment: Engineering Applications

Norman N. Hantzsche

Over the past decade major emphasis has been given to improving methods of handling wastewater discharges. This is particularly so in the case of small- to medium-sized communities that, in comparison with large communities, typically pay a disproportionate amount to achieve the same level of conventional wastewater treatment.

A wide range of alternative and innovative wastewater technologies have been investigated and promoted. These have included, for example, improved on-site sewage-disposal methods to defer or eliminate the need for central sewerage systems; wastewater reclamation through land application and reuse; and wetlands and natural biological treatment systems.

The interest in wetlands for wastewater treatment can be attributed to four basic factors:

1. Public demands for more stringent wastewater effluent standards, including removal of nutrients and trace contaminants as well as organic and suspended matter;
2. Rapidly escalating costs of construction and operation associated with conventional treatment facilities;
3. Recognition of the natural treatment functions of wetlands, particularly as nutrient sinks and buffering zones;
4. Emerging or renewed appreciation of aesthetic, wildlife, and other incidental environmental benefits associated with the preservation and enhancement of wetlands.

WETLAND CHARACTERISTICS

While the interest in wetlands for wastewater treatment is a fairly recent innovation, the term wetlands is also a relatively new expression, encompassing what for years have simply been referred to as marshes, swamps, bogs, and so on. Descriptions of the principal types of natural wetlands, according to hydrological factors, are provided in Table 1.1. System characteristics, representative vegetation types, and some of the important environmental sensitivities are also displayed. Wetlands occur in a wide range of physical settings at the interface of terrestrial and aquatic

7

Table 1.1
Classification of Natural Wetland Systems

Wetland System	System Characteristics	Vegetation Types	Sensitivities
Freshwater marshes--riverine (associated with water channels)	Water circulation distributes dissolved and suspended materials through system. Good aeration and light penetration.	Emergent plants--cattails, reeds, sedges, bulrush, watercress; floating algae.	Subject to sedimentation, scouring, and seasonally changing water levels. Pollutant loadings vary with watershed.
Freshwater marshes--lacustrine (associated with ponds and lakes)	Temperature/oxygen stratification and light attenuation can cause major differences in top, middle, and bottom layers. Circulation is often poor.	Floating plants--duckweed, water fern, water primrose, pondweeds and others. Emergent plants--see riverine system, submerged plants.	Closed or semiclosed systems. Pollutants enter food chain or accumulate in sediments.
Freshwater marshes--palustrine (not confined by channels or adjacent to lakes)	Surface layer has thick and/or porous deposits with high organic content. Marsh is fed by subsurface seepage/high ground water.	Peat bogs, cypress, mangrove, and papyrus swamps--vegetation types often specific to geographical area.	Isolation from open water bodies (streams, rivers, and lakes) limits water exchange, forming potential pollutant sink.
Brackish marshes--(salinity > 0.4 ppt)	Marsh fed by seasonal surface flows and/or seepage; may also be subject to tidal influences; can experience salinity fluctuations.	Emergent plants--sedges, bulrush, pickleweed, saltgrass, saltbush.	Evaporation can lead to salinities of 60-80 ppt and concentration of pollutants.
Salt marshes (subject to tidal influence)	(1) Wetlands near streams (2) Lower wetlands--reversing flow (3) Lower wetlands--drained only at low tides (4) Upper wetlands--inundated only at high tides	Emergent plants--pickleweed, cordgrass, sedges, saltgrass.	Salinity and sediment interactions can trap pollutants; however, low pH and oxidizing muds can re-release pollutants to system on a continuing basis.

Source: From E. Chan, T. A. Bursztynsky, N. N. Hantzsche, and Y. J. Litwin, 1981, The use of wetlands for water pollution control, U. S. EPA Grant No. R-806357, derived from Tchobanoglaus and Culp 1980; Good et al 1978; Tourbier and Pierson 1976.

ecosystems. Because of this position, some wetlands have been subjected to inadvertent municipal, industrial, agricultural, and storm-water discharges for many years. It is only in the past 10 to 15 years that attention has focused on planned use for wastewater treatment.

There are four basic functions of wetlands that make them potentially attractive for wastewater treatment (Chan et al. 1981):

1. Dispersion of surface waters over a large area through intricate channelization of flow;
2. Physical entrapment of pollutants through sorption in the surface soils and organic litter;
3. Uptake and metabolic utilization by plants;
4. Utilization and transformation of elements by microorganisms.

WETLAND TREATMENT OBJECTIVES

The types of treatment purposes that have been experimented with or proposed for wetland systems cover a broad range of water pollution concerns.

MUNICIPAL WASTEWATER

Greatest attention to date has been given to the use of wetlands, either natural or artificial, for tertiary treatment of municipal wastewater, mainly to meet nutrient-removal requirements. Achievement of wastewater treatment standards through the use of natural biological systems is generally found to offer an economically attractive alternative to conventional advanced wastewater treatment practices (Williams and Sutherland 1979). In a few instances, raw sewage and primary effluent have been applied to artificial wetlands for treatment purposes (Small 1976).

INDUSTRIAL WASTEWATERS

Wetlands and other natural biological treatment processes have been employed for renovation and disposal of certain types of industrial wastewaters, particularly those with high organic or nutrient loads. Examples include wastes from fish-rearing operations and canneries (Jackson, Bastian, and Marks 1972). The possibility of utilizing wetlands for treatment of waste fluids from geothermal drilling operations has also been investigated and found to be feasible and economically viable in certain situations.

STORMWATER RUNOFF

One outgrowth of the nationwide 208 planning studies has been the interest in making better use of wetlands for the control of non-point source pollutants contained in surface runoff. Several demonstration projects have been undertaken around the United States to determine the effectiveness of marshes, meadows, and vegetated channels for treating urban runoff pollutants (Litwin, Advani, and

Chan 1981; Morris et al. 1980). The focus in these cases has been on a wide range of pollutants, including suspended solids (SS), nutrients, and heavy metals.

AGRICULTURAL RETURN FLOWS

Pollutants carried in agricultural runoff and irrigation return flows are also considered possible candidates for wetland treatment. Pesticide residues, SS, salts, and nutrients are the major constit- uents of water quality concern. Possible wetland treatment strate- gies that have been examined include the use of vegetated waterways, wetland buffers at the edge of agricultural plots, and marsh construction projects for flow-through treatment of major agri- cultural drain waters (Blumer 1978; Miller 1981).

NATURAL WETLAND TREATMENT SYSTEMS

Experimentation with wastewater discharges to natural wetlands has occurred in many different parts of the United States and Canada. The major types of wetlands that have been examined for wastewater treatment include: northern freshwater marshes (Wisconsin, Michigan, Massachusetts, Canada); peatlands (Michigan); southern marshes and hardwood swamps (Florida); and freshwater tidal marshes (New Jersey).
Table 1.2 presents a comparative summary of some of the major investigations of wastewater treatment by natural wetlands, noting in particular system characteristics, loading rates, pollutant removal efficiencies, and other observations concerning treatment processes.

NORTHERN MARSHES

Due largely to problems of nutrient enrichment of surface waters, a significant amount of attention has been given to the use of fresh- water marshes for advanced wastewater treatment in several northern states and Canada. Some of the most extensive work has been in Wisconsin (Fetter et al. 1978). The studies of wastewater impacts on a natural cattail marsh at Brillion, for example, showed the following: (1) significant improvement in water quality as a result of passage through the marsh; (2) substantial nitrate removal (greater than 50%), but only limited reduction in phosphorus (P); and (3) strong seasonal fluctuations in nutrient removal efficiencies. Other studies have also identified distinct seasonal trends in nutrient retention. Greatest removal is generally found to occur in the spring-summer growth period. Low biological activity and high runoff flows account for decreased nutrient retention during the fall-winter period.

PEATLANDS

The work with natural peatlands at Houghton Lake, Michigan, provides some of the most detailed and thorough information on the utilization of peat wetlands for tertiary treatment of wastewater. In this particular application, a high degree of nutrient removal is

achieved after flow through a very short distance in the peatland.
This result has been attributed to the following four factors (Kadlec
and Tilton 1978): (1) high nutrient-adsorption capacity of organic
litter and peat soils; (2) slow subsurface movement of water; (3)
high denitrification rates within the water-logged soils; and (4)
nutrient uptake by some plants. In general, a high degree of contact
between surface water, peat, and the litter layer has been found to
be particularly important for successful renovation of wastewater.

SOUTHERN MARSHES AND HARDWOOD SWAMPS

Several wastewater-wetland studies have also been carried out in
the southern states, most notably in Florida. The interest here has
been mainly in the use of freshwater marshes, hardwood swamps, and
cypress domes for increased nutrient removal. High levels of
nitrogen (N) and P removal, on the order of 90%, have been demonstrated (Boyt, Bayley, and Zoltek 1977). In some studies, these
removal rates have been correlated with increased productivity of
wetland vegetation, most significantly in hardwoods.

Cypress wetlands or domes have attracted considerable attention
since they allow for much greater contact with soils than do marshes.
Studies at the University of Florida have shown rapid nutrient
removal with no noticeable effect on local groundwaters (Mitsch,
Odum, and Ewel 1976). Wastewater application to cypress domes is
severely restricted, however, by the limited occurrence of this type
of wetland ecosystem relative to potential need for wastewater application. It has been estimated, for instance, that only 3% of
currently produced wastewater in Florida could be diverted for
treatment in cypress domes (Fritz and Helle 1979).

TIDAL MARSHES

To date, relatively little use has been made of tidal marshes
for wastewater treatment, although the impacts of several long-
standing discharges have been studied (McCormick, Grant, and Patrick
1970; Simpson, Whigham, and Walker 1978). There is some indication
that tidally influenced systems may be able to process more effluent
than can nontidal wetlands. Treatment effectiveness is strongly
influenced by periodic inundation and flushing and seasonal variations (Simpson, Whigham, and Walker 1978). Similar conclusions have
been reached in regard to the potential effectiveness of salt marshes
for wastewater treatment (Valiela, Teal, and Sass 1973).

In general, what is apparent from these experiences with natural
wetland treatment systems is the following:

1. There is extreme variability in wetland characteristics, making
   it difficult, if not impossible, to translate results from one
   geographical area to another.
2. The functioning of component parts of wetland systems is complex
   and often extremely difficult to characterize. This severely
   restricts the ability to predict responses to wastewater
   application.

Table 1.2
Characteristics and Observed Performance of Natural Wetland-Wastewater Treatment Systems

System/Location (reference)	Wastewater Discharge Status	Wetland Characteristics	Hydraulic Loading	Removal Efficiencies	Comments
Brillion, Wisconsin (Fetter et al. 1978)	Existing discharge Secondary effluent	Cattail marsh 156 ha	2.2×10^3 to 4.6×10^3 m^3/month	NO$_3$ - 51.3 Total P - 13.4 BOD - 80.1 SS - 29.1	Decrease in NH$_3$-N and NO$_3$-N partially attributable to plant uptake. P shows a strong seasonal fluctuation Nutrients mobilized during periods of low biological activity (fall-winter) BOD values consistently reduced to below background levels except in winter, when marsh iced over.
Hay River, Northwest Territories, Canada (Hartland-Rowe and Wright 1974)	Existing discharge Secondary effluent	Northern swampland 32 ha	Not reported	NH$_4$-N - 96.2 Total P - 97.5 BOD - 97.7	Biological diversity and water quality reduced in initial impact area. Phosphate uptake by dead organic matter in root masses of floating swamp vegetation and also by active plant uptake.
Cootes Paradise, Ontario Canada (Murdoch and Capo-bianco 1979)	Existing discharge Secondary effluent	Lacustrine marsh 5.2 Km^2	30-40% of stream flow	Total N - 41 Total P - 33	Total N and P reduction related primarily to settling of particulate matter from effluent and some amount of biological uptake. Metal uptake also observed in wetland plants. Main supply to plants judged to be through soils rather than water.

Location (Reference)	Type of study	Wetland type / size	Loading rate	Removal efficiency (%)	Comments
Clermont, Florida (Zoltek et al. 1979)	Pilot study, Secondary effluent	Freshwater marsh, 200 m² plots	1.3, 3.8, and 10.2 cm/wk	Total N – 97.3, Total P – 97.5	Dead standing crop and belowground biomass acted as N sinks. Storage in roots, litter, and soil complex accounted for P reduction.
City of Wildwood, Florida	Existing discharge, Secondary effluent	Hardwood swamp, 202 ha	570 m³/d	Total N – 89.5, Total P – 98.1 seasonal 13 annual	High levels of N-removal during August and September attributable to plant utilization. P release during dormant period responsible for lower annual removal efficiency.
Houghton Lake, Michigan (Kadlec and Tilton 1978; Tilton and Kadlec 1979)	Pilot study, Secondary effluent	Peatland, 6.5 ha	3.9 cm/wk	NO_3-N – 99, Total P – 95	Nutrient immobilization due to denitrification and sorption by peat. P reductions varied with water depth (6-30 cm) and seasonally. Greatest release during late winter and early spring.
Bellaire, Michigan (Kadlec and Tilton 1978, 1978a)	New discharge, Secondary effluent	Freshwater marsh, 20.45 ha	550-630 m³/d	NO_3-N+ NH_4-N – 91, Total P – 97	Monthly removal rates varied in response to hydrologic conditions, plant growth, and temperature.
Great Meadows National Wildlife Refuge (Yonika and Lowry 1979)	Existing discharge, Secondary effluent	Freshwater marsh, 19 ha	780-2100 m³/d	TKN – 35, NH_4 – 58, NO_3 – 20, Total P – 47	Highest N removal in spring and early summer growing season. Denitrification most significant factor. Low removal rates and release during fall-midwinter (TKN) and winter-early spring (NO_3-N).

3. There is no question that natural wetlands act to renovate wastewater, but the extent of their treatment capacity is largely unknown. Thus, reliable design criteria cannot as yet be defined.

The need to conduct carefully controlled pilot studies prior to initiating full-scale projects using natural wetlands for wastewater treatment is widely accepted.

ARTIFICIAL WETLAND TREATMENT SYSTEMS

The extreme differences in natural wetland characteristics and the difficulty in understanding or predicting pollutant-removal capabilities has brought about interest in artificial wetlands. Such systems can be constructed to provide the same basic hydraulic and vegetative treatment functions as natural wetlands, but usually with a much greater degree of control.

Artificial wetlands are the result of establishing wetland vegetation and the required hydrological conditions in locations where they previously did not exist. Examples might include the creation of ponds and marshes for wildlife enhancement. Lowlands converted for use as permanent storm-water retention basins are another possibility. Trenches and ponds constructed specifically for physical and biological treatment of wastewaters may also take on the form of artificial wetlands. In many such cases, the substrate and vegetation established are foreign to the immediate locale. Table 1.3 provides a listing of some of the principal types of artificial wetlands that have been used or investigated for wastewater treatment purposes.

Small-scale wetlands have been created expressly for the purpose of providing wastewater treatment (DeJong 1976; Seidel 1976; Small 1976), while others on a larger scale have been implemented with multiuse objectives in mind—for example, using treated sewage effluent as a freshwater source for the creation and restoration of marshes for environmental enhancement (Demgen 1979).

The use of artificial wetlands for wastewater treatment is founded largely on the work of Kathe Seidel and her co-workers at the Max Planck Institute in Germany (Seidel 1976). Since the early 1950s they have been studying the wastewater treatment capabilities of plants established on artificial substrates. More recent developments have included the use of peat filters, floating plants (e.g., duckweed and water hyacinth), and large-scale creation of wetland habitats. Artificial wetlands have been studied for their capabilities to provide primary and secondary wastewater treatment, as well as for advanced or tertiary treatment. They afford much greater operational flexibility than do natural systems. Some systems are set up to recycle a portion of the wasteflow and to direct the final effluent into the soil for recharge purposes (Williams and Sutherland 1979). Others act as flow-through systems, discharging final effluent to receiving waters (Demgen 1979).

Possible constraints to the use of artificial wetlands for wastewater treatment include the following (Chan et al. 1981):

1. Geographical limitations of plant species, as well as the potential that a newly introduced plant species will become a nuisance or an agricultural competitor;

Table 1.3
Artificial Wetlands Used for the Treatment
of Wastewater or Storm Water

Type	Description
Marsh	Areas with impervious to semipervious bottoms planted with various wetlands plants such as reeds or rushes
Marsh-pond	Marsh wetlands followed by pond
Pond	Ponds with semipervious bottoms with embankments to contain or channel the applied water. Often, emergent wetland plants will be planted in clumps or mounds to form small subecosystems
Seepage wetlands	Wastewater irrigated fields overgrown with volunteer emergent wetland vegetation as a result of intermittent ponding and seepage of wastewater
Trench	Trenches or ditches planted with reeds or rushes. In some cases the trenches have been filled with peat
Trench (lined)	Trenches lined with an impervious membrane usually filled with gravel or sand and planted with reeds

Source: From E. Chan, T. A. Bursztynsky, N. N. Hantzsche, and Y. J.
Litwin, 1981, The use of wetlands for water pollution control,
U. S. EPA Grant No. R-806357, derived from Tchobanoglaus and Culp
1980; Good et al. 1978; Tourbier and Pierson 1976.

2. Land-based treatment systems require 4 to 10 times more land area than a conventional wastewater treatment facility;
3. Plant biomass harvesting is constrained by high plant moisture content and wetland configuration;
4. Artificial wetlands may provide breeding grounds for disease and insects and may generate odors.

Table 1.4 summarizes the characteristics and pollutant removal efficiencies for several major artificial wetland-wastewater treatment systems. These examples show a wide geographical distribution and active interest in a range of wetland vegetative systems. Many in the field are currently promoting expanded utilization of artificial, rather than natural, wetland systems (Yonika and Lowry 1979; Fetter, Sloey and Spangler 1978), mainly because of the higher degree of control and reliability. The environmental enhancement they provide is an added incentive.

WETLANDS TREATMENT OF STORM WATER

By nature, most wetland systems receive surface runoff from adjacent lands and watercourses. To varying degrees, this process provides treatment of runoff waters. In the past few years serious attention has been given to capitalizing on wetland processes as a means of providing detention storage and treatment of storm-water flows. The emerging interest may be attributable to three factors:

1. Increased knowledge of and concern regarding the control of non-point source storm-water pollutants;
2. Heightened interest and experience with natural biological treatment systems for wastewater pollutants;
3. Alarmed concern for preservation of this nation's diminishing wetland resources.

Many wetlands have been receiving inadvertent discharges of storm waters for a number of years. To date, there have been only a few instances where storm water has been specifically routed into natural or artificially created wetlands for flood control or water quality management purposes. Where the practice has been employed, consistent reduction of biochemical oxygen demand (BOD), SS, and heavy metals generally have been shown. Storm-water treatment through wetlands encompasses three categories:

1. Systems planned primarily for flood control with treatment as an incidental benefit;
2. Systems planned and operated with treatment of storm-water pollutants as a primary objective;
3. Existing wetland systems providing detention and treatment of storm-water flows as an unplanned, natural function.

The factors responsible for wetlands treatment of surface runoff are largely the same as those noted previously in reference to wastewater treatment. In principle, wetlands offer hydraulic resistance to surface runoff flowing through them, resulting in decreased velocities and increased deposition of SS. The large surface area provided by surface soils and vegetation contributes to higher levels of absorption, adsorption, microbial transformation, and biological utilization than normally occur in more channelized watercourses (Tilton and Kadlec 1979).

The utilization or creation of wetlands for storm-water treatment is limited by several factors, including:

1. Proximity of natural wetlands to sources of runoff needing treatment;
2. Seasonal and sporadic nature of storm-water runoff where reliable water supply is needed for maintenance of wetland vegetation;
3. High flows and flushing action associated with runoff events;
4. Potential for creating nuisance vector and odor problems.

A variety of management practices may be employed to overcome these limitations and enhance the overall capabilities of wetlands for treatment of storm-water runoff. Techniques might include, for example, inflow/outflow regulation, water-level manipulation, and flow distribution.

Examples of instances where wetlands have been employed, examined, or proposed for treatment of storm water include: northern peatlands (Minnesota); cypress wetland (Florida); brackish marsh (California); high-altitude meadows (California); vegetated retention basins (Maryland); and southern freshwater marsh (Florida).

HYDROLOGICAL FACTORS

A clear knowledge of the hydrogeology is crucial to understanding the wetland environment and assessing its potential utility for assimilation of waterborne pollutants. The lack of adequate hydrological information has hampered the efforts of numerous researchers in their attempt to quantify and evaluate the pollutant-removal efficiency of wetlands.

In considering the application of wastewaters to wetlands, the relationships between hydrology and ecosystem characteristics need to be recognized. Factors such as source of water, velocity, flow rate, renewal rate, and frequency of inundation have a major bearing on the chemical and physical properties of the wetland substrate, which in turn influence the character and health of the ecosystem, as reflected by species composition and richness, primary productivity, organic deposition and flux, and nutrient cycling (Gosselink and Turner 1978). In general, water movement through wetlands tends to have a positive impact on the ecosystem.

Hydrology controls pollutant removal by wetlands through its influence on the processes of sedimentation, aeration, biological transformation, and soil adsorption. Critical hydrologic factors are velocity and flow rate, water depth and fluctuation, detention time, circulation and distribution patterns, turbulence and wave action, seasonal and climatic influences, and groundwater conditions.

Various criteria and practices can be identified for hydraulic/hydrological management of wetlands to yield improved wastewater and storm-water treatment.

Flow routing: Initial introduction and subsequent distribution of flow should attempt to maximize effective contact between water and wetland soils and vegetation.

Table 1.4
Characteristics and Observed Performance of Artificial Wetland-Wastewater Treatment Systems

System/Location (reference)	Wastewater Discharge Status	Wetland Characteristics	Hydraulic Loading	Removal Efficiencies (%)	Comments
Brookhaven National Laboratory, New York (Small 1976; Small and Wurm 1977)	Experimental system Raw sewage and primary effluent	Freshwater marsh-meadow-pond systems 0.2 and 0.4 ha	80 m^3/d 60-80% recycled	Total N - 79 Total P - 77 BOD - 88-92 SS - 91.5 Heavy metals - 23-94	Reductions are for the total system (meadow-marsh-pond). The removal abilities of the individual components have not been analyzed. Meadow-marsh-pond system somewhat more effective than marsh-pond system.
Mt. View Sanitary District, Martinez, California (Demgen 1979; Demgen and Nute 1979)	Full-scale system Secondary effluent	Freshwater marsh 6.1 ha	2800 m^3/d	Total Org. N - 12 PO$_4$-P - 13	Marsh purposely managed to enhance wildlife rather than to optimize pollutant uptake. Best phosphate removal during summer months. High BOD readings in summer months due to algae. Measurably lower values during cooler months but significant increase during summer algae blooms.
Vermontville, Michigan (Williams and Sutherland, Sutherland and Benis 1979; Williams 1980)	Full-scale system Secondary effluent	Seepage wetland 4.6 ha	630 m^3/d (Jun. - Oct.)	NO^3-N - 60 Total P - 97	Reduction occurs through denitrification in the shallow wetland soil. P removal occurs through soil absorption of wastewater prior to encountering groundwater; 95% occurs in upper 3 ft of wetland soil.

Location (Reference)	System	Marsh type	Hydraulic loading	Removal efficiency (%)	Comments
Arcata, California (Gearheart and Finney 1981; Gearheart et al. 1982)	Pilot system; Secondary effluent	Freshwater marsh	107–414 m^3/d ha-d	NH_4-N – 10 NO_3-N – 42 BOD – 31 SS – 83	Seasonal variations in BOD removal efficiency; best results in October to April. Removal efficiencies appear to be related more to vegetation density than to hydraulic loading rate. Marsh effluent consistently shows low pH and no toxicity. Fecal coliform density reduced by 10 to 1000 times in experimental cells.
Listowel, Ontario, Canada (Black et al. 1981; Wile et al. 1981)	(a) Pilot system; Conventional Lagoon Effluent	Freshwater marsh	180–680 m^3/ha-d	Total N – 16–75 Total P – 13–85 BOD – 20–67 SS – 27–51	Test systems operated year-round in cold climatic zone. Pretreatment of BOD and SS recommended to prevent excessive O_2 demand and sludge buildup.
	(b) Pilot system; Complete mix Aerated Effluent	Freshwater marsh	150–500 m^3/ha-d	Total N – 34–90 Total P – 71–98 BOD – 81–96 SS – 84–98	Long-term P removal efficiencies expected to decline. Geometric configuration (length:width ratio) has significant effect on treatment performance.

Water-level maintenance: Manipulation of water levels is a useful means of enhancing pollutant removal by vegetation and soil. Regulation of levels must take into account competing ecosystem needs and the additional nuisance problem of mosquitos.

Inflow/outflow regulation: Possible techniques might include inflow and containment of the "first flush" of runoff, or retention storage during spring runoff until marsh communities are functioning at higher uptake rates.

Seasonal application: Where possible, seasonal applications of wastewaters and storm waters might be used for specific treatment or flushing purposes, taking into consideration biological activity in the wetlands, availability of dilution flows, and seasonal uses and quality of downstream receiving waters.

Infiltration: Maximum soil contact should be emphasized, with attention given to routing and/or ponding wastewaters in areas of highest soil permeability.

CONCLUSIONS

The use of wetlands for treatment of various wastewaters has attracted considerable interest and research attention during the past 10 to 15 years. What can generally be concluded on the basis of the experience to date is that:

1. Wetland systems can provide measurable renovation of wastewaters and storm waters, but the necessary understanding and criteria to take the best advantage of these processes on a routine basis do not currently exist.
2. Natural wetlands are highly variable in characteristics, making it difficult, if not impossible, to apply study results to different geographical areas.
3. The use of artificial or constructed wetlands appears to have the greatest promise for general application because of better reliability and process control.
4. There is a substantial amount of interest in creating or restoring wetlands simply for environmental enhancement. There are also strong desires to couple environmental enhancement with programs for treatment of municipal wastewaters, stormwaters, agricultural return flows, and various types of industrial wastewaters.
5. Pilot or demonstration studies are still needed before engineers can confidently proceed with design and implementation of full-scale wetland-wastewater systems.

REFERENCES

Black, S. A., I. Wile, and G. Miller, 1981, Sewage effluent treatment in artificial marshland, paper presented at Water Pollution Control Federation Conference; Detroit, Mich.

Blumer, K., 1978, The use of wetlands for treating wastes--wisdom in diversity?, in Environmental quality through wetlands utilization, proceedings from a symposium sponsored by the Coordinating Council on the Restoration of the Kissimmee River Valley and Taylor Creek-Nubbin Slough Basin, February 29-March 2, 1978, Tallahassee, Fla.

Boyt, F. L., S. E. Bayley, and J. Zoltek, Jr., 1977, Removal of nutrients from treated municipal wastewater by wetland vegetation, Water Pollut. Control Fed. J. 49(5):789.

Chan, E., T. A. Bursztynsky, N. N. Hantzsche, and Y. J. Litwin, 1981, The use of wetlands for water pollution control, U. S. EPA Grant No. R-806357.

DeJong, J., 1976, The purification of wastewater with the aid of rush or reed ponds, in Biological control of water pollution, J. Tourbier and R. W. Pierson, Jr., eds., University of Pennsylvania Press, Philadelphia.

Demgen, F. C., 1979, Wetlands creation for habitat and treatment at Mt. View Sanitary District, California, paper presented at the aquaculture systems for wastewater treatment seminar, University of California, Davis, Calif.

Demgen, F. C., and J. W. Nute, 1979, Wetlands creation using secondary treated wastewater, Proceedings of water reuse symposium,, vol. 1, March 1979, Washington, D. C.

Fetter, C. W., Jr., W. E. Sloey and F. L. Spangler, 1978, Use of a natural marsh for waste water polishing, Water Pollut. Control Fed. J. 50(2):290.

Fritz, W. R. and S. C. Helle, 1979, Cypress wetlands for tertiary treatment, in Aquaculture systems for wastewater treatment-- seminar proceedings and engineering assessment, R. K. Bastian and S. C. Reed, eds., EPA 430/9-80-006, EPA Office of Water Programs Operations, Washington, D. C., pp. 75-81.

Gearheart, R. A. and B. A. Finney, 1981, Vascular plants in wetland wastewater treatment systems--a pilot project, U. S. EPA Water Reuse Symposium II, August 1981, Washington, D. C.

Gearheart, R. A., S. Wilbur, J. Williams, D. Hull, N. Hoelper, K. Wells, S. Sundberg, S. Salinger, D. Hendrix, C. Holm, L. Dillon, G. Moritz, P. Griechaber, N. Lerner, and B. Finney, 1982, City of Arcata marsh pilot project--second annual progress report, project no. C-06-2270.

Good, R. E., D. F. Whigham, and R. L. Simpson, eds., 1978, Freshwater wetlands: Ecological processes and management potential, Academic Press, New York, 378p.

Gosselink, J. G., and R. E. Turner, 1978, The role of hydrology in freshwater wetland ecosystems, in Freshwater wetlands: Ecological processes and management potential, R. E. Good, D. F. Whigham, and R. L. Simpson, eds., Academic Press, New York.

Hartland-Rowe, R. C. B., and P. B. Wright, 1974, Swamplands for sewage effluents: Final report, Environmental-social committee northern pipelines, report no. 74-4, Information Canada (cat. no. R72-13174), QS-1553-000-E-A1, Canada.

Jackson, W. B., R. K. Bastian, and J. R. Marks, 1972, Effluent disposal in an oak woods during two decades, Climatology 25(3):20-36.

Kadlec, R. H., and D. L. Tilton, 1978a, Monitoring report on the Bellaire wastewater treatment facility, University of Michigan Wetlands Ecosystem Group, Ann Arbor, Mich.

Kadlec, R. H., and D. L., Tilton, 1978b, Waste water treatment via wetland irrigation: Nutrient dynamics, in Environmental quality through wetlands utilization, proceedings from a symposium sponsored by the Coordinating Council on the Restoration of the Kissimmee River Valley and Taylor Creek-Nubbin Slough Basin, February 29-March 2, 1978, Tallahassee, Fla.

Litwin, Y. J., R. Advani, and E. Chan, 1981, Treatment of stormwater runoff by a marsh/flood basin, U. S. EPA Grant No. R-806357.

McCormick, R., R. Grant and R. Patrick, 1970, Two studies of Tinucum Marsh, Delaware and Philadelphia Counties, Pa., The Conservation Foundation, Washington, D. C., 123p.

Miller, A. W., 1981, Evaluation of irrigation drain water for marsh management in the San Joaquin Valley--study plan, U. S. Fish and Wildlife Service, Dixon, Calif.

Mitsch, W. J., H. T. Odum, and K. C. Ewel, 1976, Ecological engineering through the disposal of wastewater into cypress wetlands in Florida, paper presented at the National Conference on Environmental Engineering Research, Development and Design, University of Washington, Seattle.

Morris, F. A., M. K. Morris, T. S. Michaud, and L. R. Williams, 1980, Meadowland natural treatment processes in the Lake Tahoe Basin: A field investigation, Environmental Monitoring Systems Laboratory, U. S. EPA, Las Vegas, Nev.

Mudroch, A., and J. A. Capobianco, 1979, Effects of treated effluent on a natural marsh, Water Pollut. Control Fed. J. 51(9).

Seidel, K., 1976, Macrophytes and water purification, in Biological control of water pollution, J. Tourbier and R. W. Pierson, Jr., eds., University of Pennsylvania Press, Philadelphia.

Simpson, R. L., D. F. Whigham, and R. Walker, 1978, Seasonal patterns of nutrient movement in a freshwater tidal marsh, in Freshwater wetlands: Ecological processes and management potential, R. E. Good, D. F. Whigham and R. L. Simpson, eds., Academic Press, New York, 378p.

Small, M. M., 1976, Marsh/pond sewage treatment plants, in Freshwater wetlands and sewage effluent disposal, D. L. Tilton, R. H. Kadlec, and C. J. Richardson, eds., University of Michigan, Ann Arbor.

Small, M. M., and C. Wurm, 1977, Data report, meadow/marsh/pond system, Brookhaven National Laboratory, Upton, N. Y.

Sutherland, J. C., and F. B. Bevis, 1979, Reuse of municipal wastewater by volunteer freshwater wetlands, Proceedings of water reuse symposium, Vol. 1, Washington, D. C., March 1979.

Tchobanoglous, G., and G. L. Culp, 1980, Wetland systems for wastewater treatment: An engineering assessment, in Aquaculture systems for wastewater treatment--an engineering assessment, S. C. Reed and R. K. Bastian, eds., EPA-430/9-80-007, pp. 13-42.

Tilton, D. L., and R. H. Kadlec, 1979, The utilization of a freshwater wetland for nutrient removal from secondarily treated wastewater effluent, J. Environ. Qual. 8(3):328.

Tourbier, J., and R. W. Pierson, Jr., eds., 1976, Biological control of water pollution, University of Pennsylvania Press, Philadelphia.

Valiela, I., J. M. Teal, and W. Sass, 1973, Nutrient retention in salt marsh plots experimentally fertilized with sewage sludge, Estuar. Coastal Mar. Sci. 1:261.

Wile, I., G. Palmateer, and G. Miller, 1981, Uses of artificial wetlands for wastewater treatment, in Proceedings of the Midwest conference on wetland values and management, St. Paul, Minnesota, pp. 255-271.

Williams, T. C., 1980, Wetlands irrigation aids man and nature, Water and Wastes Eng. 16:11.

Williams, T. C., and J. C. Sutherland, 1979, Engineering, energy, and effectiveness features of Michigan wetland tertiary wastewater treatment systems, in <u>Aquaculture</u> <u>systems</u> <u>for</u> <u>wastewater</u> <u>treatment:</u> <u>Seminar</u> <u>proceedings</u> <u>and</u> <u>engineering</u> <u>assessment</u>, R. K. Bastian and S. C. Reed, eds., EPA 430/9-80-006, EPA Office of Water Programs Operations, Washington, D.C., pp. 141-173.

Yonika, D., and D. Lowry, 1979, Feasibility study of wetland disposal of wastewater treatment plant effluent: Final report, Commonwealth of Massachusetts Water Resources Commission, Research Project 78-04.

Zoltek, J., Jr., S. E. Bayley, A. J. Hermann, A. R. Tortora, and T. J. Dolan, 1978, Removal of nutrients from treated municipal wastewater by freshwater marshes, Progress report to the City of Clermont, Florida.

DISCUSSION

<u>Berger</u>: Can you give us some idea of the area of wetland required in terms of the population served?

<u>Hantzsche</u>: I think that is still a question mark. There have been various proposals. One figure that comes to mind is 1 acre for a population of 40 people. Other estimates range from something like 30 to 60 acres, I believe, for something like a million gallons per day flow. Significant land areas are required for wetland treatment of wastewaters. Engineers, I think, are unable to be specific at the present time and are not comfortable with the design parameters, such as the application rates that might be necessary. I think this is the crux of the issue concerning pilot studies and demonstration prior to implementing full-scale projects.

<u>Zedler</u>: In the impounded wetlands, how does the water get out of the system? In Arcata Marsh, for instance.

<u>Hantzsche</u>: The Arcata wetland that has been constructed presently has water pumped into it from the adjacent storm drainage area. They will eventually need a supplementary source and are looking at wastewater as one of the sources. The wastewater will be pumped in and will flow by gravity through the marsh system and into Humbolt Arcata Bay.

<u>Zedler</u>: It just seeps through the marsh?

<u>Hantzsche</u>: Yes. It would just traverse the marsh system, going through the different sections, and exiting at an overflow. Initial pilot studies are showing something like a tenfold reduction in coliform bacteria through these marsh cells. The goal in a full-scale marsh system would be to achieve the same level of coliform reduction and then to chlorinate and dechlorinate the effluent at the end of the marsh system prior to discharge into the bay. At least this is what is being contemplated now, the thought being that the cost savings in chlorine demand would be important.

<u>Zedler</u>: Actually, my current concern is the water going out. You say you've got a spillway, so it's a point source of fresh water.

Hantzsche: Yes, it would be a point source.

Hodson: Is there a finite period of time during which a natural or artificial wetland is useful for treatment purposes and, if so, about how long?

Hantzsche: I think that's a very good question, and I don't think the engineers or I have the answer to that. What we see in these wetland treatment systems is an initial removal efficiency that takes place through the settling out of particulate matter, physical entrapment, adsorption and so forth. How those processes continue to function in the long term is, I think, somewhat in question, perhaps more so than in other types of treatment systems. One of the possible consequences for a marsh flood basin like the one in Palo Alto is more and more confinement of flow to the channels and less sedimentation in the quiescent areas, resulting in less and less contact with the soils and vegetation, which is important to the treatment efficiency. So, in that regard, I think you would expect some reduction in the treatment effectiveness of a wetland over time. What period of time that is, I think, is still in doubt.

Brinson: I noticed your title mentions "engineering applications," and I wonder just what kind of data engineers need to show that a wetland is effective in its treatment?

Hantzsche: It would depend a lot on the particular performance standards or discharge requirements that you would be seeking from that wetland system. In some of the midwestern or northern regions, interest is really in nutrient removal, principally nitrogen and phosphorus, because of the potential eutrophication problems in surface waters. So the necessary information would be what percentage nutrient reduction can be achieved by passage through soil or wetland vegetative systems.

Brinson: I guess the next question is, What is accepted [by the engineering profession] in treatment when dealing with nitrogen and phosphorus?

Hantzsche: This would relate to particular discharge standards. If the discharge standard was, for instance, a tenth of a part per million, you would want to see that level of reduction achieved through the use of the wetland treatment system as a polishing facility. The interest in pilot studies of this nature would be to estimate percent reductions basically and then relate those to the quality of the water entering the marsh and the predicted outflow quality.

Bastian: Real problem here. The engineers have an achievement goal in mind for the water quality coming out of a plant. There is no real requirement for the marsh to achieve all of the necessary treatment, since you could have all sorts of front-end treatment before the wastewater enters the wetland. When you responded to an earlier question about how long treatment plants might last, you stated that it would depend on what you're trying to achieve at the end. But if you're talking about finite ability to hold phosphorus, you're either going to have to start harvesting or make some change, whereas if its nitrogen removal and conversion of ammonia to nitrate, there's no reason why it shouldn't have an infinite capacity. There are so many permutations and combinations and they are so site specific that it's very difficult to say something in a very general sense, so you must break it down into categories as well as types of wetland. This business of mixing conventional and wetland processes together to achieve a final water quality goal allows a lot more flexibility than trying to achieve ten-ten standards exclusively through use of a wetland.

Hantzsche: Yes. Some of the artificial wetland/wastewater projects, in particular the one that I mentioned at Mountain View, did not have water quality improvement as a specific objective of that system. The effluent was fine when it left the treatment facility. They constructed the marsh system basically to provide an environmental enhancement as a trade-off for a deep-water outfall. So, that's a situation where there is some benefit from routing the wastewater through the marsh, but it's not necessarily to achieve a higher degree of treatment for the wastewater.

Pilling: I noticed that the Coyote Hills Project has a primary treatment cells-catchment basin. In treating non-point sources of runoff, have you seen any design primers or recommendations for a primary sedimentation basin prior to the wetland cells to reduce the sedimentation rate?

Hantzsche: There are no design parameters other than to say it's a good idea to collect sediment as well as other debris that may be washed with storm waters in order to keep them out of the treatment areas, and it also serves as a regulating and distribution reservoir, as I guess you might call it, to route those flows to different areas. But there may be methods for sizing a specific basin for a specific amount of sediment control. This is used a lot, for instance, where storm-waters-runoff waters are diverted for infiltration or groundwater recharge. In some cases, they actually add flocculation treatment. This occurs in the southern areas of the San Francisco Bay area, where there is a major groundwater recharge operation and it is important that solids are filtered out. Something similar has been proposed in the Coyote Hills Project.

2

Design and Use of Artificial Wetlands

I. Wile, G. Miller, and S. Black

In recent years, the use of wetlands for wastewater treatment has received increasing attention in North America. To date, attention has largely focused on the use of natural wetlands for tertiary treatment.

Artificial wetlands, however, offer greater scope for general use and are not restricted by many of the environmental concerns and user conflicts associated with natural wetlands. Unlike natural wetlands, which are confined by availability and proximity to the sewage source (Sutherland 1981), engineered marshes can be built anywhere, including lands with limited alternative uses. They also offer greater scope for design and management options and thus may provide superior performance and reliability.

Use of artificial wetlands for year-round treatment of sewage has been under study in Ontario since 1979. An experimental facility in Listowel, Ontario, consists of five separate marsh systems (Fig. 2.1) and provides flexibility in pretreatment (conventional lagoon; complete-mix aerated cell), system configuration, hydraulic loading rates, liquid depths, and detention times (Wile, Palmateer, and Miller 1981; Black, Wile, and Miller 1981). Data from this facility has been used to develop preliminary design and management guidelines.

DESIGN CONSIDERATIONS

The development of suitable design criteria and management techniques is key to the performance and reliability of a wetland sewage treatment system. The criteria discussed below are oriented primarily to artificial wetlands operating on a year-round basis in colder climatic zones and providing an effluent of better quality than secondary treatment. Data from Listowel, Ontario, indicates that effluent equivalent to tertiary quality can be produced during some of the ice-free periods from a system receiving raw aerated sewage.

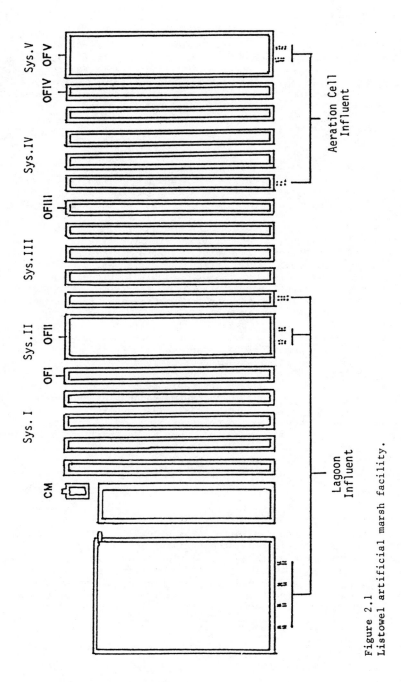

Figure 2.1
Listowel artificial marsh facility.

27

SITE SELECTION

Artificial wetlands can be constructed almost anywhere. In Ontario, experimental systems have been built in heavy clay soils (Listowel) and in an abandoned mine-tailing basin (Cobalt). Since grading and excavating represent a major cost factor, topography is an important criterion in the selection of an appropriate site. In heavy clay soils, additions of peat moss or top soil will improve soil permeability and accelerate initial plant growth.

PRETREATMENT

To reduce capital and operating costs, minimal pretreatment of wastewater prior to discharge to a wetland is desirable. However, the level of pretreatment will also influence the quality of the final marsh effluent, and therefore treatment objectives must also be considered.

Coarse screening and comminution represent minimal pretreatment requirements. However, final effluents from the artificial wetland are unlikely to meet receiving water criteria and may require additional polishing (Small 1978). Preceding wetland treatment with a conventional primary treatment plant is capital intensive and impractical unless such a facility is already in existence. Pretreatment with a conventional lagoon is land consumptive and will generate hydrogen sulphide problems in winter and algal problems in warmer weather. As illustrated in Figure 2.2, a small facultative aerated cell (5 to 10 day detention) with a continuous alum feed would reduce biochemical oxygen demand (BOD), suspended solids (SS), and phosphorus (P) without significant sludge accumulation and may represent the best pretreatment alternative. Based on studies at Listowel, some reduction of SS and BOD is desirable to reduce oxygen demand and prevent sludge accumulations in the upper reaches of the marsh. Although P removal in the Listowel marshes is substantial (up to 98%), efficiencies are expected to decline on a long-term basis, since no permanent escape mechanism, such as denitrification, exists for this element. Therefore, P reduction by chemical addition is recommended in the pretreatment step.

VEGETATION

Cattails (<u>Typha</u> spp.) are cosmopolitan in distribution, hardy, capable of thriving under diverse environmental conditions, and easy to propagate and thus represent an ideal plant species for artificial wetlands. They are also capable of producing a large annual biomass and provide greater potential for nitrogen (N) and P removal, when harvesting is practiced, than most other emergent plant species. Cattail rhizomes planted at approximately 1 m intervals will produce a dense stand within three months.

MARSH CONFIGURATION

Studies at Listowel have demonstrated the importance of the geometric configuration of an artificial wetland to performance

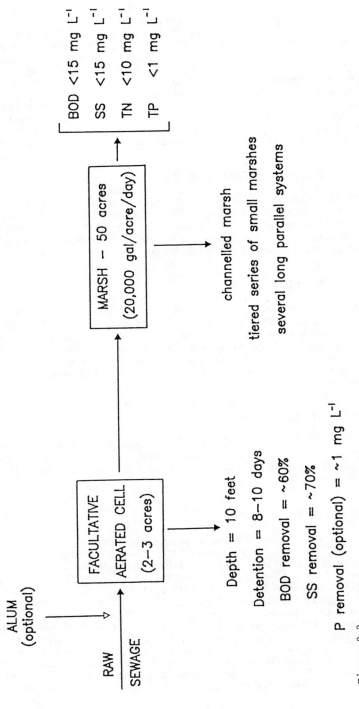

Figure 2.2
Potential design for a 1 MGD (U. S.) marsh treatment facility.

efficiency. System IV has a serpentine configuration with a length-to-width ratio of 75:1, whereas System II is an open rectangular marsh with a ratio of 4.5:1. System IV has consistently outperformed System II, although it receives raw, aerated sewage with higher concentrations of SS, BOD, total nitrogen (TN), and total phosphorus (TP) than the more dilute conventional lagoon influent used in System II (Table 2.1). Although laser grading with an accuracy of ±5 cm was employed during construction, dye studies (Fig. 2.3) have confirmed that short-circuiting of wastewater is occurring in System II. Internal flow distribution must therefore be achieved by using high length-to-width ratios or by internal berming or barriers.

HYDRAULIC LOADING/DETENTION TIME

Results obtained to date at Listowel, Ontario, suggest that hydraulic loading rates of some 200 m^3/ha/day (Tables 2.1 and 2.2) and detention times of about seven days will provide maximum treatment efficiencies. Detention times can be regulated by changing liquid depths in the marshes for summer and winter periods.

LAND REQUIREMENTS

Land requirements will depend on the hydraulic loading rates. For example, at a loading rate of 200 m^3/ha/day, a 1 MGD (3785 m^3/day) community would require a 20 ha (50-acre) wetland plus additional land for pretreatment facilities. Although land requirements appear excessive, a seasonal discharge conventional lagoon (180-day retention) would require some 40 ha (100 acres) for an equivalent sewage loading.

MANAGEMENT CONSIDERATIONS

Proper management techniques are essential for optimal system performance and reliability. Performance and reliability requirements should not be so complex that they necessitate excessive labor or skill level.

HYDRAULIC DETENTION TIME

As previously indicated, a seven-day detention time appears optimal for artificial wetlands. At a constant hydraulic loading, detention times will be influenced by two major factors: evapotranspiration in summer and ice formation in winter. High evapotranspiration rates will significantly increase detention times and cause stagnation and severe anoxia, whereas ice formation will reduce the available volume, thereby decreasing detention time. In Listowel, evapotranspiration during warm, dry weather conditions often exceeded 60% of the hydraulic loading rate, resulting in a significant increase in detention time. Coincident with increased detention times, treatment efficiencies declined, reflecting extreme reducing conditions. Maintaining low summertime liquid depths (<10 cm) in the marshes will generally circumvent these problems. Heavy rains may

Table 2.1
Removal Efficiency (%) in Systems II and IV, Listowel, Ontario

Marsh System	Pretreatment	Parameter (%)	1 (Aug.-Dec.)	2 (Dec.-Apr.)	3 (May-July)	4 (July-Sept.)	5 (Sept.-Dec.)	6 (Dec.-Apr.)
II	Conventional lagoon	TP	85	30	55	13	-	-
IV	Complete mix aeration cell		98	71	84	92	84	79
II		TN	75	16	43	20	-	-
IV			90	34	69	76	68	56
II		BOD	67	20	58	27	-	-
IV			96	75	82	81	92	88
II		SS	27	50	32	51	-	-
IV			98	84	91	94	98	97
Average water load (m^3/ha/day)			200 / 150	680 / 500	320 / 320	180 / 160	180 / 160	180 / 160

31

Table 2.2

Selected Parameter in Effluents from Marsh Systems II and V, Listowel, Ontario

Parameter	System	Hydraulic Periods					
		1	2	3	4	5	6
TP	II	0.290	0.646	0.778	2.16	–	–
	IV	0.099	0.577	1.25	0.460	0.615	0.747
TN	II	2.7	10.1	9.8	13.9	–	–
	IV	2.5	10.4	12.0	8.2	6.3	10.6
BOD	II	5.9	20.6	14.7	10.1	–	–
	IV	2.9	15.2	21.3	9.0	4.9	10.7
SS	II	16.8	16.1	24.0	10.3	–	–
	IV	3.37	8.17	21.0	8.90	3.4	4.6

32

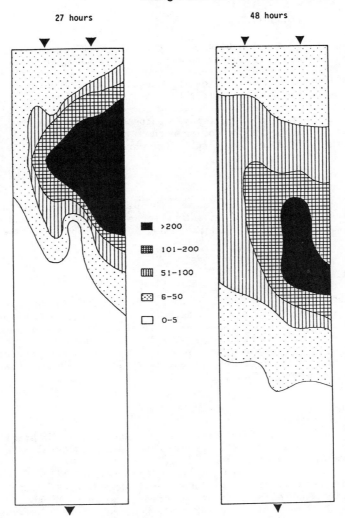

Figure 2.3
Dye intensity (relative fluorescence) after 27 h
and 48 h in System II, Listowel, Ontario.
Average water load 180 m³/ha/day; detention time
7 days.

temporarily reduce detention times; however, the resulting dilution
will mitigate any loss in treatment efficiency. Similarly, during
the winter, ice formation will reduce marsh volume and consequently
decrease detention time. Raising water levels in the marsh (>30 cm)
prior to the onset of winter conditions will remedy this problem.

HARVESTING OF VEGETATION

Harvesting of plant biomass is generally not regarded as a practical method for nutrient removal. For example, in Listowel, a single, late-season harvest removed 200 g/m^2 (dry weight) of plant material but only 8% and 10% of the annual N and P loading to the marsh, respectively. An earlier harvest, prior to translocation of nutrients by the cattails, or several harvests per season would be more effective for nutrient removal purposes. Harvesting may be desirable to reduce the excessive accumulation of litter that could shorten the life span of an artificial marsh.

RESOURCES RECOVERY

Where harvesting is practiced, the plant biomass could be utilized for production of livestock feeds (Bagnall 1979) and energy (Wolverton and McDonald 1980; Pratt and Andrews 1980) as a means of offsetting operating expenditures.

PEST CONTROL

Marshes provide an ideal breeding environment for many insect pest species, particularly mosquitoes. In Listowel, population densities of Culex pipiens were directly related to the presence of high organic loadings and inversely related to surface water coverage by dense duckweed (Lemna spp.) growths. Obviously, mosquito control by physical, chemical, or biological means is an essential component of marsh management.

REFERENCES

Bagnall, L. O., 1979, Comparison test of water hyacinth, hydrilla
    and cattail silage, Institute of Food and Agricultural Sciences,
    University of Florida, 3p.
Black, S. A., I. Wile, and G. Miller, 1981, Sewage effluent treat-
    ment in an artificial marshland, paper presented at Water
    Pollution Control Federation Conference, Detroit, Mich.
Pratt, D. C., and N. J. Andrews, 1981, Research in biomass--special
    energy crop production in wetlands, Proceedings of the Midwest
    conference on wetland values and management, B. Richardson, ed.,
    June 1981, St. Paul, Minn., pp. 71-81.
Small, M. M., 1978, Wetland wastewater treatment systems, in State of
    knowledge in land treatment of wastewater: proceedings of the
    international symposium on land treatment, vol. 2, H. L. McKim,
    coord., August 1978, Hanover, N.H.
Sutherland, J. C., 1981, Economic implications of using wetlands for
    wastewater treatment, in Proceedings of the Midwest conference on
    wetland values and management, B. Richardson, ed., June 1981,
    St. Paul, Minn., pp. 295-305.
Wile, I., G. Palmateer, and G. Miller, 1981, Use of artificial wet-
    lands for wastewater treatment, in Proceedings of the Midwest
    conference on wetland values and management, B. Richardson, ed.,
    June 1981, St. Paul, Minn., pp. 255-271.

Wolverton, B. C., and R. C. McDonald, 1980, Vascular plants for water pollution control and renewable sources of energy, paper presented at the Bio-energy World Congress and Exposition, Atlanta, Ga.

DISCUSSION

Odum: They got started in the sixties trying to recycle wastes at Calico Creek, Morehead City, North Carolina. One of the objectives was self-organization--in other words, letting the ecosystem build its own pattern. That minimizes the cost of management and construction, and so forth. For example, in many Florida situations, plants will move into locations wherever there are nutrients; so, instead of open channels developing, the system develops plant beds, making for a lot of genetic diversity. The system operates itself. You may learn much by setting up an experiment with organized channels and doing experiments, but when it comes to finding a cheap way to do it, it looks as though we've got to hang on to this self-organizing principle, which we sometimes call ecological engineering. This minimizes the engineering costs and maximizes nature's self-organizational process.

Wile: That's great if you have a system available. But if you don't have a natural wetland within a hundred miles, it becomes necessary to create one.

Odum: It seems as though it would apply also to the wetlands you are going to start. If you have a land area and some water and life, you can turn it loose with as many genetic varieties as you can to see what happens. I'm suggesting that the self-organizing principle will take over. For example, in that Arcata system, consider that mud flat out there in the marine environment--we would predict marine marsh developing there pretty soon. Why not help it along? There is an invasion of Spartina townsendii in a situation like this in New Zealand, which we studied recently. Why not encourage it in the California situation? There's always an adapted genetic variety somewhere, and it will come in eventually. This is an opposite philosophy to those who vainly try to delay exotic invasion.

Wile: I tend to agree with you that you could leave the system alone and let the plants invade. It will likely be cattails, anyway, in our climate. But for the sake of speeding up the development of the system, I don't think it hurts to initiate establishment artificially. I think the really important thing is that if you're going to design a system--create a mudflat, if you like--you may as well design it to provide optimal treatment. And as we found, if you have large, open systems, you will get short-circuiting and very poor quality treatment.

Odum: Well, I was objecting to that, because I don't think that's what happens.

Hemond: There's a tendency to think of engineering as a process of using established relationships and formulae and handbooks, and so on, so that you predict ahead of time how something you're going to build is going to perform. Are there real engineering tools for

predicting ahead of time how a given system is going to perform?
Or do we have to rely pretty much on pilot projects in any given
situation--that is, the establishment of a small pond and the
extrapolation of its performance?

Wile: With natural marshes, which are more site-specific, you
probably have to rely on pilot studies. I think the advantage of
artificial systems is that you can develop standard criteria, and
that's really the objective in our studies at Listowel: to define
design criteria that the engineers can apply anywhere in Ontario.

Kaczynski: I would agree with what she says--that it is completely
within today's state of the art to successfully design completely
managed systems with standard design criteria.

Ewel: There's a very common concern that relates to system self-
organization in all of these projects. I think it is a common
misconception that we have to worry about the organic matter accumulating, as you mentioned in the case of the cattails. It's been
shown in several projects that decomposition rates actually increase
quite a bit, in large part because of the increases in the nutrient
levels in the plants. So I would suspect that harvesting is simply
not necessary because decomposition in situ will take care of a lot
of the accumulated biomass. I think this also offers the possibility
of letting the system take care of itself, perhaps by undergoing some
kind of self-organizing successional change.

Bastian: I assume the reason that you go for harvesting is that you
want to control the depth of water flowing through the wetland. If
you let it fill up with organic matter, you would no longer be able
to push water through the system.

Wile: To add to Bob's comments, you also tend to get channelization
of the influent because of dense Typha litter in these shallow
systems. Litter removal is the main reason we are looking at
harvesting as a management tool. We're not saying it's necessary.

Day: What happens in the winter? It seems to me that the temperature
would be such that the system would stop operating?

Wile: No. As a matter of fact, all our design criteria are aimed at
a system operating year-round in a cold climate and producing better-
than-secondary-quality effluent at all times. In the winter, we do
get some decrease in removal efficiencies. Where we might have
effluent ammonia concentrations of less than 1 mg/l in summer, we
will have ammonia concentrations of 2 or 3 mg/l in the winter.
Suspended solids in the marsh effluent might increase from 5 to 10
mg/l. Phosphorus might increase from 0.5 to 0.7 or 0.8 mg/l.
Obviously, we are getting removal in the winter but I'm not entirely
sure of the mechanisms.

Reed: What kind of water depth do you deal with in the winter?
Wile: Our influent wastewater (raw sewage) is at about zero degrees,
so we do have ice formation on the surface of the marsh. The ice is
approximately 10 cm deep, with wastewater flowing underneath the ice
layer all winter long. Prior to the onset of winter, depths are

increased to at least 30 cm, so that we have at least 20 cm of water beneath the ice.

Guntenspergen: Have you tried to see what sort of effect the winter operations have on belowground structures in cattails? I'd think that if there's any sort of elevated temperature, belowground carbohydrate content utilization in winter would have a certain effect on the life history of these plants.

Wile: We haven't noticed any detrimental effect on cattails, and we have had two very rigorous winters. As a matter of fact, our first winter was the coldest on record in 50 years, with -40°F temperatures. Typha is not totally dormant during the winter, according to some recent studies by Isabel Bailey.

Guntenspergen: I was just thinking that if you elevated effluent temperatures . . .

Miller: Sewage effluent elevates temperatures at most 2 degrees--not 10 or 15 degrees.

Jackson: Have you thought at all about secondary problems, like creating an ideal redwing blackbird roost area with possible depredation of corn fields?

Wile: Yes, we have. I should point out that for pest control, we do muskrat trapping to prevent destruction of berms. We are studying mosquito populations. Redwing blackbirds also represent a problem. As a matter of fact, we have about one bird per cattail in the summer, and we are seriously considering including bird droppings as part of our nutrient loading measurements. I suspect that bird-control methods may ultimately have to be employed in wetlands where these birds present a problem for agriculture.

3

Mosquito Considerations in the Design of Wetland Systems for the Treatment of Wastewater

Rich Stowell, Scott Weber,
George Tchobanoglous, Barbara A. Wilson,
and Kenneth R. Townzen

In the Central Valley of California, water hyacinth systems have produced year-round an effluent containing consistently less than 10 mg/l of biochemical oxygen demand (BOD) and 10 mg/l of suspended solids (SS) at costs substantially lower than other technologies, including overland flow. Recently, several local hyacinth systems have been taken out of service because of mosquito production problems, the major drawback of the systems. The purposes of this paper are to present mosquito-production research data from the Roseville, California, hyacinth systems, to discuss how to minimize hyacinth-system mosquito production by design, and to suggest design strategies for hyacinth and other wetland-type systems.

ROSEVILLE, CALIFORNIA, MOSQUITO DATA

The Roseville hyacinth system consisted of three parallel 25 ft x 300 ft clay lined ponds operated at a water depth of 1.5 ft. Each pond received approximately 16 gal/min of secondary effluent with BOD and SS concentrations averaging 17 mg/l and 10 mg/l, respectively. These ponds had been in operation as research hyacinth systems for approximately one year before mosquito research was begun.

Each of the ponds was operated differently. Pond 1 was operated as a totally unmanaged hyacinth-covered pond. The hyacinths were at full stand density as evidenced by long, nonbulbous petioles. Pond 2 also had a totally unmanaged hyacinth stand but the first 200 ft of this pond were aerated via a submersed manifold bubble aeration system. Pond 3 did not contain an aeration system but had approximately 20% of its hyacinths harvested biweekly during the warm seasons. After harvesting, the remaining plants were not redistributed. Consequently, harvested areas typically remained relatively open for one to three weeks.

Mosquito research monitoring of these ponds was conducted from March through November. Mosquito larvae and larval predaceous insect populations were quantified by weekly dip samples at 20 locations in each pond. Mosquitofish (_Gambusia_ spp.) populations were assessed using fish traps.

Results of the mosquito larvae monitoring program are shown in

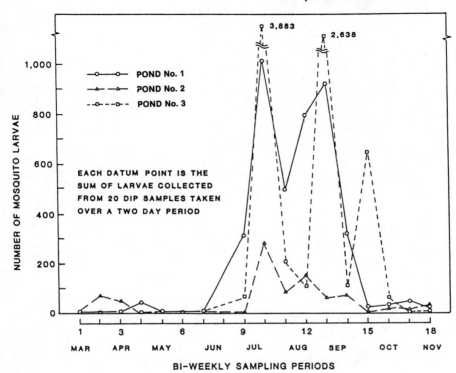

Figure 3.1
Mosquito larvae collected during each sampling.
(After E.W. Mortenson, 1982.)

Figure 3.1. Mosquito larvae populations were seasonal and sporadic.
At Roseville, larvae of the nuisance mosquito Culex peus and of the
disease vectors Culex pipiens and Culex tarsalis were dominant during
the summer (July–September). Culiseta incidens and Culiseta inornata
larvae were dominant in early spring (March–April). Culex erythro-
thorax larvae were dominant in fall (October–November).

Mosquito larvae populations tended to be lowest in Pond 2 and
highest in Pond 3. This situation is thought to be a direct consequence of the number of larvae predators (specifically, mosquitofish)
present in the respective ponds, as shown in Figure 3.2. Pond 2 had
more fish than Pond 1 probably because the former had a higher
dissolved-oxygen (DO) concentration as a result of the aeration
system. In terms of DO, Pond 3 also should have had more fish than
Pond 1, but this was not the case. The relatively low mosquitofish
population in Pond 3 is thought to be a result of inadequate
stocking.

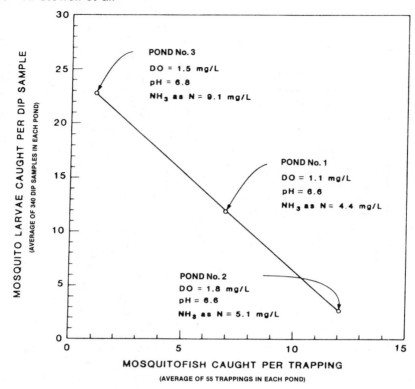

Figure 3.2
Mosquito larvae populations as a function of
mosquitofish populations. (After E.W. Mortenson,
1982.)

The Roseville hyacinth system mosquito monitoring program also
produced information about the general nature of the mosquito problem
in wetland systems and how this problem should be monitored. Because
of the seasonal and sporadic nature of mosquito larvae populations,
mosquito monitoring programs must be of long duration and entail
frequent sampling at many sampling stations to be significant. The
selection of sampling stations should reflect the various subhabitats
that unavoidably form or exist in a wetland system. As an example,
in the Roseville ponds the highest mosquito larvae populations were
found in shallow water sometimes and in the deeper water at other
times for no apparent reason. Highest concentrations of larvae also
shifted for no apparent reason between open water and the sheltered
water between petioles in the hyacinth mats. Areas dominated by
filamentous algae tended to have only Culex tarsalis in reduced
numbers. Areas dominated by duckweed had few mosquitoes of any
genus. Special sampling equipment may also have to be developed for
use in the mosquito monitoring program, as was the case at Roseville
where a pointed steel "dip sampler" was built to penetrate dense mats
of hyacinths to obtain samples of the mosquito larvae population
existing in these areas.

DESIGNING TO MINIMIZE MOSQUITO PRODUCTION

On the basis of reported experiences at Lakeland, Florida, and Roseville, California, it appears that at this time the best approach to controlling mosquito production in wetland systems is to design the system so that natural predators of mosquito larvae will thrive throughout the wetland system. Because most natural predators of mosquito larvae (such as mosquitofish, dragonfly and damselfly nymphs, and a variety of water beetles) are strict aerobes, wetland systems should be designed so that anaerobic conditions are avoided, even in the immediate vicinity of influent points where "hot spots" of mosquito production would be more likely to occur. To avoid anaerobic conditions, the wastewater BOD load to the wetland system must be kept low and well distributed over the wetland surface area to avoid creation of the aforementioned hot spots. Consequently, the mosquito-production potential of a wetland system is very much a function of: (1) wastewater pretreatment prior to its entry to a wetland system and (2) wastewater distribution within a wetland system.

In addition to designing wetland systems so that the aquatic environment is aerobic, several other wetland system design, operational, and abatement measures may reduce mosquito larvae populations. Observations at Roseville indicate that wetland systems should be designed so that the occurrence of hydraulically static areas is minimized. The design should also be such that shallow areas will not tend to form, but if (and when) they do, the operators should have the means to eliminate them. As an example, operators should have the means to remove areas of vegetation (alive or dead) that become so thick as to cause hydraulically static or anaerobic conditions.

Nonbiodegradables such as grit, plastics, and grease can plug portions of the piping system that distributes the wastewater over the wetland. Static conditions occur in the area(s) influenced by the plug(s). Overloaded conditions occur elsewhere that may lead to very low DO concentrations. Long-term accumulation of nonbiodegradables can cause shallow, stagnant areas and the accompanying problems. For these reasons it is very desirable to remove nonbiodegradables prior to wetland treatment.

Pretreatment for wetland systems beyond removal of nonbiodegradables is simply a matter of economics and local policy. Until uncertainties regarding the mosquito-production potential of wetland systems are clarified, gross BOD loading rates should be kept under 100 lb/acre/day, and the influent wastewater should be applied more or less uniformly to as much as three-fourths of the wetland surface area. These conditions could be achieved by a spray application system or an extensive submerged influent manifold piping system.

Because design considerations for treating relatively high strength wastewaters in wetlands without causing mosquito problems are ill defined, pretreatment for wetland systems may be specified by local officials. At this time in California, health officials are very hesitant to approve wetland systems of any sort if pretreatment is not a secondary process. As more information is gathered on the mosquito-production potential of wetland systems as a function of pretreatment and distribution over a wetland, this position may change.

If these and other design and operational measures do not reduce the mosquito problem to acceptable levels, then mosquito-abatement measures will be necessary. Current mosquito-abatement technology includes the use of parasitic and pathogenic organisms or hormonal substances for the control of mosquito larvae, but the applications of these control measures have not been evaluated in wetland systems. Traditional mosquito-abatement methods of applying oil and pesticides could be used only to the extent that the wetland wastewater treatment system would not be "upset" significantly. In the few trials of traditional control methods in California, mosquito larvae kills were undetectable.

SUGGESTED DESIGNS FOR WETLAND SYSTEMS

In what follows, several wetland system designs are examined as either innovative or as systems that provide useful modifications of existing treatment systems. Water quality throughout the wetland must be adequate for mosquitofish production. In all cases, it is assumed that the effluent from these systems will meet secondary requirements.

NEW SYSTEMS USING WETLANDS

Several systems incorporating a wetland for wastewater treatment are shown in Figure 3.3. Several pretreatment options are used; appropriate design parameters for these options are given in Table 3.1. Each option is considered separately below, followed by the design of a wetland sytem that is the same for all options.

Pretreatment Options

The use of an Imhoff tank as a pretreatment step (see Fig. 3.3a) is recommended for small communities and residential developments where the land required for lagoon systems may not be available. Because of the extreme flow variations that occur in the wastewater flow rates from small communities and developments, the hydraulic detention time should be about 3 h to 4 h, based on the average flow rate.

In Figure 3.3b, a facultative lagoon system is used for pretreatment. The recommended detention time is about 10 days. In warmer climates, where significant algal population shifts occur during the spring and fall turnovers, it is recommended that the available storage capacity of the lagoon be at least 21 days. This storage capacity is used to contain the flow during the period when algal populations shift in size and type from relatively large to extremely small and back to relatively large. Extremely small algae will pass through most wetlands.

A conventional sedimentation tank is used for pretreatment in Figure 3.3c. Sedimentation facilities should be used for intermediate size communities where sludge digestion is economically feasible. Appropriate design values are given in Table 3.1.

A more elaborate pretreatment option is shown in Figure 3.3d. In this flowsheet, primary effluent is filtered prior to discharge to the wetland. Primary effluent filtration would be used where it is desired to recover additional energy and to reduce the size of the

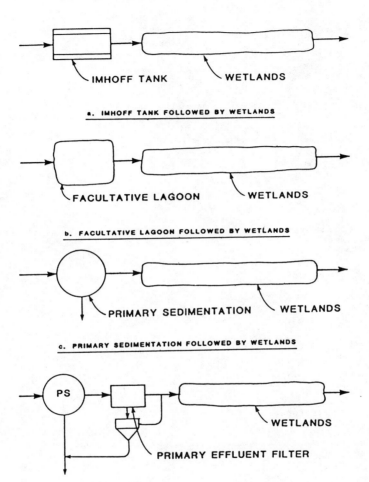

Figure 3.3
Proposed design options for new wetland/wastewater
treatment systems.

Table 3.1
Design Parameters for Pretreatment Options Used in
Conjunction with Wetland Treatment System

Pretreatment	Design parameters	
	Units	Value
Imhoff tank		
Overflow rate	$gal/ft^2/d$	600
Detention time	h	3
Sludge storage capacity	mo	6
Lagoon		
Aerated		
Detention time	d	5
Depth	ft	12
Facultative		
Detention time	d	10
Depth	ft	5–6
BOD_5 loading	lb/acre/d	50
Anaerobic		
Detention time	d	30
Depth	ft	14–16
BOD_5 loading	lb/acre/d	300
Primary sedimentation		
Overflow rate	$gal/ft^2/d$	1000
Detention time	h	2
Depth	ft	12
Primary filtration		
Surface loading rate	$gal/ft^2/min$	2

wetland treatment system.

Wetlands Options

On the basis of analysis of the data available in the literature (Black, Wile, and Miller 1981) and the results of research studies conducted at the University of California at Davis, wetlands for wastewater treatment should be designed as plug flow reactors with aspect ratios greater than 15 (length to width). Recommended design criteria include: (1) BOD_5 loading rate = 100 lb/acre/d, (2) depth = 1.0 ft, and (3) temperature coefficient = 1.10. Depending on the geographic location, either cattails or water hyacinths could be used. Cattails are preferred in all but the southern United States, where cattails, like water hyacinths, can build up excessive biomass. Under these conditions, harvesting or plant management is necessary to maintain a well-functioning biosystem. Wetland systems meeting these criteria will continue to function effectively, even when covered with a foot or more of ice, as long as detention time is maintained under the ice (Black, Wile, and Miller 1981). Depending on the temperature and the degree of pretreatment, it is anticipated that a design envelope such as the one shown in Figure 3.4 will encompass the actual operating conditions found in the field.

MODIFICATION OF EXISTING SYSTEMS

Three typical systems in which wetlands are used to replace existing treatment components are shown in Figure 3.5. In Figures 3.5a and 3.5b, the aeration tank of an existing plant is bypassed. The secondary sedimentation tank could be used to provide additional sedimentation capacity. An activated sludge plant is shown in Figure 3.5b. In Figure 3.5c, an existing lagoon system is modified by converting one or more of the lagoons to plug flow wetlands. The effluent take-off location would vary, depending on the local temperatures. As shown, effluent would be taken from pond 2 in the summer. The design of the wetland treatment system would be as previously discussed for new systems.

CONCLUSIONS

The wetlands options for new or existing wastewater treatment systems represent a departure from what has been done in the past. The documented performance and lower cost of these systems make it clear that where the conditions are suitable, these systems should be considered in any cost-effective analysis of wastewater treatment alternatives. However, before the widespread use of wetland wastewater treatment systems can be recommended routinely, research and surveillance of additional pilot projects are needed to determine the mosquito-production potential of wetland systems as a function of design, operation, and mosquito-abatement measures.

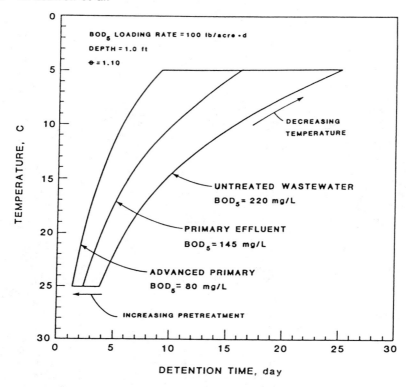

Figure 3.4
Detention time operating envelop to achieve secondary
treatment in wetlands/wastewater treatment systems.

REFERENCES

Black, S. A., I. Wile, and G. Miller, 1981, Sewage effluent treatment
    in an artificial marshland, paper presented at the 1981 Water
    Pollution Control Federation Conference, October 1981, Detroit,
    Mich.
Chan, E., T. A. Bursztynsky, N. Hantzsche, and Y. J. Litwin, 1981,
    The use of wetlands for water pollution control, U. S. EPA--
    Municipal Environmental Research Lab, Office of Research and
    Development, Cincinnati, Oh.
Mortenson, E. W., 1982, Mosquito occurrence in wastewater marshes:
    a potential new community problem, California Mosquito and Vector
    Control Association, American Mosquito Control Association, Joint
    Proc. 50:65-67, Sacramento, Calif.
Stowell, R., R. Ludwig, J. Colt, and G. Tchobanoglous, 1981, Concepts
    in aquatic treatment systems design, Amer. Soc. Civil Eng.
    Environ. Eng. Div. J. 107:EE5.
Townzen, K. R. and B. A. Wilson, 1983, Survey of mosquito production
    associated with water hyacinth, California State Water Resources
    Control Board, Sacramento.

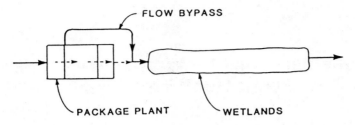

FLOW BYPASS

PACKAGE PLANT   WETLANDS

a. SMALL PACKAGE PLANT

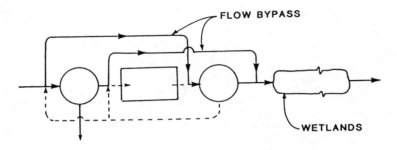

FLOW BYPASS

WETLANDS

b. CONVENTIONAL ACTIVATED SLUDGE PLANT

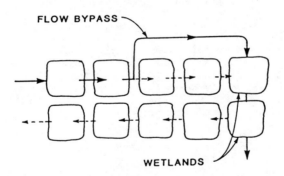

FLOW BYPASS

WETLANDS

c. LAGOON TREATMENT SYSTEM

Figure 3.5
Proposed design options for incorporating wetland
treatment in existing treatment systems.

4

Considerations for Wetland Treatment of Spent Geothermal Fluids

V. W. Kaczynski

Most wetlands treatment studies have dealt with the disposal of partially treated municipal sewage effluent; industrial applications are not common. We recently performed a generic research study that investigated the feasibility of disposing spent industrial geothermal fluids by wetlands treatment. The study was done for the U. S. Fish and Wildlife Service in close cooperation with the U. S. Environmental Protection Agency. Highlights of the study are presented below. The main objectives of this paper are: (1) to present two sets of ecological criteria for effluent application in wetlands in order to discuss their appropriateness (especially in terms of industrial effluents or municipal effluents with major industrial contributors that do not have pretreatment programs) and (2) to present some general design and ecological factors associated with selecting sites and designing wetlands for effluent treatment. Our reference studies were directed primarily toward creating and managing wetlands as alternative treatment systems, with secondary objectives of providing waterfowl and wildlife habitat. These studies were done mainly in relatively water-short western areas.

GEOTHERMAL EFFLUENT--WETLAND APPLICATION

The study of wetland treatment and disposal of geothermal effluent involved a multidisciplinary approach with each task proceeding in a dependent manner. Initially, all known geothermal fluids (with industrial potential) in the states of Montana, Idaho, New Mexico, Utah, Nevada and Oregon and in northern California were characterized to determine their physical and chemical makeup. We found patterns that allowed the individual geothermal sources to be pooled into 91 geographic areas based on the physical/chemical characteristics of the fluids. Next, we developed two levels of wetland ecological tolerance criteria for these effluent constituents: Level 1, a very conservative set, and Level 2, a less conservative set of concentrations. These levels will be discussed in more detail below. We then compared characteristics of the geothermal sources to ecological tolerances (see below) to determine whether the potential for application was general or isolated. Most effluents met at least the less conservative level, which shows good general potential for the wetlands treatment application. Within the seven states, 21 geothermal areas were directly usable (met the very conservative criteria); 47 areas were probably usable (met the less conservative criteria); and three areas would need pretreatment before wetland disposal/treatment would be feasible. Several geothermal areas could not be categorized and evaluated because their fluid data were insufficient to allow reasonable chemical characterization.

We then evaluated appropriate technologies and costs for geothermal effluent pretreatment based on the physical/chemical characteristics of the effluent and the need to meet, at least, the less conservative effluent criteria before wetland application.

Wetlands construction, operation, and maintenance costs were next developed on a general (not site-specific) basis (Figs. 4.1 and 4.2). Major factors affecting cost are quantity of effluent, application rates, land area requirements, geographic location, topography, competing land use, allowable discharge, and actual system design (internal complexity relating to multipurpose management). Costs generally are higher when new wetlands are created (as opposed to when existing wetlands are enhanced), no discharge is allowed, and effluent flow rates are higher. Costs for a zero-discharge wetland are about four times higher than those for a wetland where discharge is allowed. These higher costs are directly related to the amount of land required to obtain total water consumption (via transpiration, evaporation, and groundwater recharge).

When these costs were added to the effluent-pretreatment costs, we could then compare the total cost of a wetland disposal alternative to the cost of reinjection (the conventional disposal technique). Wetlands disposal is economically attractive when compared to fluid reinjection. Many geothermal-wetland applications would require little or no pretreatment. Low-technology pretreatment would be favorable. High-technology pretreatment would be cost prohibitive and could not compete with reinjection. A combination of high effluent flow plus a zero-discharge requirement also favors reinjection.

Legal and institutional constraints were also studied, and no serious obstacles were found. Site-specific factors can be serious constraints. These factors include potential geologic reservoir depletion, subsidence, and groundwater contamination constraints.

The remainder of this paper will deal with the development of criteria for effluent discharged to wetlands and wetland site selection and design considerations.

EFFLUENT APPLICATION CRITERIA DEVELOPMENT

If wetland communities cannot tolerate excess levels of various pollutants, they cannot function in a manner that provides effective wastewater treatment and other benefits. Effluent application criteria specific for a number of constituents (in terms of known physical and chemical tolerances) would be invaluable. These criteria should be set at levels that would allow the successful functioning of wetland systems. Where possible, these biocriteria should be developed from known toxicity tolerances or effects. We studied this problem for the constituents typically found in

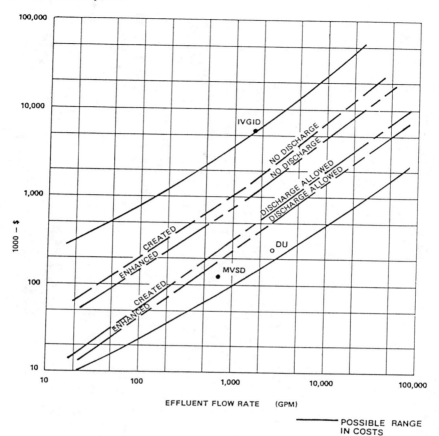

Figure 4.1
Wetlands construction cost. IVGID = Incline Village General Improvement
District; MVSD = Mountain View Sanitary District; DU = Ducks Unlimited,
Pitt Marsh (Barnes 1980).

geothermal effluent. (It would be helpful to have similar nutrient
application criteria.) Constituent concentrations that affected
principal wetland groups (flora, aquatic invertebrate, aquatic
vertebrate, and avian) are summarized in Tables 4.1 and 4.2. The
toxicity levels encountered in the literature showed considerable
variance, which reflected different testing conditions, different
combinations of compounds tested, differences in response criteria,
and different species used as "indicators" of the major wetland
groups. In cases where vegetative bioaccumulation occurs, toxic
effects generally do not occur in the concentrating plant species but
are more likely to be evident in the higher trophic levels.

Toxic levels for aquatic invertebrates are generally well known;
for aquatic vertebrates they are reasonably known; and direct
toxicities for waterfowl are least known. About half of the common
geothermal constituents could directly affect aquatic fauna or could

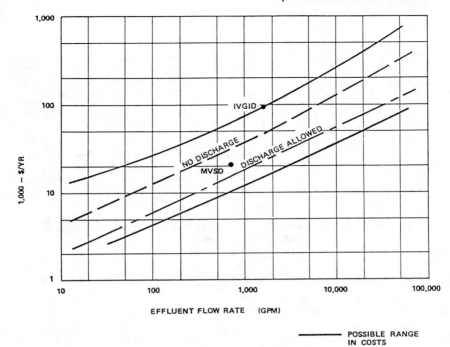

Figure 4.2
Wetlands operation and maintenance cost. IVGID = Incline Village
General Improvement District; MVSD = Mountain View Sanitary
District.

be bioaccumulated by aquatic fauna. About 30% of the geothermal
constituents have comparable effects on wetland vegetation. The same
comparison to avifauna could not be made because of insufficient
data. Aquatic invertebrates are often especially sensitive to
pollutant stress, but many of them are not critical to a wetland
designed to provide pollutant removal and waterfowl nesting habitat.
However, providing effluent criteria that make wetland waters
tolerable for sensitive invertebrates lends a level of assurance and
safety to the concept of wetland effluent treatment. It is also
reasonable to ask if such stringent criteria are really necessary.
This was a dilemma for us, and from the list of paper topics here, it
appears that this workshop shares our dilemma.

Even by using all of the available direct toxicity data (mostly
from invertebrates, some from vertebrates, fewer from waterfowl), we
were not able to derive direct geothermal effluent water quality
criteria suitable for wetlands application. The data were too
limited. We had to fill in blanks using other sources, primarily
livestock and irrigation water quality standards, as guidelines.
These data were used in combination to establish our Level 1

criteria, "Parameter-Specific Criteria for Hydrothermal Effluent To Be Used in Waterfowl Wetlands," shown in Table 4.3. The Level 1 criteria are very conservative, and any effluents meeting these limits should be very safe. Good, functional wetlands with no impairments should result from their application. We termed as direct use effluents meeting these criteria. These Level 1 criteria can probably be adjusted downward as more directly relevant wetland species (especially waterfowl) toxicity information is obtained. For example, most western wildlife refuges cannot meet these Level 1 criteria (Table 4.4).

We also determined more practical guidelines that would still result in functioning wetlands with a level of safety. For this purpose, we established the potential use, or Level 2, criteria (Table 4.3). These criteria are set at the highest concentration (or level) of each constituent that would still allow growth of desired marsh plant species (including alkali bulrush, three-square bulrush, hardstem bulrush, Olney's bulrush, sago pondweed, spikerush, and widgeongrass). Water quality information (Table 4.4) at western waterfowl refuges was also used as a guideline in setting Level 2 criteria. Natural wetlands processes appear to be sufficient to reduce the Level 2 parameter concentrations to levels that would be environmentally safe for discharge (if the wetland was sufficiently sized).

At present, the best uses of the Levels 1 and 2 effluent criteria are as interim guidelines; they should not be used as final water quality standards for effluent disposal into wetlands. Two groups may find the results useful: parties studying the geo-thermal effluent disposal/wetlands treatment alternative, and wetland researchers evaluating the appropriateness of the criteria against more directly relevant information as it is collected.

WETLAND SITE SELECTION AND DESIGN CONSIDERATIONS

Any effluent that meets appropriate guidelines (i.e., our Level 1 or Level 2 criteria for geothermal or similar industrial effluents, or advanced-primary treated municipal sewage effluent) can be used to enhance or create wetlands on almost any flat and open space. However, consideration of design criteria can minimize costs and other problems and maximize the effectiveness of the resulting wetland in terms of its wastewater treatment and other goals (such as waterfowl management).

Table 4.1
Summary of Effects of Geothermal Constituents on Flora

Constituent	Geothermal Range (ppm)	Toxicity Range (ppm)	Bioaccumulation	Comments
Aluminum (Al)	0-2	1-20	No	No data available on effects on marsh species.
Arsenic (As)	0-2.7	0.1-10 0.1 irrigation of crops[a]	No[a]	Tolerance limits of several aquatic plants in the 4-5 ppm range. Exhibits antagonisms with Zn, Fe, and Al compounds.
Barium (Ba)	0.00001-0.2	>10	No	Waterweed killed at 10 ppm soluble Ba. Most plants only exhibit toxic effects under extreme conditions.
Beryllium (Be)	0.005-0.007	0.1-2	No	No data for marsh plants. Usually unavailable in soils.
Boron (B)	0-828	0.8-2 0.75 for sensitive crops[a]	No	Raft River effluent did not harm crop species. Several aquatic species accumulated B with no toxic effects.
Cadmium (Cd)	<0.002-0.1	0.005-0.1	Yes	Marsh plants readily take up, translocate, and accumulate Cd.
Calcium (Ca)	0.28-320	_[b]	No	Essential macronutrient. No apparent direct toxic effects at levels tested.
Chromium (Cr)	<0.02-<0.04	5-20 0.1 freshwater aquatic life[a]	No	Accumulated by roots of marsh plants but not translocated.
Copper (Cu)	0.004-0.08	0.2-10	No	Submerged aquatic plants unaffected at 10 ppm Cu.
Fluorides (F)	0-24	-	Variable	No apparent toxicities from aqueous uptake. Gaseous and particulate fluorides may concentrate sufficiently to pose bioaccumulation problems.
Iron (Fe)	0-76	-	No	Extremely abundant element. Adverse effects related to deficiencies.
Lead (Pg)	0.003-0.07	5-20	Yes	Usually unavailable in soils. 10 ppm results in greatly decreased production in aquatic plants.

Table 4.1 (Continued)

Constituent	Geothermal Range (ppm)	Toxicity Range (ppm)	Bioaccumulation	Comments
Magnesium (Mg)	0–73	>25	No	Essential to plant growth. Often found at high levels in plant tissues with no toxic effects.
Manganese (Mn)	0–15	2–20	Yes	Often concentrated to high levels by aquatic plants with no toxic effects.
Mercury (Hg)	0.00007–0.02	–	Yes	Methylated form most toxic, but rarely found in soils.
Molybdenum (Mb)	Trace	–	Yes	Toxicity not a problem to plants, but can poison animals if excessive amounts (>5 ppm) occur in forage.
Nickel (Ni)	<0.02–0.08	–	No	Normally toxic at "low levels."
Phosphate (PO₄)	0.01–1.4	–	No	An essential nutrient. Aquatic plants can absorb high concentrations. Problems mainly related to eutrophication of standing waters.
Potassium (K)	0.15–660	–	No	An essential element. Commonly accumulated to high levels by aquatic vegetation with no toxic effects.
Selenium (Se)	0–0.014	5–300	Yes	Toxicity level applies to non-accumulating plants. Duckweed and many other plants can accumulate high Se levels. Livestock poisoning occurs at >5 ppm.
Sodium (Na)	5.8–3,100	Variable	No	Cattail and bulrush relatively tolerant.
Zinc (Zn)	0.005–2.32	<10 ppm	No	An essential micronutrient. Deficiencies more common than toxicities. Readily accumulated and translocated by marsh plants.

[a] EPA recommended criterion.
[b] Undetermined.

Table 4.2
Summary of Effects of Geothermal Constituents on Fauna

Constituent	Geothermal Range (ppm)	Avian Toxicity Range (ppm)	Aquatic Invertebrate Toxicity Range (ppm)	Aquatic Vertebrate Toxicity Range (ppm)	Bioaccumulation	Comments
Aluminum (Al)	0-2	_[b]	6.7	0.07-133	No	Little information available.
Antimony (Sb)	0.005-0.4	-	9-37	9-20	-	Rapidly precipitates into sediments. Probably not a problem.
Arsenic (As)	0-2.7	-	1.1-20	5-60	No	Food chain transference has low toxicity due to organic binding.
Ammonia (NH$_3$)	<0.01-4	-	90-134	0.6-24.7	No	Unionized form is by far the more toxic.
Barium (Ba)	0.00001-0.2	-	11.1-170	200-4,400	No	Sulfates and carbonates relatively insoluble. Most toxicity studies conducted on chlorides and nitrates.
Beryllium (Be)	0.005-0.007	-	-	0.15-20.3	No	Normally present in only trace amounts. Few data available.
Boron (B)	0-828	-	5.2-240	1,600-19,000	No	Generally low toxicity.
Bromine (Br)	0-3	-	10	20	Yes	Apparently bioaccumulates, but no toxicity data associated with this phenomenon.
Cadmium (Cd)	<0.002-0.1	20-200	0.005-4.9 / 0.004-0.012^a	0.017-80	Yes	Exhibits synergisms with other metals as well as direct toxicity at very low concentrations. Toxic effects in waterfowl related to kidneys, testes, and egg production and are sublethal.

55

Table 4.2 (Continued)

Constituent	Geothermal Range (ppm)	Avian Toxicity Range (ppm)	Aquatic Invertebrate Toxicity Range (ppm)	Aquatic Vertebrate Toxicity Range (ppm)	Bioaccumulation	Comments
Calcium (Ca)	0.28–320	–	920–1,730	160–56,000	No	Unlikely to present problems.
Cesium (Cs)	0.001–4.76	–	–	–	Yes (radioactive forms)	Radioactive forms bioaccumulate, but the element and its salts react similarly to potassium and are not considered toxic.
Chromium (Cr)	<0.02–<0.04	–	0.016–148 0.1[a]	2–300	No	Fish generally more tolerant than invertebrates.
Copper (Cu)	0.004–0.08	–	0.0024–32 0.1 x 96 hr LC_{50}[a]	0.009–1.25[a]	No	Extremely toxic constituent in reactive state exhibiting numerous synergisms and antagonisms. Relatively nontoxic in combined state with any degree of water hardness.
Chloride (Cl)	0.5–5.400	–	variable	2,000–10,500	No	The chloride ion readily complexes with many elements, leading to extremely variable toxicity values. Tolerances of even freshwater species are relatively high.
Fluorides (F)	0–24	–	270	419	variable	Not considered a problem in aqueous state. Deposition of particulates a potential problem.
Hydrogen Sulfide (H_2S)	0–7.4	–	0.059–1.07 0.002 undissociated H_2S[a]	0.002 H_2S[a]	No	A very toxic component that concentrates under anaerobic conditions.
Iodine (I)	0.01–0.24	–	–	28.5	–	Unlikely to present problems.

Element					Bioaccumulation	Comments
Iron (Fe)	0–76	–	0.32^a / 0.001^a	0.9	No	Generally not a problem in aerated waters due to oxidation and precipitation.
Lead (Pb)	0.003–0.07	<200 mg. (ingested metal)	0.3–64 / .01 x 96 hr LC$_{50}$^{a}	0.4–75 LC$_{50}$^{a}	Yes	Not considered a direct problem, but bioaccumulation possible.
Lithium (Li)	0.01–20	–	2.6	2.6	–	The 2.6 value is for "aquatic organisms." Li is a rare element and few data are available.
Magnesium (Mg)	0–73	–	400–3,500	100–400	No	Not considered a problem.
Manganese (Mn)	0–1.5	–	5.7–700 / 0.1^a	15–40	Yes	Not considered a direct problem but could bioaccumulate.
Mercury (Hg)	0.00007–0.02	.1–.5	0.006–2.1 / .00005^a	0.0001–.0?	Yes	Toxic at low levels. Poses bioaccumulation problems and exhibits synergisms.
Nickel (Ni)	<0.02–0.08	–	0.51–64 / 0.01 x 96 hr LC$_{50}$^{a}	4.58–9.82	No	Moderately toxic to freshwater organisms.
Nitrates (NO$_3$)	0–42	–	2670	420	No	Not considered a problem except for eutrophication potential.
Nitrites (NO$_2$)	0–0.12	–	–	1.5	No	Not generally considered a problem in natural surface waters.
Potassium (K)	0.15–660	–	127–1,000	50–2,010	No	Most common form (KCl) is least toxic.
Selenium (Se)	0–0.14	–	2.5 / 0.01 x 96 hr LC$_{50}$^{a}	2.9–40	Yes	Carcinogenic and teratogenic.
Silver (Ag)	0.004–0.02	–	0.0051–0.15 / 0.01 x 96 hr LC$_{50}$^{a}	0.003–0.0043	–	Nitrates and sulfates most toxic forms to aquatic life.

Table 4.2 (Continued)

Constituent	Geothermal Range (ppm)	Avian Toxicity Range (ppm)	Aquatic Invertebrate Toxicity Range (ppm)	Aquatic Vertebrate Toxicity Range (ppm)	Bioaccumulation	Comments
Sodium (Na)	5.8–3,100	–	1,000–2,100	1,000–2,000	No	Increasing concentrations alter community structure.
Sulfates (SO_4)	2–4,520	–	0.01–7105	12–34,000	No	Moderately toxic within range of geothermal sources.
Tin (Sn)	<0.02–<0.2	–	25–146	–	No	Not considered a problem.
Vanadium (Vn)	–	25–300	–	–	No	Some evidence of effects on lipid metabolism and interaction with calcium mobilization.
Zinc (Zn)	0.005–2.32	–	0.1–0.3 0.1 x 96 hr LC_{50}[a]	0.88–0.96	No	Relatively toxic when reactive; nontoxic when combined. Exhibits synergisms with copper and nickel. Will be in combined state with any amount of hardness present in geothermal effluent.

[a]EPA recommended criterion.
[b]Undetermined.

58

Table 4.3
Parameter Specific Criteria for Hydrothermal Effluent
to be Used in Waterfowl Wetlands

Parameter	Level 1 (Direct Use) (ppm)	Level 2 (Potential Use) (ppm)	Average of 23 Natural Geothermal Springs
TDS	1,000	15,000	6,500
pH[c]		6.0 9.0	
Ag	0.003	0.5	0.04
Al	0.01	13.0	2.5
As	0.10	2.0	0.04
B	4.0	25	4.2
Ba	1.0	13.0	1.5
Be	1.0	12.0	0.04
Bi	ISD[a]	ISD	—[b]
Br	2.0	20	—
Ca	200	500	542
Cd	0.01	0.1	0.003
Cl	500	2,500	2,400
Cr	0.05	0.5	0.07
Cu	1.0	1.0	0.1
F	5.0	100	—
Fe	1.0	1.0	0.03
H_2S	0.07	1.0	—
Hg	0.007	0.12	0.006
I	2.85	28.5	—
K	5.0	200	86
Li	0.26	2.6	1.3
Mg	25	300	13.0
Mn	0.05	1.5	0.03
Mo	0.5	1.0	0.1
Na	500	1,400	1,350
NH_4	0.13	2.0	—
Ni	0.4	4.0	0.09
NO_3	90	420	—
Pb	0.15	1.5	0.04
PO_4	20	60	0.2
Ra	ISD	ISD	—
Rb	ISD	ISD	—
Sb	1.0	10.0	0.04
Se	0.05	3.5	0.12
Sn	2.5	25	1.5
SO_4	150	1,000*	167
V	0.1	1.0	0.23
Zn	0.2	5.0*	0.04

[a]Insufficient data.　[c]Not in ppm.
[b]No data available.　*Possibly higher.

Table 4.4
Selected Water Quality Parameters Observed at Western Waterfowl Refuges

Refuge	pH	TDS	Na	K	Ca	Mg	Cl	SO$_4$	PO$_4$	Al	Cd	Cr	Cu	Fe	Pb	Mn	Zn	NH$_4$	F	CO$_3$(HCO$_3$)
								Parameters (ppm)												
Deer Flat (Lake Lowell), Idaho	8.1	129							0.08									0.08		
Coeur D'Alene Lake, Idaho	7.5	140	2		18	11	29	6	0.14		0.16		0.02	0.07	0.14	0.8	0.56	0.06		80
Blackfoot Marsh, Idaho		431	13.5	4.1	26.5		17.2	35.3				1.08					0.46			
Desert Lake, Utah	8.29	14,026	2,925	2.9	285	814	170	9,518												34.3(249)
Ouray NWR,[b] Utah	6.88	790	96.2		118	52.4	48.6	296	0.2	0	0	0	0	0.11	0	0.41	0.01		0	
Medicine Lake, Montana		14,000[a]																		
Freezout, Montana		5,480			88	480	170	2,990						0					32	0(560)
Benton Lake NWR, Montana		3,126	147.3		122	45.5	362	1.93						0.12		0.01			2.57	

60

Location									
Bowdoin NWR, Montana	7,000	2,400		40	350	220	5,400		0.5
Red Rocks NWR, Montana	100[d]								
Stillwater, Nevada	780								
Sarasota H.S., Nevada[c]	8.25	1,213	155	5.2	150	0.06	40	685	(15)
Carson River at HWY 36, Nevada	7.8	220	225	3.2	25	6.8	10.7	29	0(120)
Well 60N Incline Village, Nevada[c]	7.76	2,693	480	11.5	280	5.2	90	1,680	(190)

[a] Average of 5 measurements taken at Medicine Lake.
[b] National Wildlife Refuge.
[c] Hydrothermal data collected and analyzed by CH2M Hill (1979).
[d] Average of 13 measurements.

61

SITE SELECTIONS

The following factors should be considered when evaluating potential wetland sites.

Distance

Distance of the wetland from any effluent source can be larger for larger flows, and distances are related to efficiencies in scale of construction. The following guidelines are suggested:

    100 gpm--up to 2 mi,
    1,500 gpm--up to 3 mi,
    20,000 gpm--up to 12 mi.

Topography

It is expensive to create wetlands on moderate to steep terrain because of the need for extensive earthwork. The following slopes are suggested:

    0% to 3% slope: use for ponds.
    0% to 12% slope: use for overland flow--desirable for a mixture
                      of wetter and drier soils and vegetation.

Soils

Soil types primarily constrain the use of structural components, such as dikes and roadways, but also affect the vegetative growth, treatment capacity, and drainage characteristics of a wetland. Peat and muck soils should be avoided if structural components are needed. If they must be crossed by structural components, they present special problems. Soil characteristics have an effect on moisture flow and adsorption reactions between soil and waste constituents. Some colloidal fines are desirable for pollutant removal, but too great a percentage of fines can increase surface runoff (and possible surface water contamination). Soil depth is important for plant root development and for removal and retention of pollutants. Soil has a finite capacity to store nutrients, heavy metals, and salts. These capacities are a function of soil mass and retention rates. Soils less than 2 ft deep over an impervious layer should be avoided. Moderately slow to very slow internal drainage is desirable, and easily eroded soils should be avoided.

Groundwater

Groundwater is a site-specific consideration since the site may be a recharge or a discharge area. Groundwater flow rates and direction of movement, quality, and "downstream" utilization will all affect wetland design and operation.

Surface Water

Floodplains generally can be used if they can be economically protected, or if periodic flooding can be contained within the wetland and such flooding will not damage the wetland function and water

control structures. Seriously flood-prone areas should be avoided
because of loss of control of wetland processes (i.e., treatment) and
serious potential for structural damage and/or possible contamination
of surface waters. Existing floodplain wetlands that would benefit
from additional water supply or from a more controlled supply are
potentially suitable. Feasibility or desirability of discharge of
wetland-treated effluent into surface or ground waters should be
evaluated on a case-by-case basis. Such discharge may be feasible
and, in many arid western regions, desirable.

Land Use

High-value, developed areas are generally too expensive. Air-
port areas should be avoided because of the likely influx of birds,
and agricultural lands may also be a problem because of potential
crop depredations by birds.

ECOLOGICAL DESIGN CRITERIA

Ecological design criteria can lead to more functional wetlands.
A few points and suggestions are offered for multipurpose management
(e.g., wetlands effluent treatment and creation of waterfowl habitat).

Retention Time

Retention time is critical for the removal of pollutants in the
mixed physical/biological wetland treatment system. The only practical way to increase retention time is through determination of
proper wetland volumes and effluent flow rates. Adequate retention
time is needed to avoid surface discharge, to obtain high polishing
efficiency, and to maximize treatment efficiency during cold months
when biological activity is minimal. Care needs to be taken during
design and construction to prevent short circuits within wetland
cells, which would result in shorter retention times.

Flow Rates

The flow rate not only affects retention time but is desirable
to help maintain aerobic water conditions to minimize the potential
development of avian botulism (which occurs under anaerobic conditions), to stimulate secondary consumers, and to minimize surface
water freezing. Flow-control structures (both upstream and downstream) are highly desirable to regulate effluent supply and control
drainage. Some surface discharge is desirable.

Vegetation/Water Interface

Limnologists use an analogous term, the degree of shoreline
development, to describe the vegetation/water interface. In wetlands, an approximate surface area ratio of 1:2 (water to emergent-
marsh or meadow-type wetland) will provide quality waterfowl habitat.
The design should include open water area (through depth and soil
considerations) as a submerged macrophyte wetland rather than a true
pond or stream habitat, although the latter would certainly be
acceptable for the waterfowl needs. Designs with suitable depths for

desired submerged macrophytes will maintain high pollutant-removal efficiencies while still providing the desired open-water waterfowl habitat. A design depth of 18 in is about optimum for emergent marsh vegetation. Areas deeper than 3 ft (up to 9 ft deep) will provide habitat for submerged macrophytes, although it may be difficult to prevent floating macrophyte invasions. Very shallow areas with some slope can become effective overland flow wetted meadow wetlands.

Vegetation Heterogeneity

It is desirable to provide a diversity of microhabitats, especially for avifauna. Three points are important: (1) interspersion of uplands, emergent wetlands, and open surface water areas; (2) number of vegetation layers (vertical dimension); and (3) number of vegetative communities. Microrelief is beneficial, and very careful grading during construction is not necessary (except to prevent short circuits and still basins). A high degree of shoreline development is desirable for emergent vegetation and open-surface-water edge effect. A large area of gradually sloping, very shallow wetted meadow wetland is highly effective for pollutant removal and provides excellent nesting habitat. Riparian phreatophytes can be encouraged along water delivery areas, canals, dikes, and other areas.

Vegetation Control

It is difficult to obtain specific vegetation; however, general wetland types can be achieved and are probably reasonable functional units for both wetland effluent treatment and waterfowl management. Shallow-wetted-meadow and emergent-marsh wetland units appear most attractive for applications in the western states. These units also offer high waterfowl habitat value when interspersed with submerged wetlands in a reasonable ratio (2:1).

Desired wetland vegetation can be promoted, although not guaranteed, through design of water depths (controlled) and rate of effluent application (balanced against projected water losses). Significant water losses are necessary for high pollutant-polishing efficiencies. Water losses can be estimated through application of appropriate soil, transpiration, and evaporation factors. Conventional agricultural techniques work quite well to calculate these losses. The soil factor will change with time as organic materials quickly accumulate. If a discharge is allowed, this should also be factored in. In addition, individual wetland cells can be designed to allow seasonal applications, draining, and drying. Vegetation can be harvested as fodder, grazed by controlled numbers of livestock, or periodically burned.

DISCUSSION

Giblin: There were some attempts in the agricultural literature to take into account interactive effects by, for example, copper and zinc equivalents, when considering heavy metal pollution. Do your criteria take these effects into account?

Kaczynski: No, they do not. They are assumed to be independent, but

I do recognize that they can be additive and even multiplicative.

Valiela: In most of the states you've mentioned, water is in very short supply. Why can't those effluents that have fairly low concentrations of metals be used for irrigation?

Kaczynski: You're right, and in fact, we have run into that controversy quite often. The agricultural interests want this water, and they consider it a real waste to put it on a wetland. We came up with a couple of schemes to combine wetland and agricultural uses in various ways. Some utilize the water for agricultural purposes first and then, through agricultural return flows, discharge it to wetlands. For geothermal fluids that can meet only the Level 2 criteria, the effluent could be passed through a wetland first and then be used for agricultural purposes. If it's designed properly, this arrangement could significantly clean the effluent before use. However, you probably can't achieve the 90% polishing levels that people keep talking about.

Hodson: If the geothermal fluid is not reinjected, what effect will it have on the subsurface aquifer?

Kaczynski: This is usually a site-specific consideration. The dangers are twofold: (1) the subsurface reservoir could be depleted, and (2) subsidence could occur. We encountered some cases where the reservoir could be depleted, in which case the effluent would have to be reinjected.

SESSION II

Ecology

5

Ecological Perspectives on Wetland Systems

Glenn R. Guntenspergen and Forest Stearns

Wetlands have long been used as sites for domestic sewage discharge, but the studies needed for an ecosystems-based evaluation are few and recent (e.g., Boyt, Bayley, and Zoltek 1977; Ewel and Odum 1979; Tilton and Kadlec 1979; Valiela, this volume; Guntenspergen, Kappel, and Stearns 1980; Mudroch and Capobianco 1979; Deghi, Ewel, and Mitsch 1980; Dolan et al. 1981).

To date, most wetland studies have emphasized water quality effects to the exclusion of other impacts. The major concern has been the ability of the wetlands to improve significantly the quality of the effluent being discharged. We believe that examination of the ecological impacts of wastewater application is essential and is long overdue. It is unrealistic to assume that there will be no ecological changes. In an earlier review (Guntenspergen and Stearns 1981), we concluded that biological changes are associated with nutrient addition to wetlands and that these changes are manifested at the ecosystem level. A number of studies have indicated that nutrient addition through effluent discharge enhanced biomass production and, at the same time, certain species were eliminated by alterations in the physical environment or by competition.

At the very beginning, several concepts must be underscored:

1. Wetland is an exceedingly broad term and encompasses a great variety of systems, each of which may have its own particular idiosyncrasies.
2. Wetlands are not static systems but highly dynamic communities lying at the interface between terrestrial and aquatic systems.
3. Wetlands change naturally. Their history is one of evolution toward equilibrium with the landscape. Rates of change vary, not only with the climate, species present, and geological situation, but also with the loading rates imposed by adjacent systems.
4. Wetlands are hydrologically controlled; structure and composition vary with such hydrologic parameters as flow rate, flood frequency, and water quality. The potential for inflow and outflow regulates the degree to which they are open or closed systems.
5. As a group of ecosystems intermediate between terrestrial and aquatic environments, wetlands show individual and group characteristics related to the species present, the environment, and

past historical events. It follows, as a result of the diversity
of species and variety of hydrologic and climatic regimes, that
wetland systems may differ in many ways. Each will show a
discrete response to chronic effluent discharge.

We know something of the effects of nutrient addition on both
terrestrial and aquatic systems, as indicated in the extensive
literature on both (e.g., Guntenspergen and Stearns 1981; Schelske
1980). For example, in a variety of ecosystems, nutrient addition
will change species composition, whether of vascular plants or
plankton. But it is possible that our understanding of wetland
ecosystems may be biased by our previous experience in other ecosystems. To fully understand the responses of wetlands to chronic
sewage addition, a more precise understanding of wetland processes is
essential. A broad based understanding is necessary to interpret the
often disparate results of studies on different wetland types.
Although wetlands are diverse, the common bonds between them must
also be considered.

Recently, Larson and Loucks (1978) highlighted our ignorance of
many basic ecosystem-level processes in wetlands. Quantitative knowledge of the processes operating in wetlands is needed before we can
attempt to predict successfully the structural and functional
responses to manipulation.

This discussion of the structural and functional aspects of
wetland ecosystems is designed to provide background by which to
evaluate the ecological response of wetlands to effluent addition.
We will begin with consideration of the structure of wetlands,
reviewing their components and the way that they are organized. We
will describe the major processes (production, decomposition, and
consumption), leaving detailed consideration of nutrient dynamics and
species changes for others. Finally, we will examine the effect of
successional process and historical factors in wetland evolution.

WETLAND STRUCTURE

Ecosystems consist of a variety of components; ecosystem
structure, or the organization of these components, is critical to
their internal functions. Organizational relationships between
components will vary from one wetland type to another and, for the
purpose of this review, need not be discussed in detail. However,
recognition of the importance of interactions among components is
essential to understanding ecosystem process.

Components of the wetland ecosystem include the following:

Water: Both the physical and chemical characteristics of water
influence the kind of wetland that develops and the way it functions.
Wave action, the timing of high water or flooding and its duration,
and the complex and diverse chemistry of wetland waters all contribute to the diversity of wetland communities (Gosselink and Turner
1978).

Sediment and substrate: The nature of wetland communites is
also regulated to some extent by the nature of the substrate
(Cowardin et al. 1979). The substrate, in turn, may be influenced
relative to accumulation by the flushing action of tides, by flood-
flows in streams, and so forth. The ability of vegetation to become
established depends in part upon the nature of the substrate.

Likewise, the ability of the vegetation to obtain nutrients from a substrate is dependent upon organic matter content and particle size, as well as nutrient concentration.

Animals: The invertebrates of wetlands principally process detritus and serve to maintain the steady flow of nutrients and energy through the system. The structure of wetland plant communities has a major influence on all animal populations, whether invertebrate, fish, bird, or mammal. Birds select wetland types by height and density of vegetation for feeding, nesting, and roosting purposes (Harris et al. 1981). Fish may use specific wetland types for spawning while avoiding others. Likewise, wetland mammals choose wetlands that provide both food and building material.

Decomposers: Wetlands also support decomposer populations that differ according to the pH of the water and substrate and aeration. The availability of nitrogen (N) and other nutrients, as well as the nature of the organic detritus, influences the species composition of decomposers. Decomposers, as usual, are the least understood part of the ecosystem.

Vegetation: In gross structure, wetlands can be grouped by the major vegetation type present--that is, as wetlands consisting of (1) submergent plants; (2) floating leaved species; or (3) emergent species, either (a) herbaceous or (b) woody.

Wetland communities are often composed of various mixtures of these groups as well as free-floating algae and epiphytes (periphyton). The nature of the vegetation obviously influences the productivity of the wetland as well as nutrient-exchange flow rates and wildlife habitat.

Many environmental factors act upon the organisms in any wetland community. Relationships between the species and different environmental gradients are far from simple. Wetland structure and function differ widely from place to place, depending on the interaction of physical and biological processes.

WETLAND PROCESSES

PRODUCTION

Production in wetlands is normally high, largely as a result of ample light, water, and nutrients and the presence of plants that have developed morphological and biochemical adaptations enabling them to take advantage of these optimum conditions (Westlake 1975; Wetzel 1975a; Whittaker and Likens 1975).

Since production provides the base for ecosystem structure and creates an organic sink for nutrients, knowledge of the process is of vital concern. Relative production rates of different species have much to do with the outcome of competitive interactions and with the specific changes that occur when effluent is discharged to wetlands. Controls over production are more complex, and some of the factors involved vary from one wetland type to another.

Several recent reviews examine the productivity, biomass, and respiration of wetland plants and present specific species and community data (Brinson, Lugo, and Brown 1981; Keefe 1972; Turner 1976; Good, Whigham, and Simpson 1978; Richardson 1979).

These published reports are based on in situ biomass harvesting, or on measurement of physiological processes, using either CO_2 or O_2 fluxes. For emergent plants, aboveground production (net biomass accumulation) usually falls into the range of 400-2000 g dry wt. $m^{-2}yr^{-1}$. Estimates of belowground production are equal to or larger than aboveground values (Jervis 1969; Bernard and Gorham 1978; Gallagher and Plumley 1979; Brinson, Lugo, and Brown 1981).

Productivity in wetlands cannot be attributed solely to emergent herbaceous and woody plants. Submerged, floating, and free-floating aquatic plants, as well as phytoplankton and periphyton, may contribute substantially to marsh community production.

Production of submerged aquatic plants is generally lower than that of emergents (Adams and McCraken 1974; Westlake 1963; Grace and Wetzel 1978; Rich, Wetzel, and Thuy 1971; Washa 1971), presumably because of the nature of limiting factors in the two habitats. Submerged aquatic plants grow in an environment where light is attenuated rapidly with depth and where CO_2 may be scarce as a result of its low diffusion rate in water. On the other hand, productivity of free-floating aquatic plants is as high as or higher than that of emergents (Westlake 1963; Bock 1969; Stephenson et al. 1980).

What adaptations have developed to make possible such high rates of production? Among the morphological adaptations of the emergent plants are the long, erect, linear leaves; this leaf form reduces self-shading and maintains a favorable structure and microclimate for optimum photosynthesis (Dykyjova and Kvet 1976; Gustafson 1976). Ecophysiological studies have also demonstrated physiological adaptations that partially account for the high rates. Some of the emergent genera—for example, _Typha_ and _Sparganium_—have high light saturation levels and high temperature optima for photosynthesis (McNaughton 1973; Guntenspergen, unpublished). Their ability to fix carbon (C) approaches the upper end of the scale for C_3 plants. Plants such as these can take advantage of the optimum light and moisture conditions of their habitat. In contrast, _Spartina alterniflora_ possesses a C_4 (dicarboxylic acid) pathway for CO_2 assimilation. Possession of the C_4 pathway gives these plants the advantage of optimal CO_2 utilization with little CO_2 loss during the day. However, despite the apparent advantages of this pathway, few other macrophytes possess it (Downton 1971; Hough and Wetzel 1977). McNaughton and Fullem (1970) found that the C_3 pathway (Calvin cycle photosynthesis) is very efficient in _Typha latifolia_. It would appear that this efficiency may result from close control over the photorespiratory pathway (the movement of photosynthetic product through glycolate oxidase).

Some submerged aquatic plants appear to reutilize CO_2 respired in roots, rhizomes, and photorespiration. And, under certain conditions, they seem able to utilize the bicarbonate ion as a source of C (Wetzel and Hough 1973; Titus 1977; Hutchinson 1970).

Which primary producers in the community are most important? In discussions of wetland communities, we often neglect that percentage of primary production contributed by algal forms, whether phytoplankton or periphyton. Epiphytic algae have high photosynthetic rates and, because of their high turnover rate, probably high net biomass production.

Data on relative contributions by algae in wetlands are scanty (Allen 1971; Westlake 1963; Pieczynska 1971; Hickman 1971; Haines 1976). Correll (1975) found algal productivity in coastal estuaries

higher than that of submerged plants; together, these two groups
contributed as much as 80% of the wetland production.

What physical factors control productivity in wetlands?
Brinson, Lugo, and Brown (1981) reported that gross primary productivity (GPP) was well correlated with water movement. Moving water
or some function associated with it increased the production of woody
emergents. Westlake (1967) suggested that flowing water increased
productivity in submerged aquatics, probably by renewing materials in
the boundary layer near the leaves. Guntenspergen, Lindsley, and
Stearns (in preparation) suggest that moving water may also provide a
nutrient subsidy for plants growing in otherwise infertile outwash
sands.

Although nutrients may be supplied by moving water, the relative
amounts of different elements may be limiting (Tilman and Kilham
1976; Tilman 1981). Certainly, the anaerobic environment created by
stagnant waters (waterlogging) involves a large respiratory cost for
wetland plants (Hook and Crawford 1978). Even so, many plants appear
to have evolved physical and metabolic adaptations to cope with
anaerobic conditions (Teskey and Hinkley 1977). Other stress
factors, such as wave action or rapid streamflow, may increase the
respiratory load. Only careful, manipulative experiments under
controlled conditions can help us predict the individual species
response to changing environmental gradients.

Absolute rates of organic matter accumulation are not in themselves always important. We should also examine the temporal and
spatial allocation of fixed C. Such biomass distribution diagrams
exist for many aquatic species and communities. A recent EPA publication (Kibby, Gallagher, and Sanville 1980) shows the yearly pattern
of standing crop biomass for a number of wetland plants. These
species each exhibit a different pattern. If we were to compare
different species patterns from specific wetlands, such as McNaughton
Marsh (Lindsley 1977) or Theresa Marsh (Klopatek and Stearns 1978),
we would also find that production peaks for various species occur at
different times throughout the growing season. If we were to examine
entire wetland communities, including submerged, floating, and
emergent plants, a similar distribution would appear (Rich, Wetzel,
and Thuy 1971).

Wetland species vary considerably in the allocation of C
belowground. Root-to-shoot ratios differ among annual, perennial,
and woody plants. Whigham and Simpson (1978) examined the distribution of biomass for 15 freshwater tidal species and found that
allocation patterns in annuals were distinct from those for
perennials. The increased allocation to belowground structures is an
important life history strategy for emergent species. Gustafson
(1976) examined the production, utilization, and storage of carbohydrate reserves in Typha latifolia. He determined that much of the
early season flush of growth is supported by stored carbohydrates.

An awareness of the belowground allocation of carbon and its
fate is crucial to understanding the movement of fixed C between
ecosystem and landscape components. Smith, Good, and Good (1979)
conducted an exacting analysis of the above- and belowground
production dynamics of the short form of Spartina alterniflora.
Belowground biomass was large compared to that aboveground. The
root/shoot ratio was 4.7. The turnover time for this component was
estimated at 5.5 years. The belowground component and its decomposition products acted as a huge sink, with questionable availability

for export from the system.

It should also be remembered (as discussed below) that the net biomass accumulation of a marsh community will change in response to a fluctuating environment. Green Bay freshwater marshes that undergo developmental changes associated with fluctuating water levels in Lake Michigan were examined by Harris and Johnson (1980) and Harris et al. (1981). The contribution of different species to community production changed during the course of the cycle. Van der Valk and Davis (1978) noted drastic fluctuations in wetland plant standing crop resulting from changes in water levels in prairie potholes.

What is the effect of added effluent on freshwater marsh production? The results of several studies on a variety of wetland systems indicate that productivity will increase (Valiela, Teal, and Sass 1975; Guntenspergen, unpublished; Mulligan and Baranowski 1969; Dolan et al. 1981; Ewel and Odum 1979; Paynok 1971; Zoltek et al. 1979; Small 1976). Clearly, effluent addition will eliminate some of the limiting factors for photosynthesis. Nitrogen (N) and phosphorus (P) addition may increase photosynthetic rates and thus net production. We have grown Sparganium eurycarpum and Typha latifolia along nutrient gradients and found significant differences in biomass accumulation between treatments (Guntenspergen, unpublished). Valiela, Teal, and Persson (1976) found that N fertilization reduced root biomass in Spartina alterniflora. Morris (1982) found similar patterns when subjecting Spartina alterniflora to varying levels of inorganic N. Root/shoot ratios ranged from 0.93 to 0.35 with increasing N supply. Similar decreases in root/shoot ratios with increasing fertilization have been documented by Bonnewell (1981) for Typha glauca. Evidently, fertilization may result in reduced root biomass while total production increases, thereby altering the ratio of root-to-shoot biomass. However, Paynok (1971) found significant increases in net aboveground production for three freshwater marsh species in Louisiana, but no significant changes in root/shoot ratios. This result may be because he included all belowground materials present.

We should not expect that nutrients alone will enhance productivity. Other environmental factors may control both productivity and the uptake of nutrients (Jefferies 1977). Whigham, Simpson, and Lee (1980) found no increase in production of a freshwater tidal wetland receiving wastewater. Morris (1980) suggests that productivity gradients in Spartina appear to be correlated with gradients of sulfide concentration and aeration that interfere with or alter the N kinetics, causing N deficiency. Hydrological fluctuations can certainly influence the distribution of various aquatic species as well as their net production and, ultimately, can change the competitive balance between species. Other characteristics of the effluent may affect the productivity of aquatic plants. The uptake of heavy metals and organic materials, although little understood, may well influence metabolic activities in plants.

Wetzel (1979) presented a theoretical model outlining the potential effects of increasing fertility. Different wetland components may react in a variety of ways. These diverse reactions have important implications for wetland succession, as we will indicate later. Phillips, Eminson, and Moss (1978) noted the significance of fertilization for the Norfolk Broads in England, as did Moss (1976) for a set of experimental ponds in Ithaca, New York. A similar model for submerged macrophytes that integrates the effects of various environmental forces has been developed by Davis and Brinson (1980). In general, a progressive increase in the fertilization of submerged macrophyte communities leads to a reduction in their productivity. Concomitantly, production of epiphytic algae increases and then phytoplankton biomass increases. Mulligan, Baranowski, and Johnson (1976) and Mulligan and Baranowski (1969) noted that moderate fertilization increased the growth of certain submerged macrophytes, but higher nutrient levels shifted the competitive balance toward phytoplankton.

C fixation by the component species of wetland communities represents an important organizing process in the structure and function of these communities. We have attempted to show the relative importance of various components of the wetland system and the changes that might be brought about by effluent disposal. As the wetland community reacts differentially to effluent disposal, the pattern of C accumulation also appears to change, and this may have important effects on community organization.

CONSUMPTION

Although it is less well documented than primary production, there is considerable information on certain aspects of wetland food webs, including direct consumption, detritus utilization, and the use of wetland production for nonfood purposes. Much of this work has been reviewed recently, from different viewpoints, by Brinson, Lugo, and Brown (1981), Crow and McDonald (1979), Livingston and Loucks (1979), and Weller (1978, 1981).

Wetland food webs differ in nature, as do the wetlands themselves; each has its set of consumers adapted to the specific water regime and vegetation present. Consumption plays an important role in the transfer of mineral elements within and between systems, and evaluation of the consumptive processes is essential when considering a wetland for effluent disposal.

This discussion considers direct consumption. However, direct consumption of primary production obviously influences the detritus supply and hence the functioning of the ecosystem food web and those of associated ecosystems downstream. Increased detrital processing will also affect nutrient export directly by means of water movement. Thus, detritus is probably the major source for energy and nutrients exported from wetlands, and export is expedited by the activities of invertebrate processors and their predators, both invertebrate and vertebrate.

Primary production also serves nonconsumptive needs by supporting nests and providing cover for nests, dens, and so on. Plant material is often harvested in quantity for habitat construction by muskrats and similar mammals. Vegetation cut or trampled during habitat use often enters the detrital flow earlier than it otherwise might under normal growth patterns, again contributing to more rapid recycling or export of nutrients.

Wetlands support a great variety of invertebrates, including a large array of insects. Although most appear to be detrital consumers, some may have a direct impact on wetland structure and nutrient mass balance. Such effects have been best documented in forested wetlands, perhaps because they are most conspicuous there—but also because species adapted to consumption of emergent and submergent macrophytes appear to be relatively few in number.

In northern swamp forests, the larch sawfly (Pristiphora erichsonii) may completely defoliate stands of tamarack (Larix laricina) within a few weeks (Drooz 1960). The resultant frass or droppings contain large quantities of soluble nutrients. These nutrients may be readily exported from the wetland if flooding occurs or may be adsorbed by the peat. Heavy rains that produce pools in the bog, especially in late summer, may limit the sawfly population, since young sawfly larvae do not survive flooding. In this way, a second year of defoliation that might eliminate the forest overstory may be avoided (Lejeune, Fell, and Burbidge 1955).

Similarly, Conner and Day (1976) noted that insect grazing by the tent caterpillar results in defoliation of tupelo gum (Nyssa aquatica) in cypress/tupelo swamp forests. They estimated that the caterpillars consumed as much as 84 g/m^2 in a tupelo stand from which the cypress had been removed earlier. They suggested that bald cypress may be successful in Louisiana wetlands, in part because it is unpalatable to the insect. Grazing impact on tupelo is somewhat lessened since leafing occurs early in the season and a second crop of leaves can be produced. The larch sawfly and the tent caterpillar occur in epidemic proportions only occasionally but can have substantial short-term effects.

In most wetland communities there is little evidence that grazing insects have any major influence, confirming Teal's (1962) conclusion in reference to salt marshes that direct herbivore consumption is a minor factor and that most primary production enters the detrital pathway. Salt marshes and freshwater wetlands support diverse insect communities. At least in salt marshes, the direct consumers appear to be sap feeders, while leaf strippers and stem borers are of minor importance (Denno 1980; Vince, Valiela, and Teal 1981).

In some salt marshes, grasshoppers may be the most important herbivorous consumers, although their net effect is relatively small when compared to detrital consumption. Parsons and de la Cruz (1980) studied a Louisiana salt marsh dominated by Juncus roemerianus and found that perhaps 0.03% of the net primary production was consumed by grasshoppers feeding on the portion of the leaf between 9.6 cm and 14 cm back from the tip. However, four times as much leaf tissue as that consumed entered the detrital pool prematurely through death of leaf tips and subsequent breakage.

Occasional outbreaks of the cattail borer (Arzome obliqua, a Lepidopteran) may damage cattail stands (Forbes 1954). Again, most of the plant material is not consumed but becomes detritus as a result of stem breakage.

Although there is considerable diversity in the insect fauna, most of the organisms appear to be detritus consumers. Simpson, Whigham, and Brannigan (1979) examined insects of freshwater macrophyte communities and identified representatives of 32 families. However, they found relatively few individuals and noted little grazing on macrophytes, save for Hibiscus palustris. They concluded that

> Because of the low number of herbivorous insects, the macrophytes of the high marsh were little affected by grazing. Thus it appears that most of the vegetation produced in the high marsh enters wetland food chains via detrital pathways, where at

least part of it is available to the larval stages of several
groups, particularly the Dipterans. (Simpson, Whigham, and
Brannigan, 1979, p. 26.)

Other studies (Judd 1949, 1953, 1960; Witter and Croson 1976) have
found freshwater wetland insect fauna to be dominated by Dipteran
taxa. The conclusion one reaches is that insect herbivory is usually
of minor importance in a wide variety of wetland types.

Little is known of direct consumption by invertebrates in most
floating or submergent wetland communities. However, insects are
being recommended for control of certain weedy, exotic, aquatic
species (Spencer 1974). Alligator weed (<u>Alternanthera philoxeroides</u>)
is being controlled by an imported flea-beetle, and several insects
are being tested as controls for water hyacinth (<u>Eichhornia crassipes</u>)
and for Eurasian milfoil (<u>Myriophyllum verticillatum</u>).

Direct consumption of living plant material by fish is limited
to a few fish species. Carp consume <u>Chara</u> as well as submerged
macrophytes (King and Hunt 1967). <u>Tilapia</u> spp. have been suggested
for use as a control for submergent macrophytes. They graze heavily
on <u>Chara</u>, filamentous algae, and rooted pond weeds (Childers and
Bennett 1967). Damage to primary production by certain bottom
feeders can also be appreciable. Carp have the most evident impact
in freshwater wetlands, uprooting submergent and emergent plants and
increasing turbidity, thus reducing light needed by submergent
species.

Recent work suggests that mammalian and avian consumers are
often responsible for heavy consumption of plant material and may
occasionally influence system structure. Most of these consumers
tend to be mobile. Much of the avian activity is seasonal, but, as
with consumption of young <u>Phragmites</u> shoots by geese (Boorman and
Fuller 1981), this impact may come at a critical time in the life of
the plant and so result in a change in wetland species composition.

Forested wetlands often support seasonal concentrations of large
herbivores. Winter concentrations of whitetail deer in wet-mesic,
white cedar swamps may reach $50/km^2$ to $124/km^2$. In this case, much
of the actual feeding occurs in the adjacent uplands, with the result
that nutrients may be imported rather than exported. However, during
periods of deep snow, these animals will consume all available
palatable browse, eliminating reproduction of cedar and thus laying
the groundwork for structural change in the system.

Muskrat and nutria may cause drastic changes with concomitant
effects on nutrient transfer. Boorman and Fuller (1981) report that
coypus (nutria) were responsible for a drastic reduction in the area
of reed swamp in the Norfolk Broads in Great Britain. Although
coypus populations were greatly reduced by cold winters and trapping
on several occasions, they contributed to a decrease in reed swamp
area from 121 ha to 49 ha in 30 years. Associated with the loss of
reed swamp has been a shift in species composition from a mixture of
<u>Phragmites</u> and <u>Typha</u> to stands composed largely of <u>Typha</u>. Increased
eutrophication of the swamps and lakes was also observed. Boorman
and Fuller suggest that grazing by coypus and by waterfowl (swans and
geese) have been at least partially responsible for the nutrient-rich
conditions that result in extensive algal mats. These, in turn,
reduce light and oxygen, eliminating aquatic macrophytes and facili-
tating continued destruction of <u>Phragmites</u>. They calculate that the
peak coypus population of 70,000 may have had an effect equivalent to

that of untreated sewage from a town of from 25,000 to 30,000 people. Other studies (e.g., Kalbe 1969) have suggested similar eutrophication from grazing ducks and geese.

Around 1950, muskrat migrating from Finland successfully colonized shallow lakes and rivers in northern Sweden (Danell 1978a 1978b, 1979). Although populations were relatively low (1/ha to 2/ha) in contrast to those in the United States, muskrat consumption created open water areas in emergent communities that equaled about 1% of the total marsh area. Later, a slightly larger population (5/ha to 6/ha) opened approximately 4% of the marsh. Generally, muskrat-caused openings were small but provided substantial edge useful to waterfowl.

In these Swedish marshes, muskrat diet and house location vary seasonally, depending on water level. As the season progresses and water levels drop, utilization shifts from Carex rostrata and Carex spp. to Equisetum fluvitale and finally to Schoenplectus lacustris. Equisetum is favored for construction material. Consumption of emergents is considerably greater in other areas of Europe, where muskrat population levels may approach those in midwestern North America and where there is greater emergent primary production. Pelikan, Svoboda, and Kvet (1970) documented muskrat populations of 25/ha to 55/ha that utilized 5% to 10% of the production.

Weller (1981) described muskrat effects on midwestern wetlands on the basis of extensive studies and stated that muskrats may "eat out" a marsh. He noted that the fluctuating water levels and heavy consumption of emergents by muskrats combine to reduce muskrat populations periodically and thus begin the cycle again. The possibility of such dynamic shifts in system structure is critical in any consideration of wetlands for effluent treatment.

Grazing and other uses of wetlands by waterfowl may also be important. In many cases, waterfowl use plant production directly as seeds, buds, or shoots of submerged or emergent species. They also utilize insect consumers, detritus feeders, and other marsh organisms extensively. Like muskrat and carp, geese may also uproot plants and increase the opportunity for export (Smith and Odum 1981). Direct use of plant material varies with species and seasonal, developmental stage. Wheeler and March (1979) detailed the use of small Wisconsin wetlands by ducks for feeding, nesting, and brood rearing. Although bird densities in these marshes are usually low, duck populations in some areas can provide a ready supply of mineral elements for growth of algae. Many of these marshes show a conspicuous, seasonal growth of duckweed.

Herbivory is clearly related to increased nutrient levels in the vegetation. An increase in grazing as a result of improved nutritional content is a familiar response in terrestrial systems where it has been documented for ruminants and microtines, as well as for insects, in communities as diverse as shortgrass range, tundra, and conifer plantations.

Considerable evidence has accumulated to demonstrate that insect herbivory in salt marshes is increased by nutrient addition, especially by N (Vince, Valiela, and Teal 1981). In a red mangrove swamp, leaf N content was greatly increased in a roost area occupied by pelicans and egrets (Onuf, Teal, and Valiela 1977). Fertilization greatly stimulated insect herbivory in contrast to areas with lower N levels. There is no reason to suspect that freshwater systems will respond any differently.

Overall, an increase in herbivory will accentuate the rate of mineral release. Under natural conditions, herbivory may occasionally reach dramatic proportions. An increase in the level of herbivory is more likely to occur where nutrient additions are taking place.

However, grazing seems relatively limited in wetland systems, particularly in those consisting of emergent species. One reason may be the presence of secondary plant metabolites, which grazers cannot metabolize (Hutchinson 1975; Su and Staba 1972). The restrictions of the semi-aquatic environment and the dynamics of the system may also help to limit the number of grazers.

In attempting to remove nutrients by harvesting wetland systems, humans act as consumers. Limited studies indicate that, at least for emergent communities, timing of the harvest is critical if the community is to survive (Sloey, Spangler, and Fetter 1978; Nichols and Cottam 1972). The results can resemble those of grazing by geese; for example, early season grazing may kill the damaged plants (Boorman and Fuller 1981).

Harvesting appears to be most effective when applied to floating species. Water hyacinth (Wolverton and McDonald 1979) has been used in Florida and elsewhere to improve the water quality of the discharge from sewage treatment plants. Under natural, undisturbed conditions, production may reach 30 metric tons/ha in 105 days (Wooten and Dodd 1976). When harvested to maintain a low-density population, production in nutrient-rich water reached 154 metric tons/ha in 7 months.

DECOMPOSITION

The C fixed by primary producers in wetlands can (1) serve as an energy source for other trophic levels, (2) accumulate in the sediments, or (3) be transported to other ecosystems. The significance of detritus to the wetland system and its fate depends a great deal on which system is being considered. In peatland systems, organic material accumulates and often alters the local hydrology, leading to the development of unique peatland types (Boelter and Verry 1977; Heinselman 1970). Decomposition products associated with coastal marshes have been hypothesized to increase production of other ecosystems by outwelling of these materials, although this has been questioned (Nixon 1980). In the littoral zone of lakes, decomposition regenerates mineral nutrients rapidly for use in further production (Carpenter 1981). Klopatek (1975) viewed macrophyte production and decay acting as a "nutrient pump," effectively transferring mineral nutrients from the sediments to open water.

Recent reviews of decomposition in freshwater wetlands conclude that although much descriptive work has been accomplished, few if any studies have attempted to examine critically and quantitatively the role of decomposition in the structural or functional framework of the ecosystem (Brinson, Lugo, and Brown 1981; Gallagher 1978; Wetzel 1978).

Our approach is to suggest a conceptual model that describes the fate of atmospherically fixed C. This view emphasizes the various pools and transfers of C and is developed much along the lines of the model used by Gallagher (1978). Live aboveground material senesces and remains upright. Some leaching and translocation occurs. The physical environment largely controls the breakdown of plant material and eventually its transfer to the litter. Wind, water flow velocity and turbulence, and animal activity then influence the process and resulting material.

Once in the litter layer, material from different wetland plants will vary in its relative rate of decay (see Brinson, Lugo, and Brown 1981 for tabular data). Each species presents a different complex of substrates and chemical composition to various decomposers. Northern peatlands generally produce litter that decays slowly. In contrast, Odum and Heywood (1978) found that species of intertidal wetlands had very high decomposition rates. Godshalk and Wetzel (1978a, 1978b) found that decomposition rates could be correlated with total fiber content. Fiber content varied, with Myriophyllum heterophyllum having the least total fiber and Scirpus acutus the most of the five species they studied.

Others have determined fiber and other organic chemical constituents in aquatic plants (Wetzel 1975b; Bernard and Solsky 1977; Klopatek 1974; Boyd 1970; Davis and van der Valk 1978; de la Cruz 1975; Mason and Bryant 1975). Godshalk and Wetzel (1978c) showed that the relative rate of decomposition may be controlled by such factors as temperature, oxygen, mineral nutrients, initial substrate composition, and particle size. In a series of elaborate experiments, they demonstrated that temperature affects the rate of decay directly, while the presence and amount of oxygen greatly influences the completeness and efficiency of the process.

Soluble nutrients, including dissolved C, are typically leached out soon after death of the plant (Otsuki and Wetzel 1974; Nichols and Keeney 1973; Puriveth 1979; Vallentyne 1962). Microbes quickly colonize the material, and an increase in N occurs as microbial biomass increases. This N usually comes from the surrounding environment (de la Cruz and Gabriel 1974; Puriveth 1979). In the later stages, degradation is very slow as a result of the accumulation of resistant materials (Godshalk and Wetzel 1978c). Invertebrates may soon begin to colonize the tissue, chewing and scraping the associated microflora. It has been difficult to determine whether invertebrates are ingesting and assimilating the plant material or the colonizing microflora (Cummins et al. 1980; Barlocher and Kendrick 1975; Hargrave 1970). Gradually, with increased fragmentation, particle size decreases and material is either exported or transferred to the substrate.

Flow rate, turbulence, and water source directly affect the amount of oxygen available for decomposition. Anaerobic conditions are common in the substrate. When restricted to anaerobic respiration, a decomposing system, without a flushing mechanism, begins to accumulate intermediate respiratory compounds, reduced electron acceptors, organic acids, and complex humic compounds. Little is known about the fate of this belowground material. Most decomposition of belowground parts occurs in situ. Thus, large quantities of the net biomass production decay at less than optimal conditions. The substrate can be viewed as a column consisting of organic material in various stages of decomposition and mineral sediments. Smith, Good, and Good (1979) estimated the turnover time of the belowground biomass of Spartina alterniflora at 5.5 years. Few other studies can make such predictions. At the other extreme, the turnaround time of organic material for some northern peatlands is estimated at thousands of years (Chamie and Richardson 1978). There is little chance for export of organic material, except perhaps dissolved organics. However, major disruptions such as storms or floods may result in the export of much detritus from certain wetlands (Nixon 1980).

The production and turnover of organic material and minerals can have significant effects on wetland succession, as well as on the structure and metabolism of wetland ecosystems. Detrital particles and attendant microbial invertebrate colonizers serve as a base for much secondary production in wetlands and adjacent ecosystems (Boyd 1976; Odum 1970; Nixon and Oviatt 1973; Kirby-Smith 1976; Thayer, Wolfe, and Williams 1975; Adams 1976; Turner 1977; Coull 1973).

To link primary production and decomposition products to the detrital food web requires the export or suspension of these materials. When production greatly exceeds decomposition and hydrologic loadings are decreased, sediment accumulates. As Gosselink and Turner (1978) suggest, this may form a feedback loop that influences hydrology even more and results in vegetation changes.

It is important that we understand the relationship between sewage effluent disposal and the decomposition process. It appears that nutrients in the effluent are sorbed by the litter layer and organically based soils (Whigham, Simpson, and Lee 1980; Tilton and Kadlec 1979). This might enhance the quality of detritus, making it more attractive to microbial and invertebrate action (Sompongse and Graetz 1981) and enhancing detrital decomposition.

Coulson and Butterfield (1978) found that the rate of decomposition in blanket bogs was highly correlated with plant N and P content. They separated the effects of N and P and found that an increase in N concentration of plant material increased the decomposition rate. Surprisingly, increased P content of Sphagnum was negatively correlated with invertebrate activity.

The effects of added sewage on decomposition appear clearest in peat bogs. This might be because decomposition rates are limited, either by both the physical conditions of the environment and the refractory nature of the materials produced, or one of the two. In any case, numerous studies suggest that addition of sewage can influence the rate of decomposition.

Ivarson (1977) found that adding lime to sphagnum peat had a significant effect on the bacterial and fungal populations. Changing the pH caused an increase in genera foreign to bog environments; the endemic fungal genera generally favored relatively acidic conditions. Nitrifying bacteria appeared to require a less acidic environment to become active. Lime treatments also tended to increase humification of the peat. Richardson et al. (1976) studied decomposition of sedge and ericaceous plant parts and could find no differences among fertilized treatments and controls. However, they did not examine the peat substrate. Levesque, Jacquin, and Polo (1980) found that the addition of nutrients resulted in pH levels and C:N and C:P ratios that increased decomposition of sphagnum peat.

The nutrient content of wetland plants and the pH of the substrate influence microbial activity and decomposition. Other controlling factors (noted earlier) include temperature, oxygen, heavy metals, and physical forces. Each of these can be partially altered by effluent addition. The functional consequences of this alteration are difficult to determine. Increased decomposition and hydraulic loading may interfere with peat accumulation in palustrine wetlands and influence downstream dynamics in riverine wetlands.

The movement of more rapidly decomposed materials and subsequent decrease (perhaps) in belowground plant detritus may affect any long-term amelioration that a large C pool could provide to an ecosystem. Altered rates of decomposition and sediment accrual could influence developmental sequences in wetlands and, perhaps, also the long-term storage of nutrients in the sediment.

SUCCESSIONAL PROCESSES

A common view of wetland development is shown by the pattern of lake-basin filling (i.e., hydrarch succession). Wetzel (1979) discussed the role of the littoral zone in lake metabolism and the ontogeny of lakes. Essentially, Wetzel and others echo the long-established view (e.g., Pearsall 1920) that the succession of plant communities is positively correlated with decreasing water depth. In this view, change in wetlands is related to productivity and sedimentation and to decomposition processes.

Wetlands in the United States occur in various degrees of development and at various levels of productivity. Where palustrine or lacustrine wetlands have developed since the retreat of the last ice sheet, the rate of development can be estimated. Similarly, riverine and estuarine systems may show overall development toward a semiterrestrial state influenced by many factors, including climatic change, sea level change, and isostatic rebound.

However, Walker (1970) presents evidence from sediment cores showing that succession is not necessarily orderly or progressive. Successional events may be retrogressive. Wetland ecosystems may be stable for long periods or change rapidly under some external influence. Thus, traditional approaches to succession and climax may not apply to wetlands. Indeed, it has been evident for some time that succession in terrestrial ecosystems shows a greater degree of multidimensionality than had originally been thought (or taught) (McIntosh 1980a, 1980b, 1981).

Horn (1976, p. 202) concludes that, "The only sweeping generalization that can safely be made about succession is that it shows a bewildering variety of patterns. Succession may be rapidly convergent on a stationary vegetational composition, or it may be slow and apparently dependent upon accidents of history." This was essentially the conclusion reached by Gleason (1927), again illustrated in brief by McCormick (1968).

Livingston and Loucks (1978) concluded that traditional theories of the relationship of wetland succession and the climax state to diversity, stability, and resilience should be reconsidered. Wetlands are dynamic systems subject to the vagaries of short- and long-term climatic events and associated changes in the hydrologic regime. Internal community development (species change) as well as external impacts affect these systems (Carpenter 1981; McIntosh 1969).

In discussing his model for wetland succession, van der Valk (1981) related the cycle of climatic events affecting water level to species composition mediated by species life-history characteristics. These included seed and propagule longevity and strategies of seedling establishment. He notes that wetlands can hardly be considered in a "stationary state." This follows the view that disturbance is a major structuring force in community development (e.g., Pickett 1980).

Van der Valk does not discuss the substantial effects that autogenic mechanisms might have in the succession process. Carpenter (1981) describes the cyclic relationship between submerged macrophytes, nutrient recycling, and sediment accumulation. Others have discussed the role that competition may play on controlling succession (e.g., Egler 1954; Horn 1974; Diamond 1975). Still another view suggests that plant-animal interactions may be important organizing forces in community development (e.g., McMahon 1981; Danell 1977). During the history of a wetland, changes in community composition may leave evidence in the layers of sediment (Livingston 1975; Jankovská 1978). However, short-term change in species composition is not necessarily representative of the successional process in the long term, although such changes are included by van der Valk in his succession model. Ombrotrophic peat bogs certainly change in the sense of sediment accretion (organic and inorganic).

Wetlands represent succession in several time frames. Over the large scale and the long term, wetlands, even though dynamic and in constant flux, usually undergo change that is directed by increased nutrient supplies and accumulated biomass and/or inorganic sediments. Short-term change results as species populations respond to fluctuating environments (disturbance). The magnitude of the disturbance is important and influences the magnitude of species change. Our view of succession, for the purposes of evaluating the effects of wastewater discharge on wetlands, focuses on the changes in species composition on a site following disturbance.

Examples of wetland development range from the aggrading wetlands of the Mississippi delta (Gosselink, this volume) to the spreading blanket bogs of northern Minnesota (Heinselman 1963) and from the tidal marshes of southern California (Zedler 1980) to the riverine wetlands of Wisconsin (Klopatek 1974) or the prairie potholes of the Midwest (van der Valk 1981). Some systems fit the long-range models of succession in which plant communities slowly accumulate substrate, either through primary production and slow decomposition or by trapping sediment in rivers or lakes. Other systems may be dynamic and change continuously, or may remain stable for long periods.

Bay-mouth and freshwater estuarine wetlands of the Great Lakes show a cyclic pattern of development and destruction as the lake water rises and falls (Stearns and Keough 1982). On Lake Michigan, these lake-level fluctuations have a maximum range of 1.2 m to 1.8 m. At high water, wave action combined with species limitations to specific water depths may eliminate large areas of marsh. When the lake level drops, plant communities again reappear. During low water, the drier marshes are invaded by trees and shrubs and the marsh plants fill the bays. In many respects, this pattern fits the successional model proposed by van der Valk.

Many isolated bogs appear to have followed the classic pattern of hydrarch succession (Boelter and Verry 1977). Peat accumulates to eventually support a wetland forest. These wetlands are also subject to disturbance--by drought, fire, and insect epidemics. Development can be altered and reversed. But, since they have accumulated an organic substrate, their development is partially directed. Other peatlands are composed of a variety of juxtaposed wetland types (Heinselman 1970; Systma and Pippen 1982).

Changes in water level, whether natural or artificial, can influence the development of riverine wetlands, although this seems less frequent in nature. The species of these wetlands are adapted to seasonal, short-term flooding each year. Now the sequence often involves dam construction, producing an impoundment that eliminates the forested wetlands, and is followed by development of marsh communities in shallow water. In some cases, the cycle is repeated with destruction of the dam, development of new wetlands, and renewed dam construction, in each case leading to a different species mix.

Development of lacustrine and estuarine marshes may be related to other shoreline processes and thus show succession from open water, first to deep (intertidal) and later to shallow (high) marsh. Such changes may be influenced by isostatic readjustment, by shoreline current transport of sediment, or by peat accumulation. Such development may occur on any large body of water and is evident on the Great Lakes (Coffin 1979) as well as along the ocean shorelines. Redfield (1972) describes a classic case at Barnstable, Massachusetts, where in the course of 4000 years, a salt marsh and its sand flats and associated channnels have occupied 2146 ha. Of this area, about 1243 ha are high marsh underlain by massive peat deposits. Redfield notes that local sources of sediment are essential for a marsh to expand rapidly. Intertidal marsh will spread as rapidly as water depths permit, while the high marsh develops more slowly.

Consideration of successional change, whether restricted to species composition changes (van der Valk 1981) or involving longer-term alterations or substrate buildup (Redfield 1972), will be conditioned by the hydrologic regime and the relation of the wetland to the surrounding upland. Mitsch et al. (in press) reviewed several models of wetland ecosystems. They conclude from the models that alterations in the hydroperiod can cause severe changes in wetlands in short periods of time.

The time scale and the nature of disruptive events, whether catastrophic or periodic, require consideration in relation to the successional process and to effluent discharge in wetlands. Short-term variations in wetland history are especially critical relative to sewage disposal—and any system subject to periodic destruction is an unlikely candidate for a disposal site.

Effluent discharge to wetlands involves adding both nutrients and water. Additions of either may have a profound effect on the developmental pattern of the wetland and thus on the community structure and function.

The evidence from sediment cores suggests that, in the sequence of postglacial lake filling, there was a period characterized by higher production. This phenomenon was related to the deposition of nutrients and sediments from the upland. Production slowed as the upland stabilized. These allogenic processes, coupled with autogenic ones, may have resulted in shifts in species composition, accompanied by associated changes in lake metabolism. In other situations, sustained high water levels may result in the death of emergent species (Harris and Marshall 1963; Kadlec 1962), a reduction in species diversity, and development of monotypic stands.

Some have argued that the nutrient-processing capability of a wetland is related to the amount of sediments not saturated by minerals. Implicit in this view is that in order for wetlands to continue functioning as nutrient sinks, more sediment must be produced. While rates of organic matter accumulation (and sediment accretion) may increase, the end result will be to affect the various environmental gradients and ultimately change the successional status of the wetland and its functional properties.

A critical feature of the successional process relative to the wetland for effluent disposal lies in the potential for stimulating or accelerating change and, in that way, altering wetland function. The potential for change is always present. In wetlands that appear stable, addition of effluent can greatly accelerate community change. Recognition of the dynamic nature of the wetlands involved and the possibility that a wetland once in use may be drastically altered by climatic or biologic events is basic to any decision to use such a system.

CONCLUSIONS

Although we have attempted throughout this review to find commonalities among processes in different wetland types, we have been only partially successful. The universal presence of water-- although under a variety of hydrologic regimes--helps to generate the diversity of wetlands that we see. Wetland species have adapted in different ways to the variables that force wetland processes, with the result that each wetland requires individual consideration.

Some short-term studies have shown wetlands to be effective in improving the quality of effluent discharges with few adverse effects. However, this evidence is equivocal. In the long term, the use of wetlands for such purposes may be counterproductive. If a wetland changes as a result of effluent addition (i.e., both nutrients and water), the new steady state that is achieved may not meet the desired goal of filtering nutrients and other pollutants and/or maintaining the traditional wetland values.

If wetlands are to be maintained for a variety of uses, we must exercise caution concerning the inputs imposed on them. We have attempted to point out the effects of effluent addition on wetland function. Changes occur, but because of the obvious individuality of wetlands, it is difficult to predict the responses exactly or to generalize about them. Too little is known about these systems to permit realistic judgments to be made concerning the economic and environmental trade-offs. One way to better understand the relation- ship between wetland structure, function, and effluent discharge is to construct artificial systems and model these interactions. Such systems can be easily manipulated and different variables isolated and monitored.

ACKNOWLEDGMENTS

We gratefully acknowledge financial assistance from the National Science Foundation (DEB 791 2516), the U. S. Forest Service, and the University of Wisconsin--Milwaukee, Center for Great Lakes Studies, each of which has supported research that has given us insight into wetland processes.

REFERENCES

Adams, S. M., 1976, Feeding ecology of eelgrass fish communities, Am. Fish. Soc. Trans. 105:514–519.

Adams, M. S., and M. D. McCracken, 1974, Seasonal production of the Myriophyllum component to the littoral of Lake Wingra, Wisconsin, J. Ecol. 62:457–467.

Allen, H. L., 1971, Primary productivity, chemo-organotrophy and nutritional interactions of epiphytic algae and bacteria on macrophytes in the littoral of a lake, Ecol. Monogr. 41:97–127.

Barlocher, F., and B. Kendrick, 1975, Leaf conditioning by micro-organisms, Oecologia 20:359–362.

Bernard, J. M., and E. Gorham, 1978, Primary production in sedge wetlands, in Freshwater wetlands: Ecological processes and management potential, R. E. Good, D. F. Whigham, and R. L. Simpson, eds., Academic Press, New York, pp. 39–51.

Bernard, J. M., and B. H. Solsky, 1977, Nutrient cycling in a Carex lacustris wetland, Can. J. Bot. 55:630–638.

Bock, J. H., 1969, Productivity of the water hyacinth Eichhornia crassipes (Mart.) Solms, Ecology 50:460–464.

Boelter, D. H., and E. S. Verry, 1977, Peatland and water in the northern lake states, gen. tech. rep. NC-31, USDA Forest Service, St. Paul, Minn., 22p.

Bonnewell, V., 1981, Typha productivity, mineral nutrition, and seed germination, Ph. D. dissertation, University of Minnesota.

Boorman, L. A., and R. M. Fuller, 1981, The changing status of reedswamp swamp in the Norfolk Broads, J. Appl. Ecol. 18:241–269.

Boyd, C. E., 1970, Losses of mineral nutrients during decomposition Typha latifolia., Arch. Hydrobiol. 66:511–517.

Boyd, C. M., 1976, Selection of particle size by filter-feeding copepods: A plea for reason, Limnol. Oceanogr. 21:175–180.

Boyt, F. L., S. E. Bayley, and J. Zoltek, Jr., 1977, Removal of nutrients from treated municipal wastewater by wetland vegetation, Water Pollut. Control Fed. J. 49:789–799.

Brinson, M. M., A. E. Lugo, and S. Brown, 1981, Primary productivity, decomposition, and consumer activity in freshwater wetlands, Ann. Rev. Ecol. Syst. 12:123–162.

Carpenter, S. R., 1981, Submerged vegetation: An internal factor in lake ecosystem succession, Am. Nat. 118:372–383.

Chamie, J. P. M., and C. J. Richardson, 1978, Decomposition in northern wetlands, in Freshwater wetlands: Ecological processes and management potential, R. E. Good, D. F. Whigham, and R. L. Simpson, eds., Academic Press, New York, pp. 115–130.

Childers, W. F., and G. W. Bennett, 1967, Experimental vegetation control by largemouth bass-Tilapia combinations, J. Wildlife Manag. 31:401–407.

Coffin, B. A., 1979, Plant distribution in relation to geomorphic processes on Presque Isle Tombolo, Stockton Island, Lake Superior, in Proceedings first conference on scientific research in the national parks, R. M. Linn, ed., New Orleans, La., pp. 115–118.

Conner, W. H., and J. W. Day, Jr., 1976, Productivity and composition of a bald cypress-water tupelo site and a bottomland hardwood site in a Louisiana swamp, Am. J. Bot. 63:1354–1364.

Coull, B. C., 1973, Estuarine meiofauna: A review, in Estuarine microbial ecology, L. H. Stevenson and R. R. Colwell, eds., University of South Carolina Press, Columbia, pp. 499–512.

Correll, D. L., 1975, Estuarine productivity, BioSci. 28:646-650.

Coulson, J. C., and J. Butterfield, 1978, An investigation of the biotic factors determining the rates of plant decomposition on blanket bogs, J. Ecol. 66:631-650.

Cowardin, L. M., V. Carter, F. C. Golet, and E. T. LaRoe, 1979, Classification of wetlands and deepwater habitats of the United States, FWS/OBS-79/31, Biological Sciences Program, U. S. Dept. of Interior, Fish and Wildlife Service, Washington, D. C., 103p.

Crow, J. H., and K. B. MacDonald, 1979, Wetland values: Secondary production, in Wetland functions and values: The state of our understanding, P. E. Greason, J. R. Clark, and J. E. Clark, eds., American Water Resources Association, Minneapolis, Minn., pp. 146-161.

Cummins, K. W., G. L. Spengler, G. M. Ward, R. M. Speaker, R. W. Ovink, D. C. Mahan, and R. L. Mattingly, 1980, Processing of confined and naturally entrained leaf litter in a woodland stream ecosystem, Limnol. Oceanogr. 25:952-957.

Danell, K., 1977, Short-term plant successions following the colonization of a northern Swedish lake by the muskrat, Ondatra zibethica, J. Appl. Ecol. 14:933-947.

Danell, K., 1978a, Food habits of the muskrat, Ondatra zibethica (L.), in a Swedish lake, Ann. Zool. Fennici, 15:177-181.

Danell, K., 1978b, Ecology of the muskrat in northern Sweden, Report from the National Swedish Environment Protection Board, report SNV PM 1043, Stockholm, 157p.

Danell, K., 1979, Reduction of aquatic vegetation following the colonization of a northern Swedish lake by the muskrat, Ondatra zibethica, Oecologia 38:101-106.

Davis, C. B., and A. G. van der Valk, 1978, The decomposition of standing and fallen litter of Typha glauca and Scirpus fluviatilis, Can. J. Bot. 56:662-675.

Davis, G. J., and M. M. Brinson, 1980, Responses of submersed vascular plant communities to environmental change, FWS/OBS-79/330, U. S. Dept. of Interior, Fish and Wildlife Service.

Deghi, G. S., K. C. Ewel, and W. J. Mitsch, 1980, Effects of sewage effluent application on litter fall and litter decomposition in cypress swamps, J. Appl. Ecol. 17:397-408.

de la Cruz, A. A., 1975, Proximate nutritive value changes during decomposition of salt marsh plants, Hydrobiologia 47:475-480.

de la Cruz, A. A., and B. C. Gabriel, 1974, Caloric, elemental, and nutritive changes in decomposing Juncus roemerianus leaves, Ecology 55:882-886.

Denno, R. F., 1980, Ecotypic differentiation in a guild of sap-feeding insects on the salt marsh grass, Spartina patens, Ecology 61:702-714.

Diamond, J. M., 1975, Assembly of species communities, in Ecology and evolution of communities, M. L. Cody and J. M. Diamond, eds., Belknap Press of Harvard University Press, Cambridge, Mass., pp. 342-444.

Dolan, T. J., S. E. Bayley, J. Zoltek, Jr., and A. J. Hermann, 1981, Phosphorus dynamics of a Florida freshwater marsh receiving treated wastewater, J. Appl. Ecol. 18:205-219.

Downton, W. J. S., 1971, Checklist of C_4 species, in Photosynthesis and photorespiration, M. D. Hatch, C. B. Osmond, and R. O. Slatyer, eds., Wiley, New York, pp. 554-558.

Drooz, A. T., 1960, The larch sawfly: Its biology and control, USDA tech. bull. 1212, U. S. Forest Service, St. Paul, Minn., 52p.

Dykyjova, D., and J. Kvet, 1976, Primary productivity of freshwater wetlands, in Proceedings international conference on conservation of wetlands and waterfowl, M. Smart, ed., Heiligenhafe, BDR, Slimbridge, England.

Egler, F. E., 1954, Vegetation science concepts in initial floristics composition, a factor in old-field vegetation development, Vegetatio 4:412-417.

Ewel, K. C., and H. T. Odum, 1979, Cypress domes: Nature's tertiary treatment filter, in Utilization of municipal sewage effluent and sludge on forest and disturbed land, proceedings of a symposium, W. E. Sopper and S. V. Kerr, eds., Pennsylvania State University Press, University Park, pp. 103-114.

Forbes, W. T. M., 1954, Lepidoptera of New York and neighboring states, Noctuidae: Part III, Memoir, Cornell University Agricultural Experiment Station, Ithaca, N. Y., 329p.

Gallagher, J. L., 1978, Decomposition processes: Summary and recommendations, in Freshwater wetlands: Ecological processes and management potential, R. E. Good, D. F. Whigham, and R. L. Simpson, eds., Academic Press, New York, pp. 145-151.

Gallagher, J. L., and F. G. Plumley, 1979, Underground biomass profiles and productivity in Atlantic coastal marshes, Am. J. Bot. 66:156-161.

Gleason, H. A., 1927, Further views on the succession concept, Ecology 8:299-326.

Godshalk, G. L., and R. G. Wetzel, 1978a, Decomposition of aquatic angiosperms: I. Dissolved components, Aquat. Bot. 5:281-300.

Godshalk, G. L., and R. G. Wetzel, 1978b, Decomposition of aquatic angiosperms: II. Particulate components, Aquat. Bot. 5:301-327.

Godshalk, G. L., and R. G. Wetzel, 1978c, Decomposition of aquatic angiosperms: III, Zostera marina L. and a conceptual model of decomposition, Aquat. Bot. 5:329-354.

Good, R. E., D. F. Whigham, and R. L. Simpson, eds., 1978, Freshwater wetlands: Ecological processes and management potential, Academic Press, New York, 378p.

Gosselink, J. G., and R. E. Turner, 1978, The role of hydrology in freshwater wetland ecosystems, in Freshwater wetlands: Ecological processes and management potential, R. E. Good, D. F. Whigham, and R. L. Simpson, eds., Academic Press, New York, pp. 63-78.

Grace, J. B., and R. G. Wetzel, 1978, The production biology of Eurasian Watermilfoil (Myriophyllum spicatum L.): A review, J. Aquat. Plant Mgt. 16:1-10.

Guntenspergen, G. R., W. Kappel, and F. Stearns, 1980, Response of a bog to application of lagoon sewage: The Drummond Project--an operational trial, in Proceedings of the 6th International Peat Congress, August 17-23, 1980, Duluth, Minn., pp. 559-561.

Guntenspergen, G. R., D. Lindsley, and F. Stearns, in preparation, Wetland plant production on nutrient poor outwash sands.

Guntenspergen, G. R., and F. Stearns, 1981, Ecological limitations on wetland use for wastewater treatment, in Wetland values and management, B. Richardson, ed., Minnesota Water Planning Board, St. Paul, Minn., pp. 273-284.

Gustafson, T. D., 1976, Production, photosynthesis, and the storage and utilization of reserves in a natural stand of Typha latifolia L., Ph. D. dissertation, University of Wisconsin, Madison.

Haines, E. B., 1976, Stable carbon isotope rations in the biota, soils, and tidal water of a Georgia salt marsh, Estuar. Coastal Mar. Sci. 4:609-616.

Hargrave, R. T., 1970, The utilization of benthic microflora by Hyalella Azteca (Amphipoda), J. Animal Ecol. 39:427-437.

Harris, H. J., and W. J. Johnson, 1980, Biological production in Green Bay coastal marshes, Sea Grant progress report R/GB-6, University of Wisconsin.

Harris, H. J., G. Fewless, M. Milligan, and W. Johnson, 1981, Recovery processes and habitat quality in a freshwater coastal marsh following a natural disturbance, in Wetland values and management, B. Richardson, ed., Minnesota Water Planning Board, St. Paul, Minn., pp. 363-379.

Harris, S. W., and W. H. Marshall, 1963, Ecology of water-level manipulations on a northern marsh, Ecology 44:331-343.

Heinselman, M. L., 1963, Forest sites, bog processes, and peatland types in the Lake Agassiz region, Minnesota, Ecol. Monogr. 33:327-372.

Heinselman, M. L., 1970, Landscape evolution, peatland types, and the environment in the Agassiz Peatlands Natural Area, Minnesota, Ecol. Monogr. 40:235-261.

Hickman, M., 1971, The standing crop and primary productivity of the eiphyton attached to Equisetum fluviatile L. in Priddy Pool, North Somerset, J. Br. Phycol. 6:51-59.

Hook, D. D., and R. N. M. Crawford, eds., 1978, Plant life in anaerobic environments, Ann Arbor Science, Ann Arbor, Mich., 564p.

Horn, H. S., 1974, The ecology of secondary succession, Ann. Rev. Ecol. Syst. 5:25-37.

Horn, H. S., 1976, Succession, in Theoretical ecology, R. M. May, ed., W. B. Saunders Co., Philadelphia, Pa., pp. 187-204.

Hough, R. A., and R. G. Wetzel, 1977, Photosynthetic pathways of some aquatic plants, Aquat. Bot. 3:297-313.

Hutchinson, G. E., 1970, The chemical ecology of three species of Myriophyllum (Angiospermae, Haloragaceae), Limnol. Oceanogr. 15:1-5.

Hutchinson, G. E., 1975, Limnological Botany, Academic Press, New York, 660p.

Ivarson, K. C., 1977, Changes in decomposition rate, microbial population and carbohydrate content of an acid peat bog after liming and reclamation, Can. J. Soil Sci. 57:129-137.

Jankovská, V., 1978, Development of wetland and aquatic vegetation in the Trebon basin since the late glacial period, in Pond littoral ecosystems structure and functioning, D. Dykyjova and J. Kvet, eds., Springer-Verlag, Berlin, pp. 88-92.

Jeffries, R. L., 1977, Growth responses of coastal halophytes to inorganic nitrogen, J. Ecol. 65:847-865.

Jervis, R. A., 1969, Primary production in the freshwater marsh of Troy Meadow, N. J., Torrey Bot. Club Bull. 96:209-231.

Judd, W. W., 1949, Insects collected in the Dundas Marsh, Hamilton, Ontario, 1946-1947, with observations on their periods of emergence, Can. Entomol. 81:1-10.

Judd, W. W., 1953, A study of the population of insects emerging as adults from the Dundas Marsh, Hamilton, Ontario, during 1948, Am. Midl. Nat. 49:801-824.

Judd, W. W., 1960, Studies of the Byron Bog in southwestern Ontario: XI, Seasonal distribution of adult insects in the Chamaedaphetum calyculatae association, Can. Entomol. 92:241–251.

Kadlec, J. A., 1962, The effects of a drawdown on a waterfowl impoundment, Ecology 43:267–281.

Kalbe, L., 1969, Uber die Auswirkungen von Hausentenhaltungen auf die Wasservogelwelt, Beitr. zur Vogelkunde 14:225–230.

Keefe, C. W., 1972, Marsh production: A summary of the literature, Contr. Mar. Sci. 16:163–181.

Kibby, H. V., J. L. Gallagher, and W. D. Sanville, 1980, Field guide to evaluate net primary production of wetlands, EPA-600/8-80-037, U. S. EPA, Environmental Research Laboratory, Corvallis, Ore., 59p.

King, D. R., and G. S. Hunt, 1967, The effect of carp on vegetation in a Lake Erie marsh, J. Wildl. Manage. 31:181–188.

Kirby-Smith, W. W., 1976, The detritus problem and the feeding and digestion of an estuarine organism, in Estuarine processes, vol. 1, M. Wiley, ed., Academic Press, New York, pp. 469–479.

Klopatek, J. M., 1974, Production of emergent macrophytes and their role in mineral cycling within a freshwater marsh, M. S. thesis, University of Wisconsin-Milwaukee.

Klopatek, J. M., 1975, The role of emergent macrophytes in mineral cycling in a freshwater marsh, in Mineral cycling in southeastern ecosysems, F. G. Howell, J. B. Gentry, and M. H. Smith, eds., ERDA conf. 740513, Technical Information Center, Office of Public Affairs, U. S. ERDA, Springfield, Va., pp. 367–392.

Klopatek, J. M., and F. Stearns, 1978, Primary productivity of emergent macrophytes in a Wisconsin freshwater marsh ecosystem, Am. Midl. Nat. 100:320–332.

Larson, J. S., and O. L. Loucks, eds., 1978, Research priorities for wetland ecosystem analysis, report to the National Science Foundation, National Wetlands Technical Council, Washington, D. C., 68p.

Lejeune, R. R., W. H. Fell, and D. P. Burbidge, 1955, The effect of flooding on development and survival of the larch sawfly, Pristiphora Erichsonii (Tenthredinidae), Ecology 36:63–70.

Levesque, M., F. Jacquin, and A. Polo, 1980, Comparative biodegradability of a sphagnum and sedge peat from France, in Proceedings 6th International Peat Congress, August 17-23, 1980, Duluth, Minn., pp. 584–590.

Lindsley, D. S., 1977, Emergent macrophytes of a Wisconsin marsh: Productivity, soil-plant nutrient regimes and uptake experiments with phosphorus 32, Ph. D. dissertation, University of Wisconsin-Milwaukee.

Livingston, D. A., 1975, Late quaternary climatic change in Africa, Ann. Rev. Ecol. Syst. 6:249–280.

Livingston, R. J., and O. Loucks, 1979, Productivity, trophic interactions, and food-web relationships in wetlands and associated systems, in Wetland functions and values: The state of our understanding, P. E. Greason, J. R. Clark, and J. E. Clark, eds., American Water Resources Association, Minneapolis, Minn., pp. 101–119.

Mason, C. F., and R. J. Bryant, 1975, Production, nutrient content and decomposition of Phragmites communis Trin. and Typha angustifolia L., J. Ecol. 63:71–95.

McCormick, J., 1968, Succession, Via 1:1–16.

McIntosh, R. P., 1969, Ecological succession, Science 166:403–404.

McIntosh, R. P., 1980a, The relationship between succession and the recovery process in ecosystems, in The recovery process in damaged ecosystems, J. Cairns, ed., Ann Arbor Science Publications, Ann Arbor, Mich., pp. 11–62.

McIntosh, R. P., 1980b, The background and some current problems of theoretical ecology, Synthese 43:195–255.

McIntosh, R. P., 1981, Succession and ecological theory, in Forest succession concepts and application, D. C. West, H. H. Shugart and D. B. Botkin, eds., Springer-Verlag, New York, pp. 10–23.

McMahon, J. A., 1981, Successional processes: Comparisons among biomes with special reference to probable roles of and influences on animals, in Forest succession concepts and application, D. C. West, H. H. Shugart, and D. B. Botkin, eds., Springer-Verlag, New York, pp. 277–304.

McNaughton, S. J., 1973, Comparative photosynthesis of Quebec and California ecotypes of Typha latifolia, Ecology 54:1260–1270.

McNaughton, S. J., and L. W. Fullem, 1970, Photosynthesis and photorespiration in Typha latifolia, Plant Physiol. 45:703–707.

Mitsch, W. J., J. W. Day, Jr., J. R. Taylor, and C. Madden, 1982, Models of North American freshwater wetlands--a review, Int. J. Ecol. Envir. Sci. 8:109–140.

Morris, J. T., 1980, The nitrogen uptake kinetics of Spartina alterniflora in culture, Ecology 61:1114–1121.

Morris, J. T., 1982, A model of growth responses by Spartina alterniflora to nitrogen limitation, J. Ecol. 70:25–42.

Moss, B., 1976, The effects of fertilization and fish on community structure and biomass of aquatic macrophytes and epiphytic algal populations: An ecosystem experiment, J. Ecol. 64:313–342.

Mudroch, A., and J. A. Capobianco, 1979, Effects of treated effluent on a natural marsh, Water Pollut. Control Fed. J. 51:2243–2256.

Mulligan, H. F., and A. Baranowski, 1969, Growth of phytoplankton and vascular aquatic plants at different nutrient levels, Int. Verein. Theor. Limnol. 17:802–810.

Mulligan, H. F., A. Baranowski, and P. Johnson, 1976, Nitrogen and phosphorus fertilization of aquatic vascular plants and algae in replicated ponds: Initial response to fertilization, Hydrobiologia 48:109–116.

Nichols, D. S., and D. R. Keeney, 1973, Nitrogen and phosphorus release from decaying water milfoil, Hydrobiologia 42:509–525.

Nichols, S., and G. Cottam, 1972, Harvesting as a control for aquatic plants, Water Res. Bull. 8:1205–1210.

Nixon, S. W., 1980, Between coastal marshes and coastal waters--a review of twenty years of speculation and research on the role of salt marshes in estuarine productivity and water chemistry, in Estuarine and wetland processes with emphasis on modeling, P. Hamilton and K. B. MacDonald, eds., Plenum Press, New York, pp. 437–526.

Nixon, S. W., and C. A. Oviatt, 1973, Ecology of a New England salt marsh, Ecol. Monogr. 43:463–498.

Odum, W. E., 1970, Utilization of the direct grazing and plant detritus food chains by striped mullet, Mugil cephalus, in Marine food chains, J. H. Steele, ed., University of California Press, Berkeley, pp. 222–240.

Odum, W. E., and M. A. Heywood, 1978, Decomposition of intertidal freshwater marsh plants, in Freshwater wetlands: Ecological processes and management potential, R. E. Good,

D. F. Whigham, and R. L. Simpson, eds., Academic Press, New York, pp. 89–98.

Onuf, C. P., J. M. Teal, and I. Valiela, 1977, Interactions of nutrients, plant growth and herbivory in a mangrove ecosystem, Ecology 58:514–526.

Otsuki, A., and R. G. Wetzel, 1974, Release of dissolved organic matter by autolysis of a submerged macrophyte, Scirpus subterminalis, Limnol. Oceanogr. 19:842–845.

Parsons, K. A., and A. A. de la Cruz, 1980, Energy flow and grazing behavior of conocephaline grasshoppers in a Juncus roemerianus marsh, Ecology 61:1045–1050.

Paynok, P. I., 1971, The response of three species of marsh macrophytes to artificial enrichment at Dulac, Louisiana, M. S. thesis, Louisiana State University, Dulac.

Pearsall, W. H., 1920, The aquatic vegetation of the English lakes, J. Ecol. 8:163–201.

Pelikan, J., J. Svoboda, and J. Kvet, 1970, On some relations between the production of Typha latifolia and a muskrat population, Zool. Listy 19:303–320.

Phillips, G. L., D. Eminson, and B. Moss, 1978, A mechanism to account for macrophyte decline in progressively eutrophicated freshwaters, Aquat. Bot. 4:103–126.

Pickett, S. T. A., 1980, Non-equilibrium coexistence of plants, Torrey Bot. Club Bull. 107:238–248.

Pieczynska, E., 1971, Mass appearance of algae in the littoral of several Mazurian lakes, Mitt. Int. Ver. Limnd. 19:59–69.

Puriveth, P., 1979, Decomposition of emergent macrophytes in a Wisconsin marsh, Hydrobiologia 72:231–242.

Redfield, A. C., 1972, Development of a New England salt marsh, Ecol. Monogr. 42:201–237.

Rich, P. H., R. G. Wetzel, and N. V. Thuy, 1971, Distribution, production, and role of aquatic macrophytes in a southern Michigan Marl Lake, Freshwater Biol. 1:3–21.

Richardson, C. J., 1979, Primary productivity values in freshwater wetlands, in Wetland functions and values: The state of our understanding, P. E. Greason, J. R. Clark, and J. E. Clark, eds., American Water Resources Association, Minneapolis, Minn., pp. 131–145.

Richardson, C. J., W. A. Wentz, J. P. N. Chamie, J. A. Kadlec, and D. L. Tilton, 1976, Plant growth, nutrient accumulation, and decomposition in a central Michigan peatland used for effluent treatment, in Freshwater wetlands and sewage effluent disposal, D. L. Tilton, R. H. Kadlec, and C. J. Richardson, eds., University of Michigan, Ann Arbor, pp. 77–118.

Schelske, C. L., 1980, Dynamics of nutrient enrichment in large lakes: The Lake Michigan case, in Proceedings of a symposium on the restoration of lakes and inland waters, U. S. EPA.

Simpson, R. L., D. F. Whigham, and K. Brannigan, 1979, The mid-summer insect communities of freshwater tidal wetland macrophytes, Delaware River estuary, New Jersey, N. J. Acad. Sci. Bull. 24(1):22–28.

Sloey, W. E., F. L. Spangler, and C. W. Fetter, Jr., 1978, Management of freshwater wetlands for nutrient assimilation, in Freshwater wetlands: Ecological processes and management potential, R. E. Good, D. F. Whigham, and R. L. Simpson, eds., Academic Press, New York, pp. 321–340.

Small, M. M., 1976, Marsh/pond sewage treatment plants, in Proceedings of a national symposium on freshwater wetlands and sewage effluent disposal, D. L. Tilton, R. H. Kadlec, and C. J. Richardson, eds., University of Michigan, Ann Arbor, pp. 197-214.

Smith, K. K., R. E. Good, and N. F. Good, 1979, Production dynamics for above- and belowground components of a New Jersey Spartina alterniflora tidal marsh, Estuar. Coastal Mar. Sci. 9:189-200.

Smith, T. J., III, and W. E. Odum, 1981, The effects of grazing by snow geese on coastal salt marshes, Ecology 62:98-106.

Sompongse, D., and D. A. Graetz, 1981, Nutrient availability from decaying vegetation in wetland ecosystems, in Progress in wetlands utilization and management, proceedings of a symposium, P. M. McCaffrey, T. Bemmer, and S. E. Gatewood, eds., Coordinating Council on the Restoration of the Kissimmee River Valley and Taylor Creek-Nubbin Slough Basin, Tallahassee, Fla., pp. 243-254.

Spencer, N. R., 1974, Insect enemies of aquatic weeds, PANS 20(4):444-450.

Stearns, F., and J. R. Keough, 1982, Pattern and function in the Mink River watershed with management alternatives, report to the Wisconsin Coastal Zone Management Program.

Stephenson, M., G. Turner, P. Pope, J. Colt, A. Knight, and G. Tchobanoglous, 1980, The use and potential of aquatic species for wastewater treatment: Appendix A. The environmental requirements of aquatic plants, publ. 65, California State Water Resources Control Board, Sacramento, 655p.

Su, K. L., and E. J. Staba, 1972, Aquatic plants from Minnesota. Part 1—chemical survey, Water Resources Research Center, University of Minnesota, Minneapolis, 50p.

Systma, K. J., and R. W. Pippen, 1982, The Hampton Creek wetland complex in southwestern Michigan: IV. Fen succession, Mich. Bot. 21:105-115.

Teal, J. M., 1962, Energy flow in the salt marsh ecosystems of Georgia, Ecology 43:614-624.

Teskey, R. O., and T. M. Hinckley, 1977, Impact of water level changes on woody riparian and wetland communities. Vol. 1. Plant and soil responses to flooding, FWS/OBS-77/158, Biological Services Program, U. S. Fish and Wildlife Service.

Thayer, G. W., D. A. Wolfe, and R. B. Williams, 1975, The impact of man on seagrass systems, Am. Sci. 63:288-296.

Tilman, D., 1981, Tests of resource competition theory using four species of Lake Michigan algae, Ecology 62:802-815.

Tilman, D., and S. Kilham, 1976, Phosphate and silica growth and uptake kinetics of the diatoms Asterionella formosa and Cyclotella meneghiniana in batch and semicontinuous culture, J. Phycol. 12:375-383.

Tilton, D. L., and R. H. Kadlec, 1979, The utilization of a freshwater wetland for nutrient removal from secondarily treated wastewater effluent, J. Envir. Qual. 8:328-334.

Titus, J. E., 1977, The comparative physiological ecology of three submersed macrophytes, Ph. D. dissertation, University of Wisconsin, Madison.

Turner, R. E., 1976, Geographic variations in salt marsh macrophyte production: A review, Contr. Mar. Sci. 20:47-68.

Turner, R. E., 1977, Intertidal vegetation and commercial yields of penaeid shrimp, Am. Fisheries Soc. Trans. 106:411-416.

van der Valk, A. G., 1981, Succession in wetlands: A Gleasonian approach, Ecology 62:688-696.
van der Valk, A. G., and C. B. Davis, 1978, Primary production of prairie glacial marshes, in Freshwater wetlands: Ecological processes and management potential, R. E. Good, D. F. Whigham, and R. L. Simpson, eds., Academic Press, New York, pp. 21-37.
Valiela, I., J. M. Teal, and N. Y. Persson, 1976, Production and dynamics of experimentally enriched salt marsh vegetation: Belowground biomass, Limnol. Oceanogr. 21:245-252.
Valiela, I., J. M. Teal, and W. J. Sass, 1975, Production and dynamics of salt marsh vegetation and the effects of experimental treatment with sewage sludge, J. Appl. Ecol. 12:973-982.
Vallentyne, J. R., Jr., 1962, Solubility and the decomposition of organic matter in nature, Arch. Hydrobiol. 58:423-434.
Vince, S. W., I. Valiela, and J. M. Teal, 1981, An experimental study of herbivorous insect communities in a salt marsh, Ecology 62:1662-1678.
Walker, D., 1970, Direction and rate in some British postglacial hydroseres, in Studies in the vegetational history of the British Isles, D. Walker and R. G. West, eds., Cambridge University Press, Cambridge, England, pp. 117-139.
Washa, A. J., 1971, The seasonal variation, standing crop and primary productivity of submerged aquatic macrophytes in Theresa Marsh, M. S. thesis, University of Wisconsin-Milwaukee.
Weller, M. W., 1978, Management of freshwater marshes for wildlife, in Freshwater wetlands: Ecological processes and management potential, R. E. Good, D. F. Whigham and R. L. Simpson, eds., Academic Press, New York, pp. 267-284.
Weller, M. W., 1981, Freshwater marshes: Ecology and wildlife management, University of Minnesota Press, Minneapolis, 146p.
Westlake, D. F., 1963, Comparisons of plant productivity, Biol. Rev. 38:385-425.
Westlake, D. F., 1967, Some effects of low-velocity currents on the metabolism of aquatic macrophytes, J. Exp. Bot. 18:187-205.
Westlake, D. F., 1975, Primary production of freshwater macrophytes in Photosynthesis and productivity in different environments, J. P. Cooper, ed., Cambridge University Press, Cambridge, England, pp. 189-206.
Wetzel, R. G., 1975a, Primary production, in River Ecology, B. A. Whitton, ed., Cambridge University Press, Cambridge, England, pp. 230-247.
Wetzel, R. G., 1975b, Limnology, W. B. Saunders, Philadelphia, 743p.
Wetzel, R. G., 1978, Foreword and introduction, in Freshwater wetlands: Ecological processes and management potential, R. E. Good, D. F. Whigham, and R. L. Simpson, eds., Academic Press, New York, pp. xiii-xvii.
Wetzel, R. G., 1979, The role of the littoral zone and detritus in lake metabolism, Arch. Hydrobiol. Beih. Ergebn. Limnol. 13:145-161.
Wetzel, R. G. and R. A. Hough, 1973, Productivity and the role of aquatic macrophytes in lakes: An assessment, Pol. Arch. Hydrobiol. 20:9-19.
Wheeler, W. E., and J. R. March, 1979, Characteristics of scattered wetlands in relation to duck production in southeastern Wisconsin, tech. bull. no. 116, Wisconsin Department of Natural Resources, Madison, 61p.

Whigham, D. F., and R. L. Simpson, 1978, The relationship between
    above-ground and below-ground biomass of freshwater tidal wetland
    macrophytes, Aquat. Bot. 5:355-364.
Whigham, D. F., R. L. Simpson, and K. Lee, 1980, The effects of
    sewage effluent on the structure and function of a freshwater
    tidal marsh ecosystem, Water Resources Research Institute, Rutgers
    University, New Brunswick, N. J., 106p.
Whittaker, R. H., and G. E. Likens, 1975, The biosphere and man, in
    Primary productivity of the biosphere, H. Leith and R. H.
    Whittaker, eds., Springer-Verlag, New York, pp. 305-328.
Witter, J. A. and S. Croson, 1976, Insects and wetlands, in Fresh-
    water wetlands and sewage effluent disposal, D. L. Tilton, R. H.
    Kadlec, and C. J. Richardson, eds., University of Michigan, Ann
    Arbor, pp. 271-295.
Wolverton, B. C., and R. C. McDonald, 1979, Water hyacinth (Eich-
    hornia crassipes) productivity and harvesting studies, Econ. Bot.
    33:1-10.
Wooten, J. W., and J. D. Dodd, 1976, Growth of water hyacinths in
    treated sewage effluent, Econ. Bot. 30:29-37.
Zedler, J. B., 1980, Salt marsh productivity with natural and altered
    tidal circulation, Oecologia 44:236-240.
Zoltek, J., Jr., S. E. Bayley, A. J. Hermann, A. R. Tortora, and
    T. J. Dolan, 1979, Removal of nutrients from treated municipal
    wastewater by freshwater marshes, final report to the City of
    Clermont, Florida, Center for Wetlands, University of Florida,
    Gainesville.

DISCUSSION

Kadlec: What's happened to Drummond Bog lately?

Guntenspergen: The Drummond Bog that Dr. Kadlec is referring to is
an ombrotrophic [sphagnum] bog in northern Wisconsin. Most of the
natural nutrient loading into the system is due to precipitation.
(There is no contact with the regional groundwater table.) For the
last several years, sewage effluent has been discharged to this
system, and Forest and I, in cooperation with several others, have
been investigating the effects of the sewage effluent on the plant
community. Our original interest was to see if additional phosphorus
and other nutrient loadings would alter bog structure and also its
functioning.
    Just to summarize briefly, we haven't really seen much of an
increase in the productivity of the herbaceous and shrub vegetation
in the bog after the first two to three years of effluent discharge.
Part of the reason is that the nutrient loading to the bog is low in
comparison with normal municipal wastewater. The lagoons where the
sewage is partially treated were overdesigned and partially filled
with lake water, so any sewage was initially diluted before being
discharged to the system. We are only talking about an additional
loading from the effluent to the bog of perhaps 1 kg to 2 kg of
phosphorus (for 1979 and probably 1980--the years for which we have
data readily at hand) plus perhaps an additonal 8 kg of total
nitrogen. The things that we have seen in the Drummond Bog are
changes in plant species composition. The changes in species composition result, we believe, from characteristics of the wastewater

other than nutrients--that is, nitrogen and phosphorus. Ombrotrophic bogs like this have a pH somewhere around 3.5 to 4. The wastewater additions have a pH up around 8, which is very alkaline compared to these acidic bogs. We believe that it is the high pH and high alkalinity of the effluent that causes the disappearance of the sphagnum mosses and the other species that are adapted to the bog environment and that really contribute most to the structure of the bog. It's the sphagnum mosses that cause, in part, the large buildup of peat and isolate the system from the groundwater table. So the sphagnum seems to have a tremendous influence on the functioning and structural characteristics of the bog. Currently, there is an invasion by nontypical bog species into the areas where the sphagnum has disappeared. These species are more characteristic of the lagg area. The lagg is much higher in nutrients than the bog proper.

Odum: Do you get cattails?

Guntenspergen: We initially saw some cattails. They do not comprise a large percentage of the species that are invading. We are finding other herbaceous species, such as bidens and northern willow herb, among others. One of the reasons that we haven't seen a lot of cattail is that initially when it started to appear, someone was taking it out. When we emphasized the experimental nature of the system to everyone concerned, we did not have any more problems.

Stearns: Also, it may partly have been a little bit too shady. I expect to see more cattails.

Jackson: One more question on the bog. Do you consider the invasion of these other plant species desirable or undesirable?

Guntenspergen: That depends on what set of values you want to base that decision on. If you're talking about . . .

Stearns: I think that we would say it's an indication that there are changes in the functioning of the bog. And from a scientific point of view, that is a valuable result. We pointed out to the Forest Service in our most recent report that they will eventually have to decide whether they want a spruce/tamarack/sphagnum bog or one of a considerably different nature.

Ewel: It sounds as though this ombrotrophic bog is becoming a minerotrophic bog. Are these other . . .

Stearns: Well, the effluent that's coming in is certainly contributing.

Ewel: From the species that you described before it sounds as though this is something we've seen before, ecologically. It's not a completely different ecosystem.

Stearns: What do you mean by "completely different?"

Ewel: It's something that, if we didn't know the effluent was going in there, we might be able to say, this looks like a minerotrophic bog, rather than something we had never seen before.

Stearns: Yes, particularly within ten feet of the pipeline on the downflow side . . . [laughter].

Bastian: In the concept of an artificial wetland you attempt to even out these variations in natural fluctuations. Does that bother you? Does that seem reasonable to you? The engineer is interested in trying to control and maintain something in a more static state than what you see in your natural wetland area.

Stearns: No, it doesn't bother me. I think that the important thing--the reason we've studied this bog--was to get people to recognize bog dynamics and how it influences the various functions of the system that in turn, influence nutrient retention and nutrient absorption. Two of the marshes we've worked in--the two riverine marshes--are partially controlled by water level. We see changes resulting from flood flows, and so forth, but the systems are more static. Even so, as this chap told me one time down in southern Indiana, looking at this big new dam that had just been built and measuring the silt content of the water, "You're going to have a wetland there soon." That is, streams are not completely static no matter what you do. Unless you start dredging.

Bastian: I just wondered if you see the form and function of a wetland when you attempt to put more control and strain on its natural fluctuations. Do you see that as a means of really getting to the end, or that a more natural system with the built-in fluctuations is a benefit in the long run over something where you want to play more of an engineering control role, as far as pollution control and associated secondary benefits are concerned?

Stearns: That question is more complicated than I would want to answer at this point, but I was going to suggest that intersystem transfers probably deserve more attention than they have received. Even in your artificial wetland, as you monitor it, you may find changes occurring in the outflow. Actually, there were changes in the outflow of Drummond Bog, depending upon the amount of rainfall, and also an infestation by the larch sawfly. So these intersystem transfers are something you want to consider. In riverine wetlands, for instance, maybe in estuarine wetlands, too, in the long term we're essentially using the established principle of dilution. The nutrients eventually move downstream, sometimes to the advantage of the downstream recipients and sometimes not. If you wanted to get a little argument going, we could say categorically that you shouldn't apply effluent to any wetland that's not hydrologically isolated from other regional water bodies, either surface or groundwater. That would be an interesting restriction . . . [laughter]. It brings to mind one of the first experiments in the use of a wetland for sewage disposal that I know about. The perpetrator, who is a long-since retired sanitary engineer, decided that septic systems were by nature inefficient, and so he began using aerated systems by digging a hole in the back of his cottage and pumping air into it. It works fine, and it has cattails.

6

Water Conservation and Wetland Values

Howard T. Odum

It might be said that the hydrologic cycle is the adenosine triphosphate (ATP) of the biosphere because of its role in organizing and driving ecosystems. Wetlands are one manifestation of landscape organization by water. While photosynthetic productivity is some-times used as a basis for evaluating ecosystems, other kinds of work involving the hydrologic cycle may be more important in wetlands. What do wetlands do with water, and how is this role related to energy and to regional value? The answers to these questions may affect the way we manage wetlands, preserve them, or generate new ones in making a better mosaic of humanity and nature.

HIERARCHY OF WATER CONVERGENCE

Solar energy's work on the atmosphere, oceans, and land converges on land through the hydrological cycle. In fact, the principal path by which oceanic solar energy reaches a landscape such as Florida may be through rain (Fig. 6.1). In Florida, water falling on uplands gradually drains and converges to form wetlands (Fig. 6.2), then strands, and, finally, floodplain streams (Odum 1978, 1982b). The process of hydrologic convergence uses energy and water through the increasing organizational development of diverse and stable abiotic and biotic systems. Thus, energy convergence via the hydro-logic cycle increases its overall quality.

PRODUCTIVITY AND TRANSPIRATION ACCORDING TO POSITION IN WETLAND SERIES

The amount of nutrients available to the wetlands in Figure 6.2 depends on the amount of water passing through wetlands and its prior trajectory over the land. Since the principal source of water for upland headwater swamps is rain, they are low in nutrients. Net productivity is slight and based on recycling of nutrients. Conse-quently, transpiration can be reduced because stomata need not be open as much. These properties were observed by Brown (1978, 1981) and are suggested in the series of transpiration rates from headwater

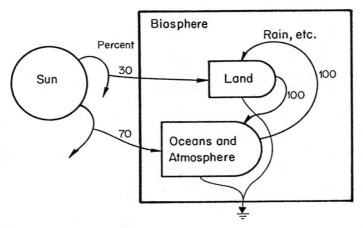

Figure 6.1
Pathways of global solar energy developing energy
flows to land. About 70% of the global solar energy
supporting land areas is indirect, being embodied in
the contributions of rain, wind, and so on, from the
ocean.

(bays) to floodplain (Table 6.1).
　　With less transpiration, leaf temperatures tend to be higher,
and increased reflectance is required to maintain a reasonable
balance. Capehart et al. (1977) and Heimburg and Vickers (1980)
observed increased reflectance for cypress relative to other vegetation. The thickened evergreen leaves may be adapted to process
less water. In the lowest nutrient situations typical of bays and
bogs (Fig. 6.3) which are supplied mainly by rainwater, evergreenness
may also be a nutrient conservation adaptation (Monk 1966). With
slightly more drainage, and consequently greater fluctuation in water
availability, the trees tend to be deciduous (cypress and gums),
thereby saving water in the dry season. Heimburg (1976) showed that
much less water evapotranspired from cypress domes than from open
water. The upland wetlands do not have high productivity, but, as
demonstrated for cypress domes (Ewel and Odum 1982), they help
maximize annual regional production by conserving water, by storing
it to help ameliorate flooding, and by recharging groundwater.

REGIONAL ROLE OF UPLAND WETLANDS

　　Many people have maintained that wetlands take water from the
ground and waste it through high transpiration. This situation may
be true of the downstream floodplains that have high productivities
and a steadier supply of water, but not for upland wetlands. In
Florida and in many other areas, most of the wetlands are upland and
perched. Notable examples are Big Cypress Swamp, Green Swamp, swamps
of the Osceola, and Okefenokee Swamp.

Table 6.1
Embodied Energy of Swamps with Increasing Water Availability

Item	Water Processed m³/m²·yr	Embodied Solar Energy Passing Through* 10⁶GSC cal/ m²·yr	Transpiration m³/m²·yr	Embodied Solar Energy Utilized† 10⁶GSC cal/ m²·yr	Dollar Equivalents of Water Used† $/ha/yr
Direct sunlight	–	1.0	–	0.85	125
Bays (rain only)	2	16.3	0.5	4.1	602
Dwarf cypress	4	32.6	0.75	6.1	897
Pond cypress	8	65.0	1.0	8.1	1191
Strand cypress	40	326.0	1.5	12.2	1794
Floodplain	100	814.0	2.0	16.2	2382

Note:

GSC = global solar calories.

† 1981 $ estimated using 6.8×10^7 global solar calories.

* Embodied energy per volume multiplied by the water passing:

Embodied energy per volume

$= (1.18 \times 10^{-3} \text{cal/g actual free energy in rain})(6.9 \times 10^3$ global solar equivalent cal per cal)

$= 8.1$ global solar cal per g water

†Embodied energy per volume multiplied by the water used in transpiration.

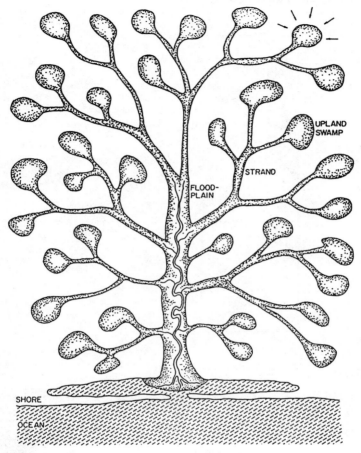

Figure 6.2
Pattern of water flow and wetlands in Florida. Rain
coverages first to small elevated swamps, next to
vegetated flowing swamps without channels, and finally
to floodplain streams.

    Rather than wasting water, upland swamps appear to save water
and thus promote increased regional production indirectly. Since
their value to the regional economy is not accurately represented
solely by consideration of their productivity, the remainder of this
paper focuses on developing a more appropriate measure, the embodied
energy of the converging inputs.

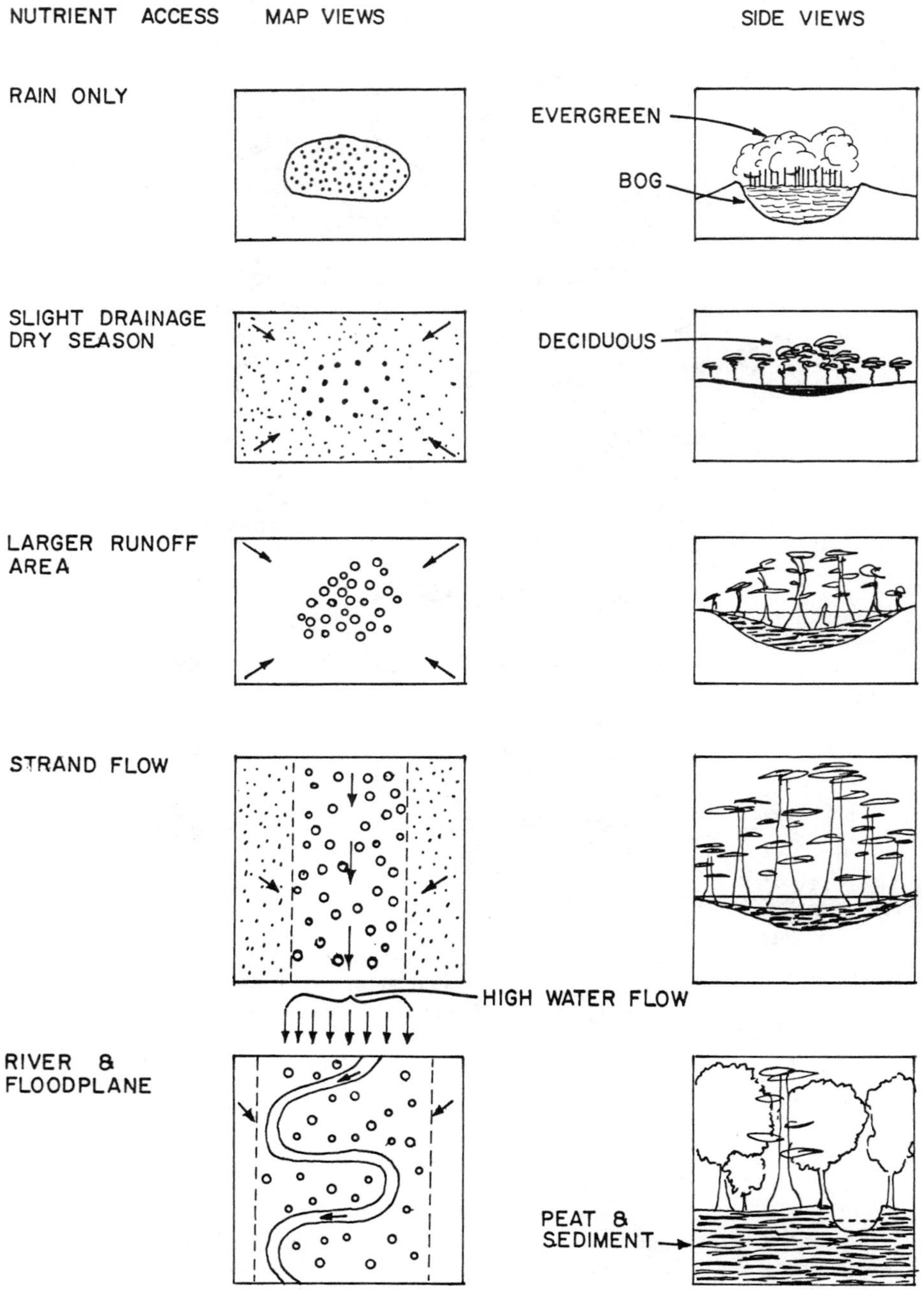

NUTRIENT ACCESS   MAP VIEWS   SIDE VIEWS

RAIN ONLY

EVERGREEN

BOG

SLIGHT DRAINAGE
DRY SEASON

DECIDUOUS

LARGER RUNOFF
AREA

STRAND FLOW

HIGH WATER FLOW

RIVER &
FLOODPLANE

PEAT &
SEDIMENT

Figure 6.3
Wetland types of Florida arranged in series order according to the flow
of water converging through them bringing nutrients.

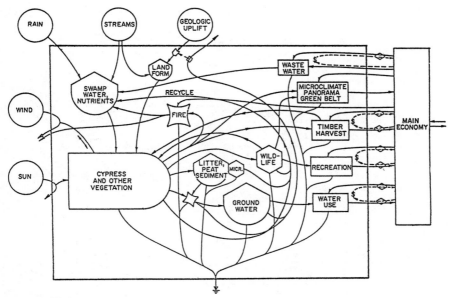

Figure 6.4
Diagram of main features of wetlands, their outside sources, and their
interfaces with human society. Symbols are those of energy circuit
language (Odum 1971, 1982). Dashed lines are dollar flows.

FUNDAMENTALS FOR ANALYSIS OF WETLAND VALUES

The water-wetland system may be summarized with an energy
circuit diagram (Fig. 6.4) that is both a kinetics model and an
energy-accounting network. The diagram includes the direct and
indirect interactions of the wetland with the economy and thus is
also an aggregated impact summary.

Water is used for photosynthesis and transpired in the process.
The work done in order to keep the water potential of plant communities from becoming too negative (too salty) as water is lost by
evapotranspiration is the Gibbs free energy change, or the difference
in the thermodynamic potential of the incoming water and that of the
plants.

Central to the valuation of wetlands is the concept of embodied
energy. As energy flows through successive compartments, the original solar energy is transformed to energy of higher organization and
quality but less caloric quantity. The increasing value of water by
virtue of its convergence to higher quality is measured by the energy
transformation ratio (global solar calories used per calorie of
higher-quality type delivered), as shown in Figure 6.5a. The
embodied energy in this process is the product of the energy transformation ratio and the Gibbs free energy change of the water used
(about 1180 kilocalories per cubic meter of fresh water). To illustrate, Figure 6.5b shows the water volume required to generate one
cubic meter of groundwater available for economic use, and Figure

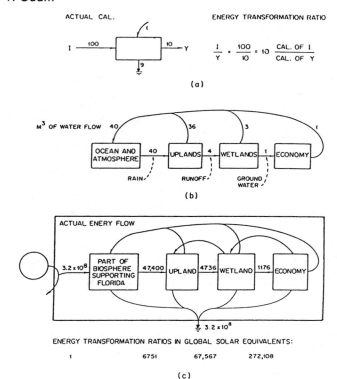

Figure 6.5
Pathways and transformations of energy that generate a cubic meter of
ground water available for economic use. (a) Definition and example of
energy transformation ratio for one transformation step; (b) typical
budgets of water flow in Florida rain passing through swamps to the
economy; (c) actual energy accompanying pathways starting with solar
energy; energy content of water pathways is the Gibbs free energy of
the water relative to salty water.

$$\underline{f} = \underline{RT} \ln \frac{c_2}{c_1}$$

cubic meter of rain water with 10 ppm salts:

$$= \frac{(2)(300)}{18} \underline{\ln} \frac{999990}{965000} \times 10^3 = 1.18 \times 10^3 \text{ kcal/m}^3$$

cubic meter of runoff water at 35 ppm salts:

$$= \frac{(2)(300)}{(18)} \underline{\ln} \frac{999965}{965000} \times 10^3 = 1.186 \times 10^3 \text{ kcal/m}^3$$

cubic meter of ground water at 350 ppm salts:

$$= \frac{(2)(300)}{18} \underline{\ln} \frac{999650}{965000} \times 10^3 = 1.176 \times 10^3 \text{ kcal/m}^3$$

6.5c evaluates the actual energy accompanying the water pathways in Figure 6.5b, starting with the portion of global solar energy involved. Below Figure 6.5c are the energy transformation ratios, in terms of global solar calories per actual calorie at each stage. Table 6.1 gives possibly representative values of water use and embodied solar energy accompanying the series of wetlands comprising the wetland compartment in Figure 6.5c. Those downstream use more water and, thus, utilize more embodied energy.

CALCULATING A DOLLAR VALUE OF WATER AND WETLAND

Figure 6.6 (Odum 1981, 1982a) indicates the way an environmental resource generates an input to the economy. With successive steps of use and value added as the environmental product is transferred and transformed through successive steps, more energy is used at each step to accomplish the next, and so on. The dollars paid to a human who first brings the free resource into the economy are a very small part of the ultimate work accomplished. The dollar circulation ultimately stimulated can be estimated by the proportion that the water's embodied energy is of the total embodied energy of the economy. This is the proportion of the Gross National Product (GNP) due to that contribution. As discussed above, comparisons must be made with all energies in equivalents of the same type. Table 6.1, last column, shows calculations appropriate for the series of swamps of Figure 6.3. The embodied energy used is multiplied by the national ratio of global solar calories to dollars. Unfortunately, the calculation has been imprecise because of uncertainties in the figure for the minimum necessary global solar energy in a calorie of coal (see revised estimates, Odum and Odum 1983).

OTHER CONCEPTS OF EMBODIED ENERGY

It may be important to clarify the difference between this method (Odum 1971; Odum et al. 1977; Odum et al. 1981) and that used by input-output approaches. Compare items in Figure 6.7b with those in Figure 6.7a. In the method used in this commentary (Figure 6.7b), every pathway of a web is regarded as a necessary by-product feedback to every other pathway, and all pathways have the embodiment of the input energies.

In the input-output method, the energies entering the web are divided up among the pathways (Figure 6.7a). The rationale for dividing up the energy is based on the idea of some carrier flowing along the pathways, such as dollars for economic systems, carbon for some biological systems, or actual energy (Herendeen 1981; Costanza 1980; Costanza and Neill 1981). But the rationale for using actual energy is really quite different, although it may have been stimulated by the former one. The input-output method seems incorrect for the purpose of evaluating necessary embodied energy for a pathway and, especially, underestimates those of small actual energy, which are often of the highest quality.

In any case, the wetlands have very high embodied energy and higher contributions to economic dollar circulation than those of many other ecosystems because of their role in converging the embodied energy of water. Water has higher dollar values to an

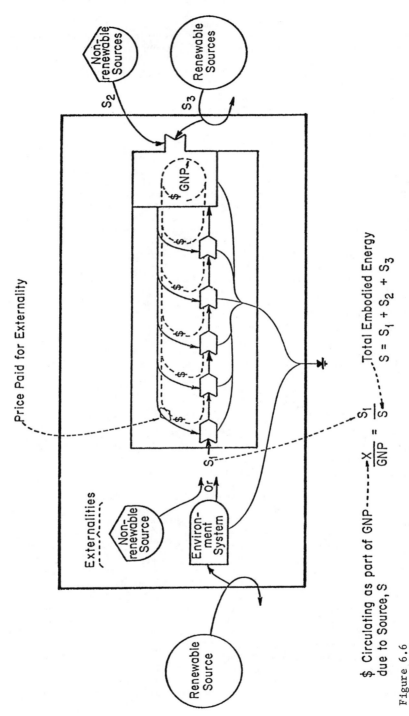

Figure 6.6
Diagram of the steps in processing an environmental resource (storage or renewable flow) into the monied economy with successive increments of value added (Odum 1983).

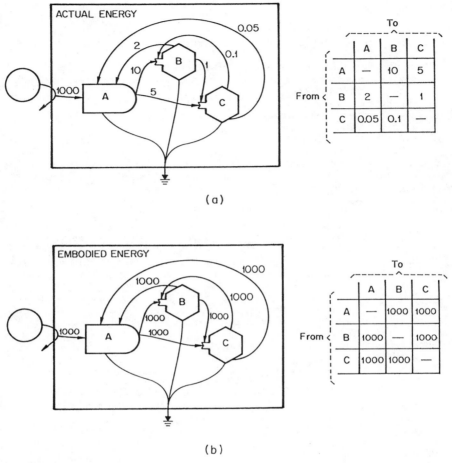

(a)

(b)

Figure 6.7
Comparison of alternative concepts of embodied energy. (<u>a</u>) The undivided concept used in this paper; (<u>b</u>) input-output concept of embodied energy, which divides the external inputs among the pathways of a web according to flows of some carrier quantity for which there are data such as dollar flows or actual energy flows.

economy than usually assigned when only human processing costs are considered. This is a larger value than the microeconomic price. Wetlands are valuable because many of them conserve water and filter recharging waters. Wetlands are especially valuable for wastewater recycling because they are solar energy driven, self-maintaining, and self-organizing. Thus they are very appropriate to the times and to the ideals of ecological engineering as first developed with the multiple seeding principle in microcosms (Odum and Hoskin 1957) and offered as a credo (Odum et al. 1963; Odum 1967).

REFERENCES

Brown, S. S., 1978, A comparison of cypress ecosystems in the landscape of Florida, Ph. D. dissertation, University of Florida, Gainesville.

Brown, S. S., 1981, A comparison of the structure, primary productivity, and transpiration of cypress ecosystems in Florida, Ecol. Monogr. 51:403-427.

Capehart, B. L., J. J. Ewel, B. R. Sedlik, and R. L. Meyers, 1977, Remote sensing of Melaleuca, Photogramm. Engin. and Remote Sensing 43(2):198-206.

Costanza, R., 1980, Embodied energy and economic evaluation, Science 210:1219-1224.

Costanza, R., and C. Neill, 1981, The energy embodied in the products of ecological systems: A linear programming approach, in Ecological modelling, W. F. Mitsch, R. W. Bosserman, and J. M. Klopatek, eds., Elsevier, Amsterdam, pp. 661-670.

Ewel, K. C., and H. T. Odum, eds., 1982, Cypress swamps, University of Florida Press, Gainesville.

Heimburg, K. F., 1976, Hydrology of some north central Florida cypress domes, M. S. thesis, University of Florida, Gainesville.

Heimburg, K. F., and C. R. Vickers, 1980, Radiation balance; preliminary results of radiation studies on an unburned dome receiving wastewater, in Cypress wetlands for water management, recycling and conservation, H. T. Odum and K. C. Ewel, eds., fifth annual report of National Science Foundation and Rockefeller Foundation, Center for Wetlands, University of Florida (NTIS), Gainesville, pp. 188-214.

Herendeen, R. A., 1981, Energy intensities in ecological and economic systems, J. Theor. Biol. 91:607-620.

Monk, C. D., 1966, An ecological study of hardwood swamps in north central Florida, Ecology 47:649-654.

Odum, H. T., 1967, Biological circuits and the marine systems of Texas, in Pollution and marine ecology, T. A. Olson and F. A. Burgess, eds., Interscience, New York, pp. 88-157.

Odum, H. T., 1971, Environment, power and society, Wiley, New York, 332p.

Odum, H. T., 1978, Principles for interfacing wetlands with development, in Environmental quality through wetlands utilization, M. Drew, ed., Coordinating Council on Restoration of the Kissimmee River Valley and Taylor Creek-Nubbin Slough Basin, State Government, Tallahassee, Fla., pp. 29-56.

Odum, H. T., 1981, Energy, economy, and environmental hierarchy, vol. 4 , Proceedings of the 3rd international conference on environment, Ministry of Environment, Paris, pp. 153-163.

Odum, H. T., 1983, Systems ecology: An introduction, Wiley, New York, 644p.

Odum, H. T., 1984, Cypress swamps and their regional role, in Cypress swamps, K. C. Ewel and H. T. Odum, eds., University of Florida Press, in press.

Odum, H. T., and C. M. Hoskin, 1957, Metabolism of a laboratory stream microcosm, Publ. Inst. Mar. Sci. Univ. Texas 4:115-133.

Odum, H. T., and E. C. Odum, eds., 1983, Energy analysis overview of nations, working paper #WP-82-83, International Institute for Applied Systems Analysis, Laxenburg, Austria, 469p.

Odum, H. T., W. L. Siler, R. J. Beyers, and N. Armstrong, 1963, Experiments with engineering of marine ecosystems, Publ. Inst. Mar. Sci. Univ. Texas 9:373–403.

Odum, H. T., W. M. Kemp, M. Sell, W. Boynton, and M. Lehman, 1977, Energy analysis and coupling of man and estuaries, Envir. Manage. 1:297–315.

Odum, H. T., M. J. Lavine, F. C. Wang, M. A. Miller, J. F. Alexander, Jr., and T. Butler, 1983, A manual for using energy analysis for power plant siting, NUREG/CR-2443, U. S. Nuclear Regulatory Commission, Washington, D. C., 242p.

DISCUSSION

<u>Odum</u>: We have a manual now for environmental procedures. It is an appendix to an NRC report [Odum, H. T., M. J. Lavine, F. C. Wang, M. A. Miller, J. F. Alexander, Jr. and T. Butler, 1983, A manual for using energy analysis for power plant siting, NUREG/CR-2443, U. S. Nuclear Regulatory Commission, Washington, D. C., 242p.].

<u>Whigham</u>: If you are adding nutrients to the wetland system, the embodied energy increases, is that correct?

<u>Odum</u>: Right.

<u>Whigham</u>: And if the embodied energy increases through time, then at some point you'll reach a crossover where the water-retention value of that wetland to the regional landscape will become a negative, right?

<u>Odum</u>: Maybe. It would be fascinating to run such a calculation, but we haven't done it. I suspect that most of our fertilization projects, like the cypress experiments, have increased transpiration, so they lose embodied energy [of water], but they may still have more energy embodied than surrounding pine lands. The bog that we were talking about may have moved from a state of relatively low transpiration to greater transpiration, but the wetland then plays a different role. It may still have as much value, but it will be doing it through another process, through a different utilization.

<u>Hemond</u>: There's something I don't understand. The concept of embodied-energy value of a nutrient-flux site, as I understand it, is equal to the solar energy input to the system that produces that flux of nutrients. Is that right? How then do you put a nutrient flux from one wetland on a par with a nutrient flux from some other entirely different system, which may have processed energy in a very different way? So an absolute nutrient flux may come up with two different embodied-energy values.

<u>Odum</u>: The embodied-energy calculation sought for nutrient delivery is that minimum energy required to deliver nutrients at maximum power. Situations with longer, larger energy requirements should be selected against. The calculation made in the presentation was of the embodied energy in the water itself, not the nutrients. To find the embodied energy in the nutrients you first calculate the actual energy, which is the Gibbs free energy, the difference between its

concentration relative to the regional one. Then you multiply by the global energy-transformation ratio. At the moment we are not talking about the geopotential energy in elevated water.

Hemond: The standard chemical state like . . . conductivity or one molar . . .

Odum: Plants, through transpiration, build up a big root pressure [water potential], 15 atmospheres or more, and so it's really the difference in water potential between the rainwater potential and that in the plants. The potential in the plant is determined by the wind, drying air, and the solar energy. These effects are integrated through the water potential. The difference in the water potential--which is really Gibbs free energy--is the actual energy that water is contributing to the system. Water is used in transpiration, maintaining the system by keeping that root potential at the same level. Whether you calculate Gibbs free energy in nutrients or water, the calculation provides actual available energy. To get the transformation ratio--which is the world solar energy that went into delivering the water to the wetland--you take the ratio of the world's solar energy and the world's water budget. We have been doing phosphorus another way.

Hemond: The water potential in the plants is invariably negative . .

Odum: Plus or minus, it is only a matter of convention.

Hemond: It gets more negative as the nutrient concentrations increase, as a matter of fact.

Odum: There is available free energy if it's negative. To obtain embodied energy from the actual energy, multiply by the global energy it took to make a calorie of transformed state.

Bastian: What kind of dollar values are you coming up with in your range of wetland types?

Odum: We're talking about $500 per acre per year.

Bastian: In which type of wetland?

Odum: Marshes or swamps without much water convergence. I guess we haven't done enough of them to generalize.

Bastian: Have you gotten down into the riverine . . .

Odum: They'd be much higher, 20 times, perhaps. See the table in my presentation. Of course, we must distinguish between the flow value and the storage value. If you have a big accumulation of peat representing many years, then you multiply the annual embodied-energy use by the number of years it takes to build it back.

Shuldiner: Since you're working over time, would you discount those future events?

Odum: Bruce Hannon [1982, in Energetics and Systems, W. J. Mitsch et al., eds., Ann Arbor Science, Ann Arbor, Mich., pp. 73-93] has a paper out on energy discounting. Discounting has the basic assumption in it that you can invest in an alternate investment and get a profit. That's based on the assumption that the economy as a whole is growing. In other words, discounting implies that having energy or money now will give you an advantage compared to having it later. Of course, the real economies of the world may be leveling soon, and the U. S. is possibly already coming down, so reverse discounting may be more appropriate.   I suggest calculations first without discounting, then discount as a separate step.

Shuldiner: We also discount because we prefer to have something today rather than waiting until tomorrow to get it.

Odum: This is correct with respect to inflation, but assuming investment will increase the real value of the money assumes growth.

7

Energy Flow in Wetlands

John T. Finn

The previous papers made it clear that wetlands are structurally and functionally complex ecosystems existing under a wide range of hydrological and chemical regimes. The ecosystem dynamics unique to each wetland pose the difficult problem of predicting the impacts of management for treatment of municipal wastewater so that undesirable changes may be avoided. Few of the available analytical tools will permit the evaluation of wetlands models or the general prediction of the consequences of wetland modification. In this paper, I will propose an analytical technique called flow analysis as a method for such evaluations and demonstrate the types of information that can be gained from analysis of energy and carbon (C) flow in two wetlands. I will also attempt to demonstrate certain properties of wetlands that must be considered in evaluating the effects of wastewater application. Specifically, I will (1) show the relative importance of direct and indirect flows in energy models, (2) demonstrate energy cycling in detritus-dominated wetlands, and (3) show the timing of flows.

I will describe the methods of flow analysis by working through two examples. More technical discussions of flow analysis can be found in Hannon (1973), Finn (1976), Patten et al. (1976), Matis and Patten (1980), Barber (1978), Finn (1980), and Finn and Leschine (1980).

The first example is the energy model of Silver Springs (Odum 1957) as modified by Hannon (1973). It provides us with all the necessary data for the analysis and, because Silver Springs is a relatively simple system, it provides an excellent opportunity to introduce the flow analysis method. Flow analysis can be used to trace flows forward and backward, but only the forward analysis will be presented here. Three questions will be asked of the Silver Springs model: (1) Given a calorie of energy starting in a particular compartment, where does that energy go? (2) How many paths are there from one compartment to another? and (3) How long does it take for the transfer of energy from one compartment to another? From the perspective of environmental impact assessment, question 1 addresses what parts of the system might be affected, question 2 addresses the complexity of the connections, and question 3 asks how long it might take for an impact to be propagated through the system.

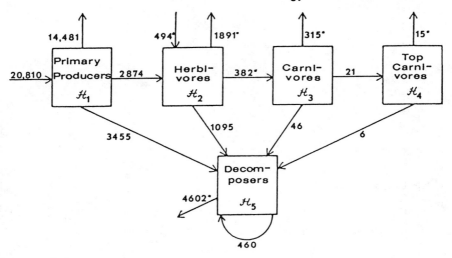

Figure 7.1
Energy flow model of Silver Springs, Florida (Odum 1957), as modified
by Hannon (1973). Flows are in kcal m^{-2}y^{-1}. Compartments are: H_1,
primary producers, including algae in aufwuchs, <u>Sagittaria lorata</u>, and
other macrophytes: H_2, herbivores, including midges, caddis flies,
snails, shrimp, mullet, turtles, and others; H_3, carnivores, including
coelenterates, miscellaneous fish, and invertebrates; H_4, top
carnivores, consisting of bass and gar; and H_5, decomposers, consisting
of bacteria and crayfish. Stumpknockers (<u>Lempomis punctatis</u>) were
divided between herbivores and carnivores. Inflows are gross production
to H_1, and an import of bread fed to fishes in H_2. Intercompartment
flows represent mortality and food ingested but not assimilated (flows
from all holons to detritus) and feeding flows. Outflows include
respiration and downstream export. Flows marked with an asterisk were
modified for balance by Hannon (1973). (After B. Hannon, 1973)

For this model, it is fairly easy to trace where flows go. For
a calorie of energy starting in primary producers (Fig. 7.1, Compartment H_1), about 10% goes directly to herbivores, 15% goes to detritus,
and 75% is respired or exported. These paths are termed one-step
paths. Of the portion that reaches herbivores (H_2), 60% is respired,
30% goes to detritus, and 10% goes to carnivores. These are two-step
paths from primary producers. The same analysis can be conducted
manually for every other pathway in the model. The results of this
process can be summarized as a set of histograms for each compartment.
These histograms show the portion of the original calorie of energy
passing through a given compartment after passing over various
numbers of paths. Thus, for flow from primary producers (H_1) to top
carnivores (H_4) (Fig. 7.2), there is no direct (one-step) flow, no
two-step flow, and only 0.1% of the original energy flow after three
steps.
    The number of paths from one compartment to another can likewise
be determined manually. Flow from primary producers to top carnivores has only one way to get there--along a three-step path.

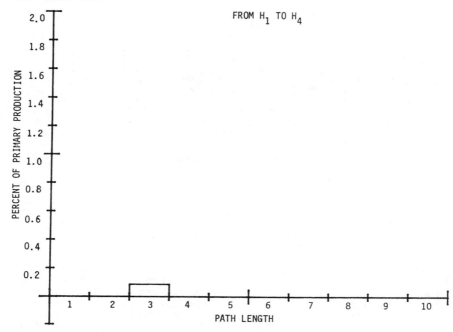

Figure 7.2
Percent of primary production ($\underline{H}_1$) in the Silver Springs system
flowing to top carnivores ($\underline{H}_4$) versus path length.

Matrix methods are available to speed the calculations for
simple models and to make them possible for complex ones. It is
assumed for this discussion that the system is in steady state (i.e.,
all compartment storages are constant). Non-steady state formu-
lations are presented in Barber, Patten, and Finn (1979). To trace
flow forward through a compartment model, a transition matrix $\underline{Q}^{**}$ is
constructed. Each element $\underline{q}^{**}$ is given by:

$$\underline{q}^{**}_{ij} = \underline{f}_{ij}/\underline{T}_j, \qquad (7.1)$$

where $\underline{f}_{ij}$ is the flow (in kcal m^{-2}yr^{-1}) from compartment j to
compartment $\underline{i}$, and $\underline{T}_j$ is the sum of all flow into (or out of)
compartment $\underline{j}$. The element $\underline{q}_{ij}$ is the portion of flow in j that goes
to $\underline{i}$. The $\underline{Q}^{**}$ matrix represents the direct or one-step flow between
compartments. Two-step flows are given by $(\underline{Q}^{**})^2$ and, in general, $\underline{n}$-
step flows are given by $(\underline{Q}^{**})^n$. The total direct and indirect flow
matrix is given by the series:

$$\underline{I} + \underline{Q}^{**} + (\underline{Q}^{**})^2 + (\underline{Q}^{**})^3 + \ldots + (\underline{Q}^{**})^n + \ldots = (\underline{I} - \underline{Q}^{**})^{-1}, \qquad (7.2)$$

where $\underline{I}$ is the identity matrix representing the starting position of
flow. This series converges (i.e., sums to a finite matrix) as long

as there is some outflow from the system. The ij element in $(I - Q^{**})^{-1}$ can also be interpreted as the number of visits a particle starting in j can be expected to make to i.

To calculate the number of paths at each step, first an adjacency matrix, A, must be constructed from the Q^{**} matrix. To do this a "1" is substituted for each non-zero element in Q^{**}. Each element a_{ij} represents the number of one-step paths from compartment j to compartment i. Multiplying A by itself gives A^2, which represents the number of two-step paths between compartments, and in general A^n represents the number of n-step paths. The total number of paths is represented by the series:

$$I + A + A^2 + A^3 + \ldots + A^n \ldots, \qquad (7.3)$$

but this series may not converge. The condition for convergence of this series is that there are no cycles in the flow model. A set of histograms may be constructed to display the number of different path lengths for each compartment.

Finally, timing can be calculated using matrices. It is usually not possible for timing to be calculated manually. A Markov chain model can be constructed to solve this problem. The probability (P_{ij}) that a calorie of energy will go from compartment j to compartment i during a length of time h is given by:

$$P_{ij} = (f_{ij} {}^*h) / ((T_j {}^*h) + x_j) \qquad (7.4)$$

where x_j is storage in compartment j. The probability, P_{ij}, that a quantum of energy will stay in compartment i is given by:

$$P_{ii} = ((f_{ii} {}^*h) + x_i) / ((T_i {}^*h) + x_i) \qquad (7.5)$$

The P matrix represents the portion of flow that travels between compartments in one-time step h. The portion of flow that travels in two time steps is given by P^2. The portion flowing in n time steps is P^n. The infinite series:

$$P^0 + P + P^2 + P^3 + P^4 + \ldots + P^n + \ldots \qquad (7.6)$$

will converge to $(I - P)^{-1}$ as long as there is some flow out of the system. An i,j element of this inverse matrix, when multiplied by h, represents the time that flow starting in j spent in i.

Flow from compartment 1 to compartment 4 is distributed in time as shown in Figure 7.3, with h equal to one week. This figure expresses flow as the average time spent in compartment 4 during each year after the unit of flow starts in compartment 1. The majority of flow from 1 to 4 occurs during the first year, but there are still traces after four years.

Now examine flow from primary producers to detritus (Fig. 7.1). Relative flow versus path length (Fig. 7.4) shows that over 16% of a calorie of primary production goes to detritus directly, 6% in two steps, 1% in three steps, 0.01% in four steps, and less thereafter. The total flow from 1 to 5 amounts to 23.4%. From compartment 1 to

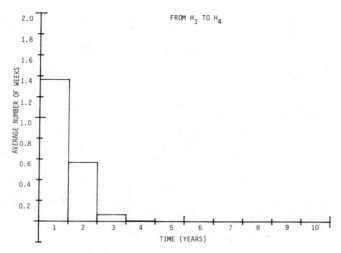

Figure 7.3
Average number of weeks that a unit of energy starting
in primary producers (H_1) spends in top carnivores
(H_4) for each year after beginning in primary
producers.

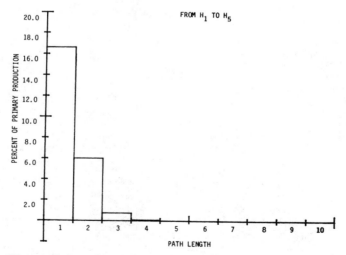

Figure 7.4
Percent of primary production in the Silver Springs
system that flows from primary producers (H_1) to
detritus (H_5) versus path length.

compartment 5 there are 1 one-step path, 2 two-step paths, 3 three-step paths, and 4 four-step paths (Fig. 7.5). Each length path from 5 to infinity has four separate paths. The total number of paths is infinite. Almost all of the flow from primary producers to detritus in this system occurs in the first year (Fig. 7.6).

A more complicated model is that of C flow in Barataria Bay (Fig. 7.7), which was presented by Hopkinson and Day (1977). Although there are a few flows of inorganic C in the model that carry no energy, the model can be used to represent energy transfers.

First examine the flow of material from marsh flora (compartment X_1) to marsh detritus (compartment X_3, Fig. 7.8). There is no direct path; 18% goes in two steps, 2% in four steps, and little after that. Total flow from 1 to 3 is 20.17% of the original unit starting in marsh flora. The number of paths (Fig. 7.9) increases exponentially with path length, there being over 800 ten-step paths. This keeps on increasing with path length forever. It takes over three years for the flow in marsh flora to be reduced to a small portion of its original (Fig. 7.10), and traces are still identifiable seven years after it started. Thus, an impact on primary production in year one would still affect marsh detritus three years later.

Flow of C from marsh flora (X_1) to aquatic fauna (X_7) is of particular interest to wetland managers because of the economic benefits accruing from fish and shellfish. There is no flow until 2% reaches aquatic fauna at four steps (Fig. 7.11), 0.5% at six steps, 0.1% at eight steps, and traces thereafter. Only 2.72% of gross primary production in marsh flora reaches aquatic fauna. There are 1 length-3 paths, 3 length-4 paths, and over 600 length-10 paths (Fig. 7.12). Flow increases during year one, peaks in year two, and decreases to a trace in year five (Fig. 7.13).

CONCLUSIONS

Energy flow through Barataria Bay is extremely complex, as indicated by the large number of pathways through the system. The model presented here is a simplified version of one originally presented by Day et al. (1973), which is in turn a simplified version of the real system. In the two paths examined, there were no direct flows; that is, all flow from marsh flora to marsh detritus and aquatic fauna was indirect. Flow analysis allows dissection of complex flow networks, so that flows from a particular compartment to any other compartment can be analyzed. As an aid to understanding complex models and their behavior it is invaluable.

In analyzing wetlands energy-flow models, indirect flows are often more important than direct flows. Moreover, energy cycling can be an important component of indirect energy flow. Energy cycling does not violate any thermodynamic laws; chemical energy (in the form of protein, carbohydrates, fats, and so forth) can be ingested many times by various organisms and even assimilated more than once before being transformed into work or waste heat (i.e., burned). Low assimilation rates and low metabolisms for many detritivores contribute to this situation.

Flow analysis can also provide estimates of the timing of flows--that is, how long it takes for energy to get from organism A to organism B. In the wetland models examined, it sometimes takes years for flow to move through a system. This procedure has direct

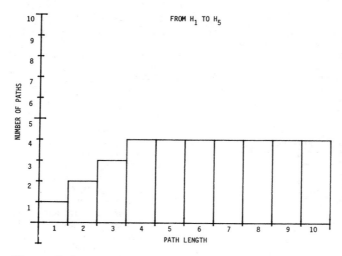

Figure 7.5
Number of paths from primary producers ($\underline{H}_1$) to
detritus ($\underline{H}_5$) versus path length for the Silver
Springs energy model.

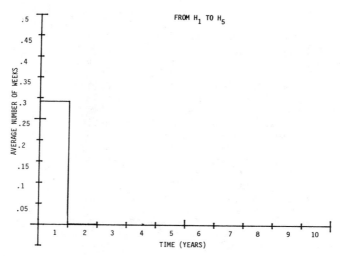

Figure 7.6
Average number of weeks that a unit of energy
starting in primary producers ($\underline{H}_1$) spends in the
detritus compartment ($\underline{H}_5$) each year.

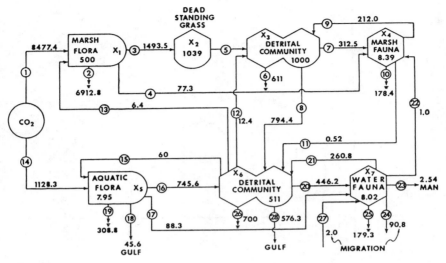

Figure 7.7
Carbon (C) flow model of the Barataria Bay salt marsh ecosystem
(Hopkinson and Day 1977). Steady-state values are shown for carbon
storages (g dry wt m^{-2}) and flows (g dry wt $m^{-2}y^{-1}$). Compartments
are: (1) live marsh plants, mainly S. alterniflora; (2) dead standing
S. alterniflora; (3) marsh microbe-detritus complex; (4) marsh
macrofauna; (5) water flora; (6) water microbe-detritus complex; and
(7) water fauna. Carbon enters the system via compartments 1 and 5
through photosynthesis, and via 7 through immigration of aquatic
fauna. Carbon leaves the system as respiration from all compartments
but 2, and via tidal exchange, fishing, and emigration from 5, 6,
and 7. Intercompartmental flows represent feeding (F_{41}, F_{43}, F_{75},
F_{76}, F_{47}), and death (F_{34}, F_{67}, F_{21}, F_{65}, F_{64}), uptake of dissolved
organic matter by flora (F_{16}, F_{56}, and settling (F_{36}). (From
Hopkinson and Day, 1977)

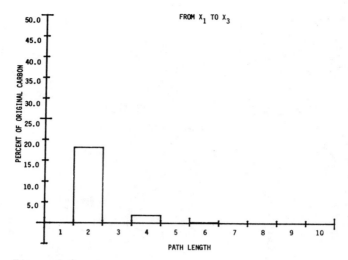

Figure 7.8
Percent of carbon entering marsh flora ($\underline{X}_1$) flowing
to marsh detritus ($\underline{X}_3$) versus path length for the
Barataria Bay carbon model.

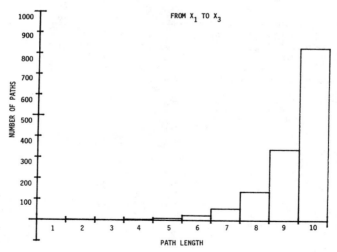

Figure 7.9
Number of paths from marsh flora ($\underline{X}_1$) to marsh
detritus ($\underline{X}_3$) versus path length for the Barataria
Bay carbon model.

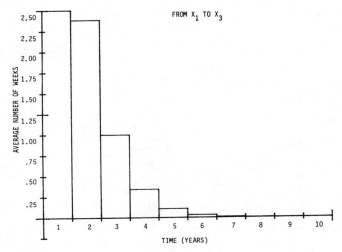

Figure 7.10
Average number of weeks that a unit of carbon
entering marsh flora (X_1) spends in marsh detritus
(X_3) each year for the Barataria Bay carbon model.

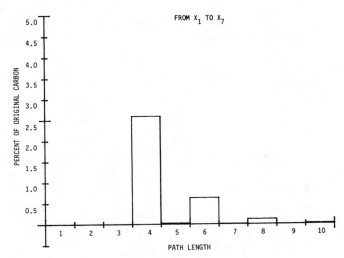

Figure 7.11
Percent of carbon entering marsh flora (X_1) flowing
to water fauna (X_7) versus path length in the
Barataria Bay carbon model.

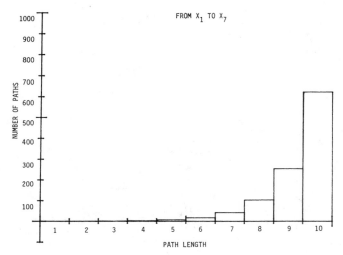

Figure 7.12
Number of paths from marsh flora (X_1) to water fauna
(X_7) versus path length for the Barataria Bay carbon
model.

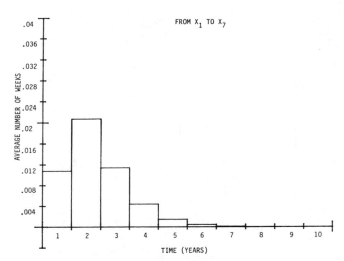

Figure 7.13
Average number of weeks that a unit of carbon
entering marsh flora (X_1) spends in water fauna
(X_7) each year for the Barataria Bay carbon model.

applicability to determining potential impacts of wastewater disposal in wetlands. In Barataria Bay, for example, an increase in primary production would still have large effects on marsh detritus three years later and on aquatic fauna four years later. Maximum effect on marsh detritus would be expected in year one, and on aquatic fauna in year two. The exact timing of impact effects cannot be determined without a time series of measurements or a simulation model because flow structure will change, but these approximations will be close.

Finally, flow analysis can also be used to address questions of elemental cycling and toxin fate. For both elements and toxins, indirect effects are much more important than for energy. Toxins can cycle through a system, visiting a single compartment many times before finally leaving the system. Exposure times of each compartment to a toxin can be calculated exactly, counting cycling and multiple visits, thus enabling more precise estimates of toxic effects. A good dynamic toxin-fate-and-effects model is better suited to this purpose, but flow analysis requires less information.

Flow analysis is heavily dependent on the compartmentalization of the model, assumptions of net flow, internalization of flows, and other major modeling decisions (Finn 1977). This dependence of model structure makes comparisons of different systems more difficult because differences in flow measures may be due to model structure as well as to ecosystem structure. However, flow analysis can be used to assess the suitablility of different model structures for particular purposes. The assumptions of the model can be examined and a better model structure chosen as a result. Flow analysis also requires a model with all flows known. For timing calculations, all storages must also be known. Often this information is not available and may be unreasonably expensive to obtain for a particular site. However, as research projects in wetland wastewater treatment progress, flow models will become available for different types of marshes and different climates. These flow models may be able to provide an indication of how flows will be distributed within a certain marsh type in a certain geographical area.

ACKNOWLEDGMENTS

I would like to thank Dr. B. C. Patten, whose presentations at several meetings during the last year inspired this talk. Dr. Patten has been advocating the importance of indirect effects and the utility of flow analysis in evaluating those effects since the mid-1970s.

REFERENCES

Barber, M. C., 1978, A Markovian model for ecosystem flow analysis, Ecol. Modelling 5:193-206.
Barber, M. C., B. C. Patten, and J. T. Finn, 1979, Review and evaluation of input-output flow analysis for ecological applications, in Compartmental analysis of ecosystem models, J. H. Matis, B. C. Patten, and G. C. White, eds., International Cooperative Publishing House, Fairland, Md., pp. 43-72.

Day, J. W., Jr., W. G. Smith, P. R. Wagner, and W. C. Stone, 1973, Community structure and carbon budget of a salt marsh and shallow bay estuarine system in Louisiana, Center for Wetland Resources, Louisiana State University, Baton Rouge, 79p.

Finn, J. T., 1976, Measures of ecosystem structure and function derived from analysis of flows, J. Theor. Biol. 56:363-380.

Finn, J. T., 1977, Flow analysis: A method for tracing flows through ecosystem models, Ph. D. dissertation, University of Georgia, Athens.

Finn, J. T., 1980, Flow analysis of models of the Hubbard Brook ecosystem, Ecology 61:562-571.

Finn, J. T., and T. M. Leschine, 1980, Does salt marsh fertilization enhance shellfish production? An application of flow analysis, Envir. Manage. 4:193-203.

Hannon, B., 1973, The structure of ecosystems, J. Theor. Biol. 41:535-546.

Hopkinson, C. S., Jr., and J. W. Day, Jr., 1977, A model of the Barataria Bay salt marsh ecosystem, in Ecosystem modeling in theory and practice: An introduction with case histories, C. A. S. Hall and J. W. Day, Jr., eds., Wiley, New York, pp. 235-265.

Matis, J. H., and B. C. Patten, 1980, Environ analysis of linear compartmental systems: The static, time invariant case, Proceedings of the 42nd session, International Statistics Institute, Dec. 4-14, 1979, Manila, Philippines, in press.

Odum, H. T., 1957, Trophic structure and productivity of Silver Springs, Florida, Ecol. Monogr. 27:55-112.

Patten, B. C., R. W. Bosserman, J. T. Finn, and W. G. Cale, 1976, Propagation of cause in ecosystems, in Systems analysis and simulation in ecology, vol. 4, B. C. Patten, ed., Academic Press, New York, pp. 457-579.

DISCUSSION

Hemond: You have a decomposer box and you have a loop on it. Is this loop really detrital-feeding organisms? Is that what you intend, or are you thinking in terms of chemical intermediates and of the prevalence of reduced intermediates in decomposition and in the low oxygen concentrations of peat environments?

Finn: Well, fortunately, Dr. Odum is here and it's his model.

Odum: If we could see more and we knew more, we would find as many pathways in the detritus subsystem as elsewhere. The number of pathways depends on the aggregation detail of the person modeling. When we know more, we will need to simulate these webs.

Hemond: You could easily have another dozen or more pathways by including hydrogen and acetate and ammonium and sulfides and a whole suite of things.

Finn: You give me a model and I'll analyze it.

Guntenspergen: I'd like to make a point here that if you're going to synthesize organic-flow models, and if you're going to use flow

analysis correctly to do these sorts of things—trying to find out
the implications of changing energy, nutrient, or toxin inputs—you
have to have an understanding of the system that you are looking at.
If you predict that most of your energy is detrital or particulate,
and later you find out that 8% or 10% of the carbon that's fixed is
released as dissolved organic carbon and you didn't account for that
in your model, that's a big gap in your model. This all really
depends on how well developed a model you're working with.

Finn: That's really right. I think one of the nice things about
this is that the technique isn't any better than the model you
analyze. It can help you figure out whether you know the system as
well as you think you know it.

Ewel: Can you tell us whether there are special features about
energy processing in a wetland as opposed to a terrestrial system, if
you accept both the forest and the swamp as being heavily detritus
based? Would you expect to find differences in those relative rates
of processing?

Finn: I don't think the data are available for terrestrial systems
to answer that. However, if you could find places where decomposition on land is more or less the same as decomposition in a particular wetland, then the two could be compared. In general, I would
speculate that you are probably going to find terrestrial decomposition rates slowing as you get colder or as you get drier. You
might then begin to find the same sorts of detrital-based communities
as in wetlands. I looked at a Hubbard Brook model of energy.
There's no cycling of energy in the detritus component at all. It
goes straight through. The reason it goes straight through is that
they know very little or nothing about cycling of energy within the
detrital compartment on land. There's some knowledge about the
insect interactions on the top of the litter, but what happens to
energy belowground—the nematodes, and so on—is really totally
unknown.

Ewel: Do we know any more about wetlands?

Finn: I think we know a lot more about the wetlands, and I think
it's at least partially because a lot of the organisms that are
detritus feeders are relatively big. In particular, in salt marshes
you've got oysters, clams, shrimp, and other benthic invertebrates,
and, in the case of freshwater wetlands, you have a lot of spinners
and scrapers, other benthic invertebrates that really are a major
part of the system.

Odum: I would expect energy processing in the two to be different.
The bigger the time, the bigger the space. So, the bigger the
territory—and the wetland has a bigger territory in the sense that
it's a unit that makes sense only in terms of its entire drainage
area—the bigger its time constant in regard to the things it stores
and the services it feeds back. This might imply that wetlands
should have bigger accumulations in storage and would deposit peat
and other material over a longer period of time.

Finn: Right, they are collectors or centers that concentrate organic matter, as you said before, and that increases their importance to the energy processing of the watershed as a whole.

Odum: They store more and tend to have even-aged populations of trees, so it will take a longer time to understand them.

Finn: The point I was making about the detritus community is that the detritus community is better understood in wetlands than it is in forests, perhaps because people like to get wet more than they like to get dirty in soils.

Hemond: I just want to argue again that the decomposition in wetlands is really qualitatively a much different process from what you have in terrestrial systems. You think of it as oxidation of carbon, and you generally have oxygen present in terrestrial detritus. But the very nature of transport limitations in wetlands sets up this whole, very long, complex set of intermediates before the carbon ultimately is oxidized. So you have a whole drawn-out string of chemical intermediates that you don't see in a terrestrial system. I think that makes it a lot different.

8

Wastewater Input to Coastal Wetlands: Management Concerns

Joy Zedler

(Note: The management concerns that I express here have developed
from experiences in an area that differs dramatically from most of
the nation's coastlines. I recognize that my concerns are regional,
and I present them in order to emphasize that the potential problems
of wastewater recycling may differ greatly from place to place.)

Southern California has a warm, dry climate, a varied landscape,
and a magnificent coastline. From the human standpoint, only one
natural amenity is lacking--freshwater. In California, virtually all
the freshwater is in the northern end of the state and most of the
people are in the south. The demands for water are met by the
nation's most ambitious water relocation system, and about 65% of
southern California's water is brought in from outside the region.
Most of it then goes to municipal uses and is treated and dumped
directly into the Pacific Ocean. For a region so heavily dependent
on imported water, there is clearly a need to recycle wastewater.

But the fact that most of the water available for recycling is
imported means that any major recycling program will modify the
region's hydrological regime. Under natural conditions, freshwater
input is highly seasonal. Along the coast, rainfall averages 20 cm
to 40 cm (8 in to 16 in) per year, with most falling between October
and March. Any wetlands used directly for wastewater treatment would
be subjected to abnormal conditions of large and continuous freshwater input. And any wetlands downstream from wastewater recycling
systems would be subjected to the indirect effects of increased
stream discharge.

Under natural conditions, most streams and some rivers flow
intermittently. Dams have been built on nearly every river to store
water and control floods, and, for the most part, dams have reduced
stream discharges. However, reservoir managers occasionally release
stored water at times when stream flow would otherwise be nil.
Recycling of wastewater for various desired uses--irrigation, groundwater recharge, artificial recreational lakes, live streams--would
increase the amount of stream flow and make the discharge continuous.

At the end of most coastal watersheds are coastal wetlands,
which are highly valued for their natural resources and are protected

by one of the strongest coastal protection acts in the nation. These wetlands owe their biological characteristics to seasonal stream discharges and persistent influence of saline water. The impact of changing their hydrology as a result of recycling wastewater is potentially negative. Yet, wastewater could play a positive role that would improve, rather than destroy, coastal wetlands. Management decisions will determine which of the effects--negative or positive--will predominate. This paper outlines my reasons for caution and suggests how to turn the problem of wastewater release into some solutions for coastal wetland management.

DISCHARGE TO EXISTING COASTAL WETLANDS

My predictions of how increased stream discharge will influence coastal wetlands follow from the responses of salt marshes to recent unusual hydrological events. The usual condition of our coastal wetlands (Fig. 8.1) is for saline ocean waters to dominate the system, for evaporation to concentrate salts in the intertidal portion of the system, and for freshwater inflow to occur in fall and winter. When the volume of flow is increased, as in years of heavy flooding, freshwater dominates the system during the normal rainy season, and the entire wetland is briefly desalinized (Fig. 8.2). However, because seawater quickly resumes its influence, the system returns to its preflood condition very rapidly--by fall of the same year. Increasing the volume of winter flow produces a brief change in the system's hydrological and salinity regimes. Such changes probably fall within the natural range of conditions to which arid-region wetlands are subjected.

Altering the period of freshwater input has different effects. Following the 1980 floods, managers released reservoir water into the San Diego River for a two-month period beyond the rainy season. Prolonged stream discharge leached salt-marsh soils and maintained the influence of freshwater for several months (Fig. 8.3).

The ecological effects of these unusual hydrological events were substantial (Table 8.1). Following the extraordinary floods of 1980, the salt-marsh vegetation at Tijuana Estuary showed improved growth, amounting to a 40% increase in August biomass compared with nonflood years. Halophytes responded to the reduced salinities (and related floodwater effects) by increasing both heights and densities, even though soil salinities returned to hypersaline conditions by the fall. The response was a functional one, and return to preflood conditions was rapid. By 1981, the low-marsh vegetation had returned to conditions present in 1979 (Zedler 1983), and I conclude that increased freshwater discharge during the normal wet season is not a major threat to the salt-marsh ecosystem.

In contrast, prolonged freshwater input has a lasting effect. In the San Diego River, the four-month period of flood flow and reservoir discharge caused heavy mortality of salt-marsh vegetation (probably due to increased inundation). Seeds of upstream marsh plants were brought in, and they readily germinated in the newly desalinated soils. Within a few months, a nearly monotypic Salicornia virginica salt marsh was replaced by freshwater marsh vegetation dominated by Typha dominguensis (Zedler 1981). Recovery of this system has been very slow. There are few seedlings in the marsh; cattails persist vegetatively; and halophytes are slowly

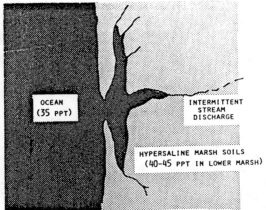

OCEAN
(35 PPT)

INTERMITTENT
STREAM
DISCHARGE

HYPERSALINE MARSH SOILS
(40-45 PPT IN LOWER MARSH)

Figure 8.1. Visual
condition: Rainfall
averages 25 cm/yr

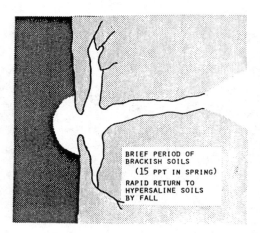

BRIEF PERIOD OF
BRACKISH SOILS
(15 PPT IN SPRING)
RAPID RETURN TO
HYPERSALINE SOILS
BY FALL

Figure 8.2. Heavy flooding:
Large volume of stream
discharge during usual rainy
season.

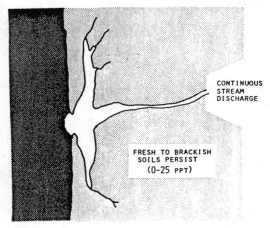

CONTINUOUS
STREAM
DISCHARGE

FRESH TO BRACKISH
SOILS PERSIST
(0-25 PPT)

Figure 8.3. Reservoir
drawdown: Prolonged stream
discharge.

Table 8.1
Response of Coastal Wetland to Different Hydrological Regimes

Usual Conditions	Heavy Flooding		Prolonged Discharge
	Spring	Fall	
Soils: hypersaline	Briefly brackish		Persistently fresh to brackish
Marshes: salt marsh with open canopy, lush algae	Salt marsh with tall, dense canopy (functional change)		Invasion of Typha, Scirpus, and forbs from upstream marshes (structural and functional change)
Ecosystem recovery	Rapid return to preflood canopy type		Slow recovery; Typha and Scirpus persist vegetatively

invading from refugia around the edge of the marsh. The preflood dominant (Salicornia virginica) is not reinvading readily.

In summary (Table 8.1), the nature of ecosystem change depends on the type of hydrological disturbance. Coastal salt marshes are quite resilient to excessive stream discharges that occur during the normal wet season. Changes are functional and recovery is rapid. However, coastal marshes may be permanently altered by persistent freshwater input. Such changes are inconsistent with California's Coastal Act, which seeks to protect the natural features of coastal wetlands as a native ecosystem:

Habitat for endangered birds (e.g., light-footed clapper rail, Belding's Savannah sparrow);

Habitat for rare and endangered plants (e.g., Cordylanthus maritimus ssp. maritimus Frankenia palmeri);

Habitat and food production for wetland consumers.

As a result, continuous discharge brought about by the release of recycled wastewater may have lasting detrimental effects on coastal wetlands.

WETLAND RESTORATION

On the other hand, there is considerable potential for making beneficial use of recycled wastewater in several of the region's coastal wetland restoration projects. Because of southern California's long history of large-scale coastal modification, there are many degraded habitats in need of what is loosely called restoration (Zedler, Josselyn, and Onuf, 1982): (1) diked areas lacking tidal flow, (2) areas denuded by off-road vehicles; (3) dredge spoil deposits; and (4) artificial habitats constructed as mitigation for altering natural wetlands.

Ideas for making these habitats more valuable ecologically focus on improving wildlife habitat by: (1) aiding recovery of endangered species (plants and animals); (2) maintaining the natural variety of habitats by restoring intertidal marshes and mudflats and by restoring brackish and freshwater marshes (where they've been lost); and (3) maximizing habitat diversity. However, the best configuration of coastal habitats (i.e., the most desirable plant and animal communities) is controversial. Whatever landscape plans are approved, there is probably a beneficial role for wastewater. Some possibilities are:

1. Seasonally controlled discharge of freshwater could increase salt-marsh vegetation height and density, thereby improving habitat for endangered species. The light-footed clapper rail is most dense in the most vigorous stands of Spartina foliosa (Barbara Massey, personal communication), and the most vigorous stands of Spartina foliosa occur where or when soil salinities are somewhat below the norm (Zedler 1983).

2. Marsh restoration efforts can probably benefit from freshwater discharge because vegetation establishment is difficult under hypersaline conditions. Our most successful transplantation occurred in the San Diego River and in seasonally brackish soils. Controlled input of freshwater to planting sites should improve survival and spread of transplants.

3. Brackish and freshwater marshes are currently rare along southern California's coast. Whether they were more abundant before the rivers were dammed is a matter of debate. Creation of brackish and freshwater marshes using wastewater may foster the goal of restoring these habitats, or it may simply be a means of increasing habitat diversity by adding a community type that is lacking.

CONCLUSION

In a region where coastal wetlands have declined to 10% to 25% of their natural extent, I am deeply concerned that freshwater-dominated ecosystems not replace the remaining native salt-marsh communities. With large-scale recycling of imported water, we could see major changes in the region's hydrology, and coastal wetlands would ultimately receive substantially larger volumes of freshwater throughout the year. This potential threat to coastal ecosystem structure and functioning could be reduced if the location, amounts, and timing of discharge were properly controlled. It could be made advantageous to coastal-wetland restoration efforts if the discharges were directed toward establishing marsh vegetation, restoring brackish and freshwater marshes, and increasing habitat diversity.

Creative management, at the watershed level, should turn a potential problem into a solution for southern California's endangered coastal wetlands.

ACKNOWLEDGMENTS

I thank Charles F. Cooper for his constructive criticism of the paper, P. Eastman and J. DeWald for help with the figures, and J. Boland, J. Covin, P. Dunn, C. Nordby, and P. Williams for help with fieldwork.

REFERENCES

Zedler, J., 1981, The San Diego River marsh: Before and after the 1980 flood, Environ. Southwest 495:20-22.
Zedler, J., 1983, Freshwater impacts in normally hypersaline marshes, Estuaries 6:346-355.
Zedler, J., M. Josselyn, and C. Onuf, 1982, Restoration techniques, research and monitoring: Vegetation, in Wetland restoration and enhancement in California, M. Josselyn, ed., California Sea Grant College Program Report No. T-CSGCP-007, La Jolla, Calif.

DISCUSSION

Kubiak: I just wanted to have you qualify "increased diversity"—when you were talking about adding on these other types adjacent to the salt marshes, was that done at the expense of the salt marsh?

Zedler: I would never want to see that.

Kubiak: Would you qualify what habitat was utilized?

Zedler: In that case, it was adjacent to a river but it's upland chaparral-type habitat. There is a lot of controversy now in southern California about whether or not we should convert areas that were once wetlands, and now are highly degraded, into fresh- and brackish-water environments using wastewater. Some people argue that these would have been the natural system there, although with our low rainfall and our small watersheds, I doubt that very much. I think that any area that could be returned to intertidal habitat ought to be, and that fresh- and brackish-water habitats should be put in other areas that cannot be made into intertidal marshes—simply because we have lost 75% to 90% (estimates vary) of our native intertidal wetlands.

Pilling: Has the proposition been put forth of augmenting the wastewater flows with a saline input to reach your desired saline content?

Zedler: In one project in the Los Angeles area (most projects are in the Los Angeles area), there's a project that would pump groundwater to provide some of this other habitat. Groundwater in that area has a salinity of 8-20 parts per thousand. So it would be brackish but with evaporation, unless you maintained a good through-flow, you would very quickly end up with a saline habitat. I guess one could engineer it so that the wastewater is diluted by seawater (it would be more expensive that way), but I think it could be engineered. Maybe if we keyed in to the brine discharges from oil wells that are very often right there in the degraded wetland, we could mix the brine and the wastewater together and come up with a nice saline wastewater.

Kaczynski: I would suggest that you not do that.

Zedler: Oh, thank you. What's the reason? Are there other toxins in the brine from the oil wells?

Kaczynski: Yes.

Zedler: Such as?

Kaczynski: Napthenic acid, which is a strong tin salt acid.

Zedler: I withdraw my suggestion!

Odum: In the mangrove marshes, transpiration alone can produce a very salty soil. In Florida, this may lead to dwarf mangroves, so if you don't have freshwater input, the only way you can get the salt out is if there is very good saltwater exchange. And another thing about adding some freshwater to the intertidal saltwater system is that overall water loss by transpiration is reduced, because the salt-water marsh plant transpiration isn't as much as in other types. At least that may be true for mangroves. So you can say, in a sense, the water goes further.

Zedler: It's not going to be used downstream from the coastal wetland, though.

Odum: The ecosystem may get more out of such effluent.

Zedler: I see what you're saying.

Odum: In Florida, one meter of freshwater is needed per year to prevent dwarfing of mangroves.

Zedler: I'd like to test that in southern California. We've got an awful lot of evapotranspiration. I'm not sure how efficient those plants are.

SESSION III

Ecosystem Dynamics

9

Comparisons of the Processing of Elements by Ecosystems, I: Nutrients

John R. Kelly and Mark A. Harwell

The environmental consequences of human activities involve problems that continue to increase in importance because of both the increasing scope and diversity of human actions and the cumulative nature of many anthropogenic perturbations affecting natural systems. Since natural systems operate over widely divergent scales of time, space, and organization, human perturbations result in an equally broad variety of consequences, ranging from effects at the scale of metabolic processes to effects operating on global biogeochemical cycles. Because immediate effects are likely to be more discernible for individual organisms, and also because scientists tend to reduce complex problems to their constituent parts, the lower level of biological organization has received the most attention with respect to human impacts. In a related sense, ecosystem-level effects are innately less tractable because of the increased complexity and reduced replicability of phenomena. Consequently, human impacts on whole ecosystems have been less well explored.

In this paper and an associated one (Giblin, this volume), we report on initial efforts to examine ecosystem responses to anthropogenic stresses based upon a look across the ecological landscape. The intent is to search for patterns of ecosystem responses in order to develop a reasonable basis for certain types of environmental management. In particular, we are trying to identify the pertinent properties of ecosystems that determine the nature of responses to exogenous forces, so that eventually a basis for predictions can be developed. But an essential element of predictability is that there be some degree of reproducibility of phenomena across systems and across time. If each system were unique and there were changes in the nature of its dynamics over time, there would be no basis for predictions. In the other extreme, if systems were to have exact replicates and little dynamic dependency upon past events, a high degree of predictability would be attainable. Ecosystems fall closer to the first case, in that no two ecosystems separated by either time or space are precisely identical. Hence, much of the ecological literature is anecdotal, and extrapolations are to be avoided. But, insofar as different ecosystems have certain common elements of structure and/or function that result in similar responses to external influences, then a scheme can be designed that will allow a reasoned basis for managing human impacts on natural systems. A key question is, What is the functional classification scheme for ecosystems with respect to responses to stress?

To establish such a classification, it is necessary to consider that the structure of ecosystems is maintained by a continual flow-through of energy and by the movement of elements. Of course, the living biota constitute the key aspect of this process for eco-systems; in fact, to a large degree they determine the nature of the nutrient and energy processes, and, in turn, have evolved within and in response to these processes. Therefore, the processes that comprise the functioning of ecosystems are intimately linked with the component communities, populations, and individuals within the eco-systems. Responses of ecosystems to stress, then, relate substan-tially to this linkage.

Through a literature and data review, we are addressing the anthropogenic inputs to ecosystems that affect (1) the ecosystem processes directly, (2) the biotic constituents directly, and (3) the linkages of these two, with concomitant indirect effects. The addi-tion of sewage effluents to natural ecosystems is of particular interest since it involves direct changes in ecosystem processes (e.g., phosphorus (P) cycling is directly affected by P additions), changes in the community structure supported by and effecting the processes, and movement through the ecosystems of materials that have little ecotoxicological impact but potentially significant human health impacts. The first two aspects relate to the nutrient addi-tion component of sewage inputs; the latter involves the ecosystem processing of heavy metals (Giblin, this volume), pathogens, and other toxicants.

ELEMENT PROCESSING BY ECOSYSTEMS: CONCEPTUAL
COMPARATIVE APPROACHES

Comparisons of the manner in which ecosystems process elements may be approached in several ways, varying in the degree of research effort entailed and the amount of structural complexity of the conceptual model necessary for the ecosystem to be characterized. Historically, there are few studies that comprehensively assess the ultimate fate or effects of elements entering natural ecosystems. Most frequently, ecosystem research has focused on several important processes operating within an ecosystem (such as production/decompo-sition); however, these processes are too infrequently related to the functioning of the whole ecosystem and its relationship to surround-ing systems. In addition to internal process studies, there have been a number of efforts to describe ecosystems on the basis of inputs and outputs, but it is rare that all external exchanges across an ecosystem's boundary have been quantitatively addressed. Studies of both input/output and internal dynamics are necessary to be able to compare the processing characteristics of different ecosystems. Our intent is to present some basic conceptual approaches (Fig. 9.1 and Table 9.1) (for which various amounts of data already exist) in order to explore what questions may be addressed by each, to what degree a functional classification of ecosystems may be accomplished, and, most important, to what extent each approach may provide an ability to predict the response of an ecosystem to external inputs.

MODEL

I.

II.

III.

G

IV.

SUBTYPE   A.   B.

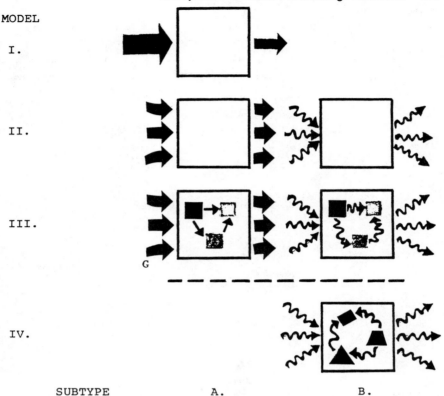

Figure 9.1
Comparative conceptual models for the processing of elements by
ecosystems. Models increase in complexity in two ways: (1) vertically,
from consideration of a novel input (I), to all inputs--that is, from
atmospheric, hydrologic, and geologic sources--(II), to the addition
of internal compartments (III), to (below dotted line) adaptable
internal structure, thus accounting for internal compartment changes
that cause process changes (IV); and (2) horizontally, from a static
(snap-shot) consideration of flow rates (A) to a dynamic (time-varying)
consideration of flow rates (B).

These conceptual models represent differing degrees of understanding ecosystems, ranging from treating them as black boxes that
somehow alter materials before releasing them to developing complex
models that are functional homologues of the real ecosystems. While
the potential predictability may increase significantly with complexity, the data requirements grow enormously. The desired goal is the
development of a minimally complex and data-consuming model that
adequately describes the ecosystem response to external forces.

Table 9.1
Comparative Approaches to the Processing of Elements by Ecosystems

Conceptual Model	Elements	Predictability
Black box (I)	Considers single input/output; site specific; static; simple, requires few data	Not a generic representation; no extrapolation possible
Mixing black box/static flow (IIa)	All inputs/outputs; site-specific; static	May allow generic classification; little extrapolation possible
Mixing black box/dynamic flow (IIb)	All inputs/outputs; time-line relationships	Can develop empirical performance characteristics; predictability within range of experience (good interpolation)
Internally structured system/static flow (IIIa)	All inputs/outputs; system state variables quantified; some mechanistic detail	Validity of predictions relates to correctness of model relationships; this cannot be verified
Internally structured system/dynamic flow (IIIb)	All inputs/outputs; system state variables quantified over time; empirical and mechanistic relationships	Very predictive within range of available data and possibly beyond, depending on validity of assumed mechanisms; verifiable
Adaptive internally structured system/dynamic flow (IVb)	All inputs/outputs/state variables over time and over differing conditions	Highly predictive; verifiable; may be more generically applicable

The predictive capabilities also depend on both the nature of the stresses upon, and the responses by, the ecosystem. For the case of the nutrient-addition aspect of sewage inputs, there are, minimally, three responses that need to be understood: (1) the nutrient outputs from the ecosystem, (2) the changes in magnitude of the ecosystem components, and (3) functional/structural changes in the ecosystem itself. The first ideally could be predicted, at least as interpolations, from the first two conceptual models; the second requires conceptual model III (Table 9.1); and only the last conceptual model can address changes in the functioning of the ecosystem itself.

The lurking problem with ecosystem response predictions relates to the third category—that is, when an ecosystem no longer responds in a way that can be extrapolated from the existing system's dynamics. For many different types of stresses, there is the potential for significant changes in the nature of the ecosystem itself, so that a single ecosystem conceptualization is no longer appropriate. For the actual ecosystem, such a shift is almost certain to correspond to a significant change in the biotic component. But the converse, that all community structural changes result in fundamental ecosystem processes alterations, is not necessarily true; indeed, biotic fluctuations may be the mechanism for a stable ecosystem response to a stress. Therefore, to a very substantial degree, the problem of predictions is determined by the limits of extrapolation from the prestressed system, and we need to determine how extensively the ecosystem can be perturbed before it no longer can be adequately represented by the same conceptual model.

To this point we have discussed the conceptual framework for use in understanding ecosystem responses to stresses. The following sections present the initial phase of our ecosystem review—that is, the survey of information available for the simpler conceptual models (Fig. 9.1) for wetlands and other ecosystems.

ECOSYSTEM PROCESSING OF NUTRIENTS

The initial look at the processing of elements by ecosystems concerns those elements that are nutrients for the system. Patterns of normal flow of these elements may form a basis against which to compare the patterns under conditions of nutrient additions. In this way, we can begin to look for the aspects of the ecosystem and external loadings that result in changes in the way the ecosystem functions.

Nutrient loading rates to various ecosystems range over four to five orders of magnitude (Fig. 9.2). Depicted are both relatively pristine ecosystems and those subject to large anthropogenic (mostly sewage) inputs; thus, there is a large variation within ecosystem type. Loading rates are presented to give a perspective on (1) wetlands in comparison with other ecosystems; (2) the extent to which aquatic and estuarine ecosystems are current recipients of wastewater; and (3) the amount of nutrients in effluents in relation to "natural" inputs (e.g., mostly from rain and runoff inputs; wetlands may receive the nutrient equivalent of approximately 1 cm/week of secondary effluent). There appears to be a progressive trend of increasing input fluxes from terrestrial to fully aquatic systems, partially since the number of sources increases (from atmospheric

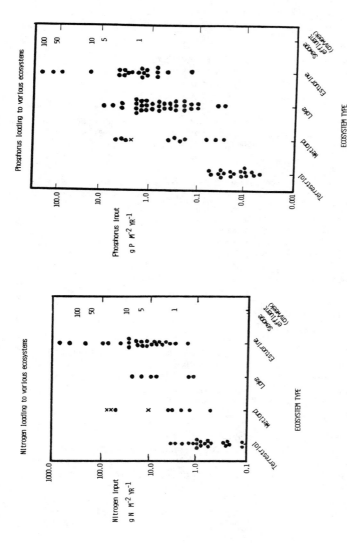

Figure 9.2

Nitrogen and phosphorous loading to various ecosystems. For wetlands, X = salt marshes, ● = freshwater ecosystems. Data are compiled from the literature, and if not specifically cited in text or legend are listed in Appendix. Data, in many cases, do not include organic forms of N or P, nor do they comprehensively assess all inputs to the ecosystem; thus, points must be considered minimum estimates, but there would not be an appreciable change in pattern even if true values for some systems were double those reported (due to log scale). Approximate values for nutrients in secondary effluent are taken from Culp, Wesner, and Culp (1979).

inputs only in forests to atmospheric, hydrologic, geologic, and anthropogenic sources in estuaries). This trend also reflects the downstream nature of ecosystem coupling, indicating the continual accumulation of nutrients into ecosystems as the hydrologic cycle runs from terrestrial to aquatic to estuarine ecosystems and, thus, as the effective nutrient collection area increases relative to the recipient ecosystem size.

SOURCES OF NUTRIENT INPUTS TO WETLAND ECOSYSTEMS

Although some studies have attempted to quantify a specific external supply of nutrients to a wetland ecosystem, comprehensive assessments of the total inputs to a given wetland, via the array of potential input pathways, are rare. Comparisons between wetlands are also limited since not all forms of the nutrients (e.g., organic and inorganic) have routinely been measured. Thus, only a preliminary impression of the relative magnitudes of various sources of nutrient inflows to wetlands may be gathered from a compilation of input rates to a variety of freshwater and saltwater wetlands (Fig. 9.3). In general, for both P and nitrogen (N), there is a considerable range of loading rates, even within a particular source category.

N and P inputs to wetlands, which have been quantified over an annual cycle, include precipitation, surface runoff, groundwater, and (saltwater) tidal exchange; additionally, fixation of atmospheric N has been estimated, at least for parts of several salt marshes. N fixation is an important process to measure since it may constitute an N input larger than precipitation by an order of magnitude, and comparable to input by runoff. Total N input to salt marshes may be an order of magnitude larger than to freshwater wetlands because of a considerable tidal gross input. A large groundwater input has also been recorded for one salt marsh (Valiela and Teal 1979). As with N, P loading rates into wetlands vary tremendously. Anthropogenic sources included in runoff can easily dictate the magnitude of the total input. Precipitation inputs are generally smaller than runoff, whereas groundwater input of P to freshwater wetlands may be much less than precipitation. The upstream source to wetlands from groundwater inflow (i.e., output from terrestrial ecosystems) may reduce P inputs via this pathway; retention of P added to terrestrial ecosystems will be demonstrated below.

OUTPUTS OF NUTRIENTS FROM WETLANDS
TO OTHER ECOSYSTEMS

Few data are available for N or P outputs from wetlands (Fig. 9.4). Denitrification, with concomitant gaseous loss of N_2 to the atmosphere, is commonly mentioned (but rarely quantified) as a major process by which wetlands remove N dissolved in water. "Estimates" available for only several ecosystems (methodological problems notwithstanding) indicate that a large output of N_2 to the atmosphere may be possible. There are surprisingly too few data for N outputs, in general, to form any basis for comparison. Output of P to adjacent ecosystems via surface runoff may be significant, but again, more data on runoff, groundwater and, for salt marshes, tidal outputs are needed.

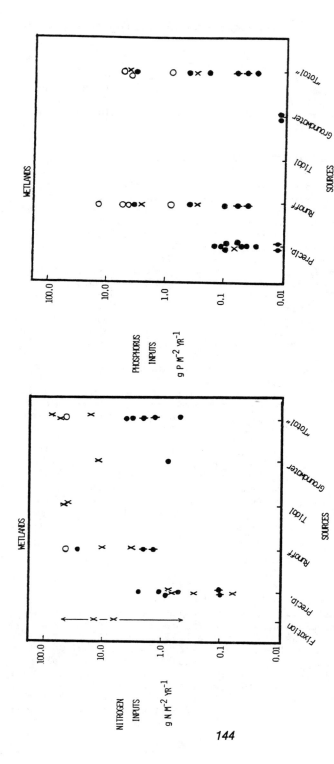

Figure 9.3
Nitrogen and phosphorus inputs to wetland ecosystems. Dark circles are for freshwater wetlands; those with vertical line are only for the growing season (Kadlec 1981, 1982). Open circles represent natural freshwater wetlands receiving an anthropogenic input (but not an experimental plot). Crosses represent salt marshes; the range for nitrogen fixation includes values for selected portions of the marsh (e.g., pans, creek banks, Spartina stands) (Nixon 1980). "Total" indicates only that at the least, runoff and precipitation are given for a site; most studies have not included all forms of nutrients nor quantified all sources. Other sources of data are given in references and Appendix (for this and the following graphs).

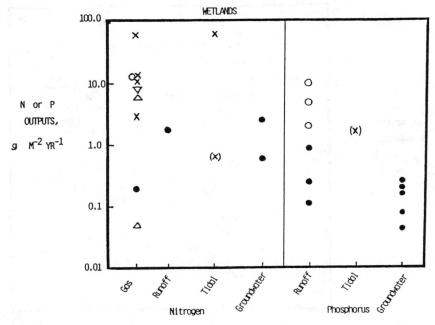

Figure 9.4
Nitrogen and phosphorus outputs from wetland ecosystems. Same notes
and symbols as for Figure 9.3, with the addition of a triangle for an
estuary (Seitzinger 1982) and inverted triangles for lakes (Likens and
Loucks 1978). Parentheses indicate value is for net tidal output only
(Correll 1981).

*INPUT/OUTPUT ANALYSIS: ECOSYSTEMS WITH AND
WITHOUT SEWAGE EFFLUENT ADDITIONS*

A comparison of nutrient-processing characteristics of ecosystems can be made by looking at input/output relationships (conceptual model IIa) for N and P (Figs. 9.5-9.8). Looking first at N, one
appreciates the range in magnitude of outputs and inputs across the
categories of ecosystems. Remembering that, for the most part,
gaseous N exchanges have not been quantified (see Fig. 9.5 notes), a
large number of "natural" sites, especially terrestrial forests,
apppear from the data that are available to retain at least 50% of
the N input. However, other ecosystems appear to pass at least 70%
of their input (those close to diagonal line, Fig. 9.5), and one
ecosystem that was surveyed, an eroding peatland (Crisp 1966),
actually had outputs greater than inputs.

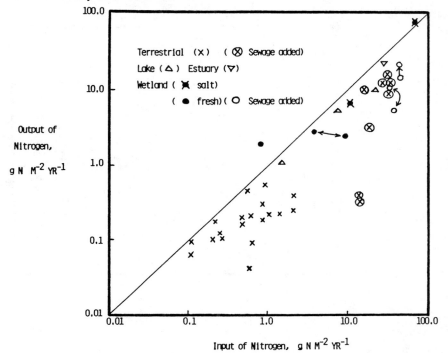

Figure 9.5
Nitrogen input/output analysis for ecosystems with or without sewage
(or nutrient) additions. Data include natural systems and experimental
field plots. Diagonal line indicates that output equals input. Total
N was considered only in studies of Woodwell et al. (1976) (terrestrial),
Zoltek et al. (1979) and Brinson, Bradshaw and Kane (1981) (freshwater
wetlands), Valiela and Teal (1979) and Correll (1981) (saltwater
wetlands), and Nixon (1981) (estuary). Single-headed arrow between two
points (freshwater wetland plot, nutrient added), shows increase in
output if denitrification is included (Brinson, Bradshaw and Kane 1981;
data are for 10 months only). Output includes only water transports
for other studies, with exception of two lakes (Likens and Loucks
1978), an estuary (Nixon 1981), and one salt marsh (Valiela and Teal
1979), all of which considered atmospheric outputs by direct measurement
or by the manner in which the budget was constructed. Double-headed
arrows between two points (freshwater wetland plots, with and without
sewage added) show variations at same site for two consecutive study
years (Zoltek et al. 1979).

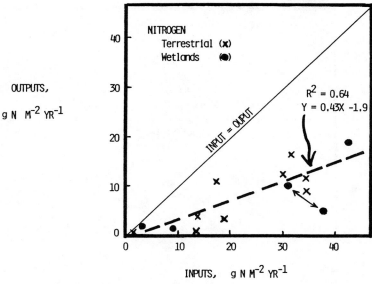

INPUTS, g N M^{-2} YR^{-1}

Figure 9.6
Nitrogen input/output analysis for terrestrial and wetland
ecosystems with sewage (or nutrients) added. Linear plot
includes natural and enriched field sites from studies of
Woodwell et al. (1976) (terrestrial), Brinson, Bradshaw, and
Kane (1981), and Zoltek et al. (1979) (wetlands). Data are
for total N, and do not include loss of gas to atmosphere.
Double-headed arrows are as for Figure 9.5, showing
variations at the same site for two consecutive study years
(Zoltek et al. 1979). Dotted line represents functional
regression (Ricker 1973) calculated for all points (n = 15).

Addition of a considerable amount of N with sewage or other
nutrient applications to terrestrial systems (Woodwell et al. 1976)
and wetlands (Zoltek et al. 1979; Brinson, Bradshaw, and Kane 1981)
does not change the gross pattern of input/output relationships
observed. Added inputs to these ecosystems are accompanied by added
outputs, resulting in a shift to parallel the ecosystems receiving
inputs of an equal magnitude. Gross N-processing characteristics of
terrestrial and wetland ecosystems in response to increasing inputs,
therefore, do not differ greatly (Fig. 9.6). In spite of a scarcity
of data, a general relationship exists: the output of N via water
flow (runoff and groundwater) is expected to be about 43% of the
input. However, there needs to be better definition at low loading
rates (1–10 g N m^{-2}yr^{-1}), since a significantly non-zero intercept
point can influence the prediction of percent passage through the
system at lower input rates. Heuristically accepting the accuracy of
the calculated regression line for purposes of illustration only, we
would predict 0% output of N up to an input rate of 4.4 g N m^{-2}yr^{-1}
(about 1 cm/week of secondary effluent), 21% output at 8.8 g N m^{-2}yr^{-1},
and asymptotically approaching 43% output (the slope of the line) at
rates exceeding 30 g N m^{-2}yr^{-1}. Richardson (this volume) has

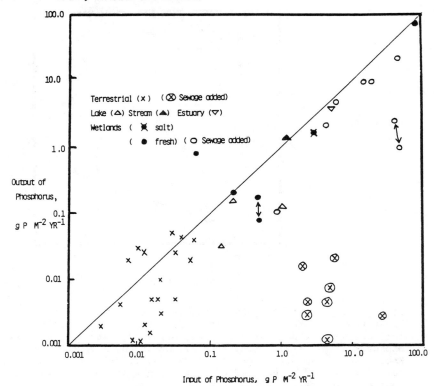

Figure 9.7
Phosphorus input/output analysis for ecosystems with or without sewage
(or nutrient) additions. Data include natural systems that are
relatively pristine and those subject to anthropogenic input, and
experimental field plots. Total P was considered in wetland, lake,
stream, and estuary studies; only PO_4 was measured in sewage additions
to terrestrial ecosystems (Woodwell et al. 1976). Wetland point at
upper right of graph is from Mitsch, Dorge and, Wiemhoff (1979) and
includes input/output due to flooding of a forested swamp to demonstrate
flow-through nature of flooded system. Double-headed arrows between two
points (freshwater wetland plots, with and without sewage added) show
variations at same site for two consecutive study years (Zoltek et al.
1979).

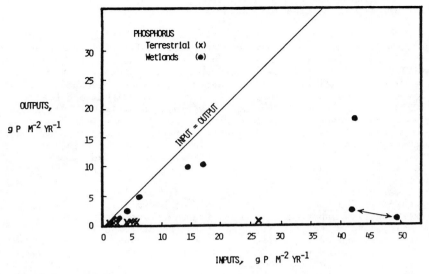

Figure 9.8
Phosphorus input/output analysis for terrestrial and wetland ecosystems
with sewage (or nutrients) added. Linear plot includes natural and
enriched field sites from studies of Woodwell et al. (1976)
(terrestrial), and (for wetlands) Zoltek et al. (1979), Ewel and Odum
(1979), Brinson, Bradshaw, and Kane (1981) (10 months only, nutrient
addition), Fetter, Sloey, and Spangler (1978), Prentki, Gustafson, and
Adams (1978) (with Prentki et al. 1977, and Loucks et al. 1977), Nessel
(1978), Boyt, Bailey, and Zoltek (1977). Double-headed arrows are as
for Figure 9.7.

indicated that retention may indeed be an exponential decay function
of the input rate, higher retention at lower inputs having been
reported for Houghton Lake studies. This is a critical area for
research. As a final caution, we must note that, although a general
relationship is evident, predictability for any particular site is
poor.

    P input/output relationships differ somewhat from the patterns
for N (Figs. 9.7 and 9.8). Many studies involve PO_4 estimation,
rather than total P (see Fig. 9.7 notes); both are reported here.
While a number of ecosystems retain a large proportion of P inputs,
many appear to be nearly flow-through systems. But the striking
aspect of P processing relates to studies of added inputs with sewage.
Terrestrial and wetland ecosystems do not respond similarly, nor does
a linear relationship of output as a function of increasing input
exist for either ecosystem type (Fig. 9.8). Terrestrial systems
exhibit quite thorough removal of added PO_4 prior to water outflow,
dramatically demonstrated by the horizontal shift along the input
axis from natural to enriched sites. This terrestrial retention
would not have been predicted simply from knowledge of the natural
ecosystem's processing capability. Finally, wetlands vary broadly in
the processing efficiency of added P, ranging across an order of
magnitude in percent retention.

INTERNAL ECOSYSTEM DYNAMICS AND NUTRIENT PROCESSING

Although improvements in the predictive capabilities regarding nutrient processing by ecosystems may ultimately entail a mechanistic coupling of external exchanges with internal dynamics, there are still few studies of wetland ecosystems where inputs, outputs, and the major internal flows and storages of nutrients have been simultaneously measured, as previously indicated by Whigham and Bayley (1978). When internal flow rates have been investigated, efforts have often concentrated on the annual accrual of nutrients in the aboveground biomass of the dominant vegetation. However, this single mechanism does not commonly account for the majority of the ecosystem's removal of nutrients from sewage. Enrichment experiments mostly indicate that although vegetative growth is stimulated greatly, only a small portion of added nutrients appear in plant biomass (e.g., salt marshes: Chalmers [1979]--2.6% of N from sludge addition appeared in Spartina; Tyler [1967]--5% of N and 2% of P from fertilizer addition appeared in Juncus; freshwater wetlands: Sloey, Spangler, and Fetter [1978]--6.8% of P from effluent appeared in Typha; Ewel and Odum [1979]--7% of P from effluent was retained in biota, only 5% in cypress wood; Nessel [1978]--about 18% of P from sewage appeared in tree parts (corrected for litterfall), only 2% in wood; Hermann [1980]--5.7% of N in sewage appeared in live aboveground marsh vegetation over two years, about 12% in live and dead aboveground vegetation combined).

Vegetative uptake in wetlands may not always be tightly coupled to external inputs. This situation can occur if internal recycling of nutrients plays a relatively large role in providing the nutrition of plants. The annual accumulation of N in aboveground vegetation in freshwater wetlands can commonly be 10-20 g N m^{-2}yr^{-1} (Whigham and Bayley 1978), which may exceed "total" (measured) N inputs to these ecosystems (Fig. 9.3). Perhaps plant uptake is derived from a nutrient pool built up in the soil or, alternatively, perhaps there is a large, unmeasured input of N by fixation from the atmosphere. Rather than attempting some definitive statement, we offer this observation simply to highlight the necessity of quantifying all inputs to, and transformations within, the ecosystem in order to begin to understand wetland processes. In any event, vegetative removal is but a portion of the information needed to couple internal processes with external exchanges. Other general factors that may add some predictability regarding ecosystem processing characteristics include those mentioned in Table 9.2, especially if the relationships with removal pathways most directly affected (Fig. 9.9) can be defined.

A final note on internal dynamics considers transformations, in this case, of N applied in wastewater discharged to a terrestrial system; this example may dampen enthusiasm for any preliminary observations made without comprehensive analyses of nutrient flows. The N cycle is complicated by the number of reactive forms dissolved in water and the degree of conversion possible between forms (Fig. 9.9). Total N input/output relationships for a sewage enrichment study (Woodwell et al. 1976) have already been discussed (Fig. 9.10). However, inorganic transformations dominate the internal dynamics. Ammonium, which constitutes the major inorganic effluent input, would appear to be completely retained over a range of loading rates. But nitrification of the ammonium within the soil has produced a nitrate

Table 9.2
Factors Affecting Internal Processes

Factor	Loss Affected (refers to Fig. 9.9)
Water residence time	Surface, atmosphere, groundwater, internal
Environment in aqueous and sediment phases	Atmosphere, internal, groundwater
"Sorptive" surfaces	Internal
Hydraulic loading	Surface, groundwater
Development of refractory organics	Internal
Migration, harvest	"Particulate" removal

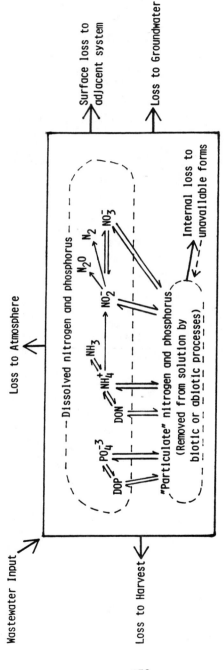

Figure 9.9
Internal ecosystem nutrient dynamics and general removal pathways of nutrients from ecosystems. Diagram shows forms of nitrogen and phosphorus to illustrate some of the possible transformations nutrients may undergo before being passed out of the ecosystem.

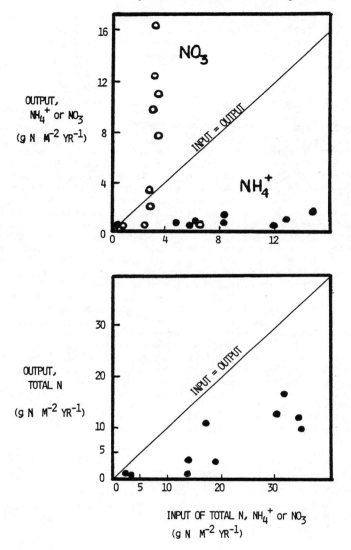

Figure 9.10
Input/output analysis of several nitrogen forms added in sewage to
terrestrial ecosystems. Lower graph shows total N (as for terrestrial
in Fig. 9.6). Upper graph shows inorganic N forms, ammonium (closed
circles), and nitrate (open circles). Ordinate and abcissa refer to
same N form for each set of points--for example, NH_4^+ input in sewage
(primary or secondary effluent)--and precipitation is plotted versus
NH_4^+ output from the ecosystem via water transport. Data are
calculated from Woodwell et al. (1976).

output in groundwater, the loss from the system that is least acceptable. Thus, while in the long run, the object may be to maximize
retention of total "reactive" forms of the nutrient, the form of
nutrient and the pathway of loss may be more critical in specific
situations.

SUMMARY AND RECOMMENDATIONS

Although many studies have included some aspect of nutrient
cycling, few comprehensively detail exchanges with other ecosystems
or consider internal dynamics in relation to external inputs. As a
result, it is premature to present many valid conclusions about
nutrient processing by whole ecosystems. Even though we have
suggested some semblance of predictive capability, results are
preliminary. Moreover, our intent has been to present a conceptual
framework by which to develop general models of ecosystem processes
in order to organize our efforts toward prediction of the fates and
effects of anthropogenic inputs upon ecosystems. Future studies
should be conscious of some of the combinations of information (i.e.,
Fig. 9.1) that may most parsimoniously coordinate data-collection
efforts to enhance the understanding of the dynamics of wetlands, and
other ecosystems, in response to human intervention.

A critical area of research, amenable to field experimentation,
relates to structural changes imposed upon the ecosystem (i.e.,
pertaining to the last conceptual model). We need to determine what
kinds of structural changes may influence element processing to a
degree that a different processing character is also induced upon the
ecosystem. We also need to determine the range of external inputs,
both in character and magnitude, that will cause these critical
structural changes. If post-structural change responses cannot be
extrapolated from previously existing dynamics, this may negate many
of our efforts to understand functional aspects by way of undisturbed
ecosystems and, hence, seriously impair our ability to predict
ecosystem responses to anthropogenic stresses.

ACKNOWLEDGMENTS

Primary funding for this research was provided by the Office of
Research and Development, U. S. Environmental Protection Agency,
under Cooperative Agreement No. CR 807856 01. Additional funding was
provided by Cornell University.

REFERENCES

Boyt, F. L., S. E. Bayley, and J. Zoltek, Jr., 1977, Removal of
    nutrients from treated municipal wastewater by wetland vegetation,
    Water Pollut. Contr. Fed. J. 49:789-799.
Brinson, M. M., H. D. Bradshaw, and E. S. Kane, 1981, Nitrogen
    cycling and assimilative capacity of nitrogen and phosphorus by
    riverine wetland forests, report no. 167, project no. B-114-NC,
    Water Resources Research Institute, University of North Carolina,
    Chapel Hill.

Chalmers, A. G., 1979, The effects of fertilization on nitrogen distribution in a Spartina alterniflora salt marsh, Estuar. Coastal Mar. Sci. 8:327-337.

Correll, D. L., 1981, Nutrient mass balances for the watershed, headwaters intertidal zone, and basin of the Rhode River Estuary, Limnol. Oceanogr. 26(6):1142-1149.

Crisp, D. T., 1966, Input and output of minerals for an area of Pennine Moorland: The importance of precipitation, drainage, peat erosion and animals, J. Appl. Ecol. 3:327-348.

Culp, R. L., G. M. Wesner, and G. L. Culp, 1979, Water reuse and recycling, vol. 2. Evaluation of treatment technology, report no. OWRT/RV-79/2, Office of Water Research and Technology, U. S. Department of the Interior, Washington, D. C., 714p.

Ewel, K. C., and H. T. Odum, 1979, Cypress domes: Nature's tertiary treatment filter, in Utilization of municipal sewage effluent and sludge on forest and disturbed land, W. E. Sopper and S. N. Kerr, eds., Pennsylvania State University Press, University Park, pp. 103-114.

Fetter, C. W., Jr., W. E. Sloey, and F. L. Spangler, 1978, Use of a natural marsh for wastewater polishing, Water Pollut. Contr. Fed. J. 50:290-307.

Hermann, A. J., 1980, Nitrogen cycling in a freshwater marsh receiving secondary sewage effluent, Ph. D. dissertation, University of Florida, Gainesville.

Kadlec, R. H., 1981, 1982, Monitoring report on the Bellaire wastewater treatment facility, 1980 and 1981, University of Michigan Wetlands Ecosystem Group, College of Engineering, University of Michigan, Ann Arbor.

Likens, G. E., and O. L. Loucks, 1978, Analysis of five North American lake ecosystems, III. Sources, loading, and fate of nitrogen and phosphorus, Verh. Internat. Verein. Limnol. 20:568-573.

Loucks, O. L., R. T. Prentki, V. J. Watson, B. J. Reynolds, P. R. Wieler, S. M. Bartell, and A. B. D'Alessio, 1977, Studies of the Lake Wingra watershed: An interim report, IES report no. 78, Center for Biotic Systems, Institute for Environmental Studies, University of Wisconsin-Madison, 45p.

Mitsch, W. J., C. L. Dorge, and J. R. Wiemhoff, 1979, Ecosystem dynamics and a phosphorus budget of an alluvial cypress swamp in southern Illinois, Ecology 60(6):1116-1124.

Nessel, J. K., 1978, Distribution and dynamics of organic matter and phosphorus in a sewage enriched cypress swamp, M. S. thesis, University of Florida, Gainesville.

Nixon, S. W., 1980, Between coastal marshes and coastal waters--a review of twenty years of speculation and research on the role of salt marshes in estuarine productivity and water chemistry, in Estuarine and wetland processes, P. Hamilton and K. B. MacDonald, eds., Plenum Publishing Corp., New York, pp. 437-525.

Nixon, S. W., 1981, Remineralization and nutrient cycling in coastal marine ecosystems, in Nutrient enrichment in estuaries, B. Nielson and L. E. Cronin, eds., Humana Press, Clifton, N. J., pp. 111-138.

Prentki, R. T., T. D. Gustafson, and M. S. Adams, 1978, Nutrient movements in lakeshore marshes, in Freshwater wetlands: Ecological processes and management potential, R. E. Good, D. F. Whigham, and R. L. Simpson, eds., Academic Press, New York, pp. 169-194.

Prentki, R. T., D. S. Rogers, V. J. Watson, P. R. Weiler, and O. L. Loucks, 1977, Summary tables of Lake Wingra basin area, IES report no. 85, Center for Biotic Systems, Institute for Environmental Studies, University of Wisconsin-Madison, 89p.

Ricker, W. E., 1973, Linear regressions in fisheries research, Fish. Res. Bd. Can. J. 30:409-434.

Seitzinger, S., 1982, The importance of denitrification and N$_2$O production in sediments of Narragansett Bay, Rhode Island, Ph. D. dissertation, University of Rhode Island, Kingston.

Sloey, W. E., F. L. Spangler, and C. W. Fetter, Jr., 1978, Management of freshwater wetlands for nutrient assimilation, in Freshwater wetlands: Ecological processes and management potential, R. E. Good, D. F. Whigham, and R. L. Simpson, eds., Academic Press, New York, pp. 321-340.

Tyler, G., 1967, On the effects of phosphorus and nitrogen, supplied to Baltic shore meadow vegetation, Bot. Not. 120:3433-447.

Valiela, I., and J. M. Teal, 1979, The nitrogen budget of a salt marsh ecosystem, Nature 280:652-656.

Whigham, D. F., and S. E. Bayley, 1978, Nutrient dynamics in freshwater wetlands, in Wetland functions and values: The state of our understanding, P. E. Greeson, J. R. Clark, and J. E. Clark, eds., proceedings of the National Symposium on Wetlands, American Water Resources Association, Minneapolis, Minn., pp. 468-478.

Woodwell, G. M., J. T. Ballard, J. Clinton, and E. V. Pecan, 1976, Nutrients, toxins, and water in terrestrial and aquatic ecosystems treated with sewage plant effluents, report BNL 50513, Brookhaven National Laboratory, Upton, N. Y., 39p.

Zoltek, J., Jr., S. E. Bayley, A. J. Hermann, L. R. Tortora, T. J. Dolan, D. A. Graetz, and N. L. Erickson, 1979, Removal of nutrients from treated municipal wastewater by freshwater marshes, final report to the City of Clermont, Florida, Center for Wetlands, University of Florida, Gainesville, 325p.

APPENDIX: OTHER WORKS CONSULTED

The following papers, not individually cited in the text or figure legends, were also consulted for data on loading rates to various ecosystems, sources of inputs and outputs, internal wetland dynamics, and nutrient transformations.

Bowman, M. J. 1977, Nutrient distributions and transport in Long Island Sound, Estuar. Coastal Mar. Sci. 5:531-548.

Dillon, P. J., 1975, The phosphorus budget of Cameron Lake, Ontario: The importance of flushing rate to the degree of entrophy of lakes, Limnol. Oceanogr. 20(1):28-39.

Jaworski, N. A., 1981, Sources of nutrients and the scale of eutrophication problems in estuaries, in Nutrient enrichment in estuaries, B. Neilson and L. E. Cronin, eds., Humana Press, Clifton, N. J., pp. 83-110.

Likens, G. E., 1975, Nutrient flux and cycling in freshwater ecosystems, in Mineral cycling in southeastern ecosystems, G. G. Howell, J. B. Gentry, and M. H. Smith, eds., ERDA symposium series, National Technical Information Service, U. S. Department of Commerce, Springfield, Va., pp. 314-348.

Likens, G. E., F. H. Bormann, R. S. Pierce, J. S. Eaton, and N. M. Johnson, 1977, Biogeochemistry of a forested ecosystem, Springer-Verlag, New York, 146p.

Meyer, J. L., and G. E. Likens, 1979, Transport and transformation of phosphorus in a forest stream ecosystem, Ecology 60(6):1255-1269.

Peterson, D. H., 1979, Sources and sinks of biologically reactive oxygen, carbon, nitrogen, and silica in northern San Francisco Bay, in San Francisco Bay: The urbanized estuary, T. J. Conomos, ed., Pacific Division of AAAS, San Francisco, Ca., pp. 175-194.

Pomeroy, L. R., and R. G. Wiegert, 1981, The ecology of a salt marsh, New York ecological studies 38, Springer-Verlag, New York, 271p.

Richardson, C. J., D. L. Tilton, J. A. Kadlec, J. P. M. Chamie, and W. A. Wentz, 1978, Nutrient dynamics of northern wetland eco-systems, in Freshwater wetlands: Ecological processes and management potential, R. E. Good, D. F. Whigham, and R. L. Simpson, eds., Academic Press, New York, pp. 217-241.

Simpson, H. J., D. E. Hammond, B. L. Deck, and S. C. Williams, 1975, Nutrient budgets in the Hudson River estuary, in Marine chemistry in the coastal environment, T. M. Church, ed., American Chemical Society, New York, pp. 618-635.

Smith, S. V., 1981, Responses of Kaneohe Bay, Hawaii, to relaxation of sewage stress, in Nutrient enrichment in estuaries, B. Neilson and L. E. Cronin, eds., Humana Press, Clifton, N. J., pp. 391-410.

Steward, K. K. and W. H. Ornes, 1975, Assessing a marsh environment for wastewater renovation, Water Pollut. Contr. Fed. J. 47(7):1880-1891.

Woodmansee, R. G., and D. A. Duncan, 1980, Nitrogen and phosphorus dynamics and budgets in annual grasslands, Ecology 61(4):893-904.

10

Comparisons of the Processing of Elements by Ecosystems, II: Metals

Anne E. Giblin

In the search for safe, cost-effective ways to dispose of sewage, increasing attention has been focused on disposal into natural ecosystems. It has been suggested that wetland ecosystems may function naturally as nutrient traps and may be capable of removing nutrients and suspended solids (SS) from wastewater. Metals are important contaminants in sewage and sludge, and it is necessary to understand their movement in these ecosystems before the relative merits of wetland disposal schemes can be fully evaluated. By understanding the internal cycling and the movement of metals through contaminated and natural ecosystems, it is possible to examine the potential of wetlands to remove metals from wastewater.

The retention and bioavailability of metals in ecosystems is strongly influenced by many factors, including hydrology, accretion and erosion rates, and the biogeochemistry of the soils or sediments. Wetlands are ecosystems dominated by saturated soil conditions for at least part of the year, and they differ from upland and permanently submerged ecosystems in their hydrology and in some aspects of their sediment chemistry. When evaluating the potential for natural or artificial wetlands to remove metals from wastewater or sewage, it is important to consider the similarities and differences in these parameters among wetland habitats and other ecosystems.

In most wetlands, due to the presence of saturated soil conditions and large amounts of organic matter, the biochemical oxygen demand (BOD) exceeds the rate of oxygen diffusion into the sediments (for review see Ponnamperuma 1972; Gambrell and Patrick 1978). Without oxygen, the sediments become highly reduced as microbial respiration proceeds, using electron acceptors such as nitrate and sulfate. Because sulfate is much more abundant in sea water than in fresh water, sulfate reduction is more important in sediment metabolism of salt- and brackish-water systems. However, some sulfate reduction can occur in freshwater systems using sulfate from rain, surface- and groundwater, and organic sulfur compounds. The occurrence of sulfate reduction in wetlands can influence the ability of the sediments to retain metals since most metal sulfides are extremely insoluble (Stumm and Morgan 1981).

Sediments from both fresh- and saltwater wetlands tend to be higher in organic content than sediments from areas that are always

completely submerged (Fig. 10.1). There is a strong tendency for
sediments from freshwater wetlands to show a negative correlation
between pH and percent organic matter (Gorham 1953) (Fig. 10.1). The
lowest pH's are found in bogs and are caused primarily by the presence of large amounts of organic acids (Hemond 1980). Salt- and
brackish-water wetlands tend to be considerably more acid than sub-
tidal marine systems. But in contrast to freshwater systems, there
is no trend of pH with organic matter (Fig. 10.1), because the low
pH's in these sediments are caused by the oxidation of metal sulfides
(such as FeS and FeS_2) present in the sediments (Gardner 1973; Lord
1980; Giblin 1982). The amount of oxidation occurring is highly
dependent upon the hydrology, sediment permeability, and macrophyte
production in the area (Howes et al. 1981), and within any wetland, a
wide range of pH's may be observed.

Sediments that are high in organic matter have been shown to
absorb more heavy metals from water than those that are high in
mineral matter (Vestergaard 1979). Since wetland sediments are
usually high in organic matter and frequently contain sulfides, it
has been suggested that sediments in these ecosystems should be
expected to retain heavy metals, possibly sequestering them in a form
unavailable to plants or animals. Factors that might be expected to
reduce metal retention in wetland ecosystems are often overlooked.
Low pH's would tend to increase the solubility of metals compared to
submerged areas where the pH is higher (Stumm and Morgan 1981).
Sediment and water chemistry in wetlands varies strongly with the
hydroperiod (Figs. 10.2a and 10.2b). Changes in pH and oxygen in the
overlying water or sediment caused by changes in water depth or
velocity can have a strong effect on the solubility of metals in
wetland ecosystems (Figure 10.2c). Therefore, to better evaluate the
potential of wetlands to remove metals from wastewater, it is
instructive to look at metal budgets from natural and contaminated
ecosystems.

Surface waters, tidal waters, groundwater, and the atmosphere
are all possible sources of metal or pathways for metal losses from
wetlands (Fig. 10.3). Because of the analytical and logistical
constraints posed by measuring all these fluxes, estimations of metal
uptake and losses from wetlands have been difficult. One approach
has been measurement of metal burial by accretion in the sediments.
When metal inputs are known or can be estimated, the accumulation of
metals by the sediment can be used to calculate the retention of
metals by the ecosystem (Windom 1975). This method relies on a
fairly accurate estimate of the sediment accumulation rate and
assumes that postdepositional migration of metals is minimal. It
must also be assumed that the ecosystem is in steady state with
regard to metal fluxes.

The role of metal uptake by the vegetation, and the subsequent
export of metal-containing detritus by tidal flushing, has been
another approach to measuring metal budgets, especially in salt
marshes (Dunstan, Windom, and McIntire 1975; Bourg, Valiela, and Teal
1979). More recently, attempts have been made to measure metal
fluxes directly from the sediment (Lord 1980). Attention is also
being given to the role of the surface microlayer in trace metal
transport into and out of ecosystems (Pellenbarg and Church 1979;
Lion and Leckie 1982). To evaluate the evidence that wetlands behave
as sinks for heavy metals, I will summarize the available information.

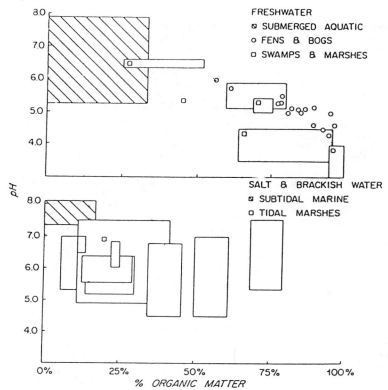

Figure 10.1
The relationship between pH and organic matter content in sediments from freshwater habitats, and salt- and brackish-water habitats. Boxes indicate the range of values for a particular location. Data for the freshwater systems comes primarily from: Gorham (1953); Nessel (1978); Richardson et al. (1976); Klopatek (1975); Malmer and Sjors (1955); Walsh and Barry (1958); and Heal and Smith (1978). Data for brackish- and saltwater marshes is taken primarily from Swanson, Love, and Frost (1972); Lord (1980); Sholkovitz (1973); Murray, Grumandis, and Smethie (1978); Lindberg, Andren, and Harris (1975); Howarth and Giblin (1982); and Breteler, Teal, and Valiela (1981b).

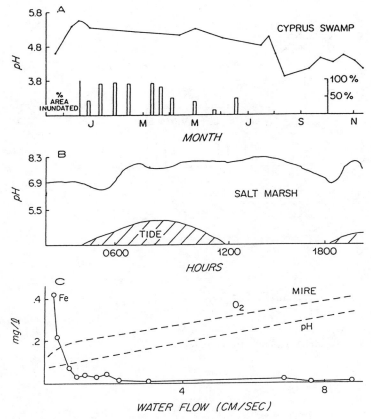

Figure 10.2
The effect of hydrology on chemical parameters in wetlands.
(A) Changes in the percent of area in a cypress swamp that
was inundated during 1980 and the change in soil pH (Day
1982). (B) Changes in the pH of the overlying water in a
salt marsh in response to changes in light and tidal
inundation (Phleger and Bradshaw 1966). (C) The
concentration of Fe, O_2, and pH in water flowing at
different velocities through a mire (Sparling 1966).

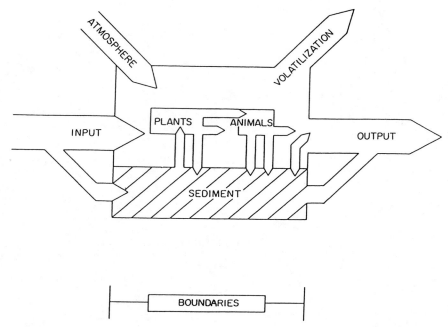

Figure 10.3
Pathways for metal fluxes in ecosystems.

SEDIMENT ACCUMULATION

Early measurements of metal concentrations in marsh sediments
suggested that salt-marsh sediments act as a sink for some metals.
Lead (Pb) profiles (Fig. 10.4) from a salt marsh in Connecticut
showed a dramatic increase in metal concentrations near the surface,
reflecting an increased anthropogenic release of Pb into the environment (Siccama and Porter 1972) Similar results were obtained from
cores taken in Great Sippewisset Marsh, Massachusetts (Banus, Valiela,
and Teal 1974), and Farm Creek, Connecticut (McCaffrey 1977).   The
source of Pb to the sediments was primarily atmospheric (McCaffrey
1977).   Since the ^{210}Pb flux McCaffrey (1977) measured in salt-marsh
cores was indistinguishable from the current rate of ^{210}Pb deposition
measured nearby, McCaffrey concluded that the marsh quantitatively
retained Pb deposited from the atmosphere.   His results for other
metals indicated that zinc (Zn) and copper (Cu) were also quantitatively retained.   As has been pointed out by Nixon (1980), the
sediment data indicate that the flux of Pb into Farm Creek Salt Marsh
is nearly eight times greater than that calculated for Great Sippewissett Marsh on Cape Cod, only 200 miles away.   Nixon suggested that
this may represent differences in the amount of time the marshes were
submerged, since McCaffrey's work was done in a high marsh area while
Banus's was done in a low marsh area.   Studies of the Pb deposition
on bryophytes have suggested that metal deposition rates on Cape Cod

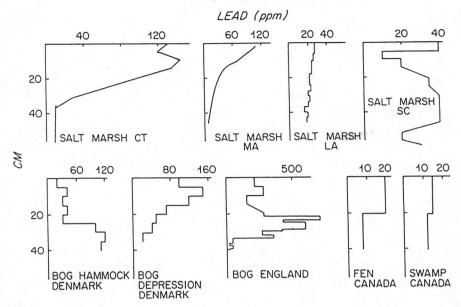

Figure 10.4
The concentration of lead in the sediments of wetlands from different areas. Data from salt marshes (top row): CT--Siccama and Porter (1972); MA--Banus, Valiela, and Teal (1974); LA--DeLaune, Reddy, and Patrick (1981); SC--Gardner (1976). Bog data (bottom row) from: Denmark--Damman (1978); Bog, England--Clymo (1978); Fen and Swamp, Canada--Glooschenko and Capoblanco (1982).

may be substantially lower than at other sites on the New England coast (Groet 1976), and this probably also accounts for some of the difference. The differences in the sediment metal-accumulation rates between these two studies do point out the difficulty of calculating metal retention from regional metal-deposition rates, without taking into account local atmospheric metal flux differences or exposure time of the sediments.

Studies of some other marshes have tended to support the idea that Pb is well retained by salt-marsh sediments. Measurements of ^{210}Pb in high-marsh sediments from Delaware indicated that the marsh retains all of the Pb deposited from the atmosphere and may extract some Pb from the tidal water (Church, Lord, and Somayajulu 1981). Salt marshes where the sediments are physically disturbed by bio-turbation or tidal action may not reflect historical metal inputs accurately. In Louisiana, a rapidly accreting salt marsh showed only a slight increase in the Pb content of sediments near the surface, and the streamside sites showed no increase (DeLaune, Reddy, and Patrick 1981). Cores taken from Dill Creek and North Inlet, South Carolina, showed that the Pb concentration with depth is highly erratic, and the majority of the cores did not show a progressive, upward increase in trace-metal concentration (Gardner 1976).

Measuring the accumulation rate of metals in wetlands is further complicated by compaction and decomposition of organic matter in the sediment. These processes can cause metal concentrations to exhibit a subsurface maxima in highly organic wetlands, such as bogs (Fig. 10.4). Pb is, apparently, quite immobile in at least the aerobic sections of bog peat (Hemond 1980; Damman 1978). Damman (1978) has suggested that Pb is removed below the water table, but Hemond (1980) found that Pb is strongly bound even in this region since both the total and the ^{210}Pb maxima occur below the water table. Cores taken from a fen in Canada showed an increase in Pb concentration in the top 20 cm of the sediment, but cores from a nearby swamp did not show a significant change in the Pb concentration with depth (Glooschenko and Capoblanco 1982) (Fig. 10.4). This data may indicate that there has been some remobilization of Pb from the swamp sediments.

Although Pb may be fairly immobile in wetland sediments, there is growing evidence that other metals are not. Iron (Fe) and manganese (Mn) are remobilized in wetland sediments, as seen both from sediment profiles (McCaffrey 1977; Damman 1978) and from direct flux measurements (Lord 1980). Zn may also be remobilized in anaerobic sediments in some freshwater wetlands (Damman 1978; Clymo 1978).

Metal-accretion rates in salt marshes vary considerably, reflecting differences in sedimentation and riverine influences (Table 10.1). Differences in the accretion rates in inland and streamside sites in Louisiana are almost solely due to differences in the mineral accretion rates (DeLaune, Reddy, and Patrick 1981), while the differences between high and low marsh sites in Massachusetts are due to differences in exposure time to the atmosphere (Banus, Valiela, and Teal 1974). Fe and Mn accumulation in Delaware differ by a factor of two, depending on how deep in the core the concentration of metals is measured, illustrating the potential importance of taking migration into account when computing metal budgets in wetlands. The atmosphere appears to be almost the only source of metals in bogs studied so far. Deposition rates are apparently much lower in bogs than in marshes. However, deposition rates calculated from age of peat in bogs or swamps such as the Okefenokee represent an average of several thousand years, and even these figures may have been underestimated due to the loss of peat by fires.

METAL UPTAKE AND EXPORT BY THE VEGETATION

Studies of metals in salt marshes have tended to focus on the role of the grasses in removing metals from the sediment. Since some of the aboveground <u>Spartina</u> production in marshes may be exported, this pathway is potentially important for metal removal from marsh systems. Attempts have been made to couple estimates of grass export with grass production and metal concentration data to calculate metal export from marshes (Banus, Valiela, and Teal 1975). Accurate measurements of fluxes by this method are complicated by the difficulty of determining whether or not the metal concentration of the grasses at harvest accurately reflects the metal concentration of the detritus when it is exported. Large changes in metal concentration take place in wetland vegetation as it decomposes, and the detritus may become highly enriched in some metals, such as Fe (Brinson 1977; Davis and van der Valk 1978; Breteler et al. 1981) (Fig. 10.5). The amount of detritus exported from salt-marsh systems is also difficult

Table 10.1
The Metal Accretion Rate of Some Salt-and Freshwater Wetlands $(mg/m^{-2}/y^{-1})$

	Accretion Rate (mm/y^{-1})	Cu	Cd	Cr	Mn	Fe	Pb	Zn	Hg
Salt Marshes									
Louisiana[a] (0-3 cm)									
Streamside	13.5	60	6		3,700	60,000	90	210	
Inland	7.5	20	3		1,300	20,000	35	70	
Georgia[b] (0-3)	1	16	2		300	45,000			0.1
S.C.[c] (0-3 cm)									
North Inlet	1.5	15					7	52	
Dill Creek	1.5	20					23	73	
Mass.[d] (0-2 cm)									
Low marsh	1	2	0.4	2.5	10	1,300	8	7	0.02
High marsh	1	3	0.4	2.7	11	940	15	4	0.02
Del.[e]									
(0-2.2 cm)	4.7				72	57,826			
(2.2-4.4 cm)	4.7				43	22,956			
Bogs									
Moor House[f]	0.1	8	1.1			490	63	42	
Sweden[g]		3	0.7			600	43	60	
Mass.[h]	0.4						45		
Swamp									
Okefenokee[i]		0.9			0.4	70		0.6	

Note: The depths of the core from which the accretion rate was calculated are shown in parentheses.

[a]DeLaune, Reddy, and Patrick 1981.

[b]Windom 1975; Windom et al. 1976.

[c]Gardner 1976.

[d]Giblin et al. 1980; Valiela 1982.

[e]Lord 1980.

[f]Clymo 1978.

[g]Calculated by Clymo (1978).

[h]Hemond 1980.

[i]Schlesinger 1978.

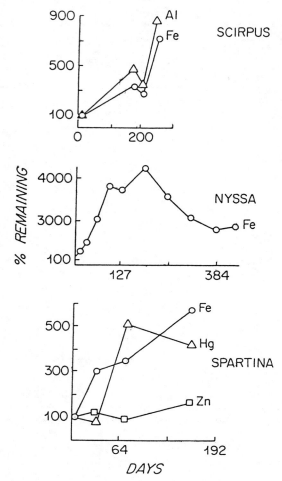

Figure 10.5
The percent of metal remaining in decomposing wetland vegetation through time. Values have been corrected for weight loss by decomposition and represent real increases in the metal content of the detritus. Data from: <u>Scirpus</u>--Davis and van der Valk 1978; <u>Nyassa</u>--Brinson 1977; and <u>Spartina</u>--Breteler et al. 1981.

to measure (for discussion see Nixon 1980).

The large changes in the metal concentration of detritus itself are an indication that metals may be released from the sediments. Pellenbarg (1978) showed that Spartina litter suspended in the water column of tidal creeks became enriched 2.5-fold in Cu and Zn and 1.5-fold in Fe within 12 hours. The source of metal enrichment for litter suspended in the creeks is probably the marsh sediments, since there is a net export of Cu, Zn, and Fe in bulk tidal water from this marsh (Pellenbarg and Church 1979). Studies of this marsh have also shown that there are complex interactions between the Spartina and the surface microlayer. As they decompose, the grasses release organic compounds that tend to chelate metals in the surface micro-layer. These metal-organic compounds are then trapped by Spartina detritus in contact with the microlayer (Pellenbarg 1978). This work points to a dynamic interaction between metals in the sediment and the biota and indicates that at least some metal remobilization from the sediments is taking place.

The actual uptake of metals by the grasses can be a significant portion of the total yearly metal input when the watershed input of metals is small, such as in the Newport estuary in North Carolina (Table 10.2); however, uptake by the grasses still represents a small portion of the total sediment inventory (Wolfe, Cross, and Jennings 1973). In the Georgia salt-marsh/river/estuary system, where river inputs of metals are larger, the grasses take up only a small portion of the annual metal input, with the exception of mercury (Hg), where about 17% of the yearly input cycles through the grasses (Windom 1975; Windom et al. 1976). In experimental plots in Great Sippe-wissett Marsh, where metals were being added at a high rate by the addition of a sewage sludge fertilizer, the grasses took up less than 3% of the annual metal additions to the plots (Bourg, Valiela, and Teal 1979).

Even though metal uptake by the vegetation probably does not represent a major flux of metals out of the ecosystem, it is relevant to metal cycling within wetland ecosystems since it represents a probable mechanism for the transfer of metals through the food chain (Williams and Murcock 1969). Uptake of metals by wetland vegetation is dependent upon sediment conditions such as pH, Eh, and percent organic matter (Windom 1977; Gambrell et al. 1977). Wetland vege-tation from contaminated sediment may have severalfold higher metal concentrations than vegetation from nearby control areas (Fig. 10.6). Some metals have a greater tendency to be accumulated than others, but there is a large variation in the degree of metal accumulation shown in different studies. These variations in enrichment most likely reflect differences in sediment chemistry and loading rates. They also imply that by manipulating sediment conditions in arti-ficial wetland ecosystems, it may be possible to minimize metal uptake by the macrophytes.

METAL BUDGETS FOR WETLAND ECOSYSTEMS

Metal budgets have been constructed for only a small number of wetland ecosystems. Most studies have used measured inputs of metals, or estimates of atmospheric fluxes, and the accumulation of metals in the sediments to calculate the flux of metals through the ecosystem. In wetlands not receiving sewage sludge, Pb and Fe showed the best

Table 10.2
The Percent of the Annual Metal Input Taken Up by the
Aboveground Primary Producers

Salt Marshes	Cu	Cd	Cr	Mn	Fe	Pb	Zn	Hg
				(%)				
Georgia[a]	3	3	NM	8	3	NM	NM	17
North Carolina[b]	NM	NM	NM	43	72	NM	100	NM
Massachusetts[c] (receiving sewage)	0.3	0.3	0.1	3	0.8	0.1	1	NM

Note: NM denotes not measured.

[a]Windom 1975; Windom et al. 1976.
[b]Wolfe, Cross, and Jennings 1973.
[c]Bourg, Valiela, and Teal 1979.

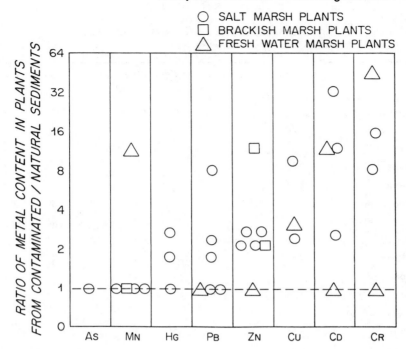

Figure 10.6
The ratio of metal concentration in plants from contaminated versus
natural sediments. Only data from field studies where there was no
significant reduction in the yield of the vegetation (contaminated,
90% of controls) were used. Data adapted from Newell et al. (1982);
Banus, Valiela, and Teal (1974, 1975); Windom et al. (1976); Seidel
(1976); Breteler, Teal, and Valiela (1981a); Drifmeyer and Odum (1975);
Broug, Valiela, and Teal (1979); Giblin et al. (1980); Edwards and
Davis (1975); Nicholas and Thomas (1978); and Kobayashi (1970).

retention by the ecosystem, while less than half of the Cu, Cd, and
Zn entering the study areas was retained (Table 10.3). Hemond (1980)
measured the input and output of lead for a bog and calculated the
yearly sediment-accumulation rate. The flux calculated by the input
data indicated that 98% of the lead entering the bog was retained,
while the sediment accumulation indicated 85% was retained. These
numbers are in fairly good agreement, but indicate that more work is
needed to determine if wetlands are in equilibrium with respect to
trace-metal fluxes. The data from Great Sippewissett Marsh, where
sewage sludge was applied, showed fairly high fluxes of metals
through the ecosystem with the exception of Hg, which was almost
completely retained. The flux of metal from experimental salt-marsh
raceways receiving dredge-spoil effluents was also as high or higher
than natural marsh sediments (Table 10.3). In this experiment,

Table 10.3
The Percentage of the Metal Coming into a Wetland That
Subsequently Leaves the System

	Method	Cu	Cd	Cr	Mn (%)	Fe	Pb	Zn	Hg
Papyrus swamp									
Africa[a]	I/O				6	3			
Bogs									
Mass.[b]	I/O						2		
Mass.[b]	I/SA						15		
Moor House[c]	I/SA	60				0	0	50	
Salt marshes									
N.C.[d]	I/O				41	30		100	
Del.[e]	I/SA						0		
Conn.[f]	SP				50				
Georgia[g]	I/SA	78	83		36	0			88
Sewage added,									
Mass.[h]	I/SA	51	85	55	73	76	40	72	0
Dredge spoil effluent,									
Georgia[i]	I/O	52	85		53	51		81	

Note: The methods used to construct the budget were I/O = input/output, I/SA = input/sediment accumulation, and SP = sediment profiles.

[a]Gaudet 1976.
[b]Hemond 1980.
[c]Clymo 1978.
[d]Wolfe, Cross, and Jennings 1973.
[e]Church, Lord, and Somayajulu 1981.
[f]McCaffrey 1977.
[g]Windom 1975; Windom et al. 1976.
[h]Giblin, Valiela, and Teal 1983, and Valiela 1981 a.
[i]Windom 1977.

Windom (1977) found no relationship between metal removal and the metal loading rate.

Another approach that has been taken to determine if marshes are "sinks" or "sources" of metals in coastal areas has been to measure the concentration of metals entering and leaving wetlands in tidal waters. Since metal inputs into the marshes were not measured in most of these studies, it is not yet possible to construct budgets for these marshes, but some interesting trends emerge. Settlemyre and Gardner (1975) found evidence that there is an export of iron from the marsh to coastal waters; they estimated the magnitude of the flux to be around 1.0 g m^{-2}yr^{-1}. Their data also indicated that Pb, Cu, and Zn were neither imported nor exported by tidal flux.

More recently, attention has been focused on the importance of the surface microlayer, since this region may be highly enriched in metals. Pellenbarg and Church (1979) found that there was a net export of Cu, Zn, and Fe from the Canary Creek Marsh into Delaware Bay. The percentage of the total metal load carried by the micro-layer ranged from 9% to 27%. A study done in San Francisco Bay found a much smaller percentage, typically less than 1% (Lion and Leckie 1982). In this study there was a net export of cadmium (Cd), Cu, and Pb from the marsh to coastal waters.

INTERACTIVE EFFECTS

Retention of metals by wetlands may decrease under higher loading rates as the sorptive characteristics of the sediments become saturated, but there is also evidence that interactive effects with nutrients may decrease metal retention by salt-marsh sediments. Experimental salt-marsh plots receiving sewage sludge fertilizer had fairly low retention of metals such as Pb and Fe when compared with other wetlands (Table 10.3). The addition of nutrients to the marsh increased the oxidation of the sediments by stimulating grass production (Howes et al. 1981). This increased oxidation decreased the concentration of sulfides and increased the solubility of metals in the pore water (Fig. 10.7, Giblin 1982).

Other effects of added nutrients or sewage that could poten-tially alter metal mobility in the ecosystem have been observed in wetlands. Chalmers (1979) found that when grass production was stimulated by nutrients, the increased water loss from the sediment by evapotranspiration caused an increase in the salinity of the sediment. When wastewater was applied to cypress swamps in Florida, the standing water in the swamp became anoxic, due to the increase in biochemical oxygen demand (BOD) and a duckweed bloom over the surface of the water.

SUMMARY AND RECOMMENDATIONS

Metal-budget and metal-flux data for wetland ecosystems show that the percentage of metal removed by passage through the ecosystem varies widely between metals and among wetlands. While some metals, such as Pb, may be well retained by wetlands under conditions of low loading rates, the majority of metals, such as Zn and Cd, may pass through the ecosystem. Although in a geochemical sense, wetlands are "sinks" for some metals, these studies indicate that they may not

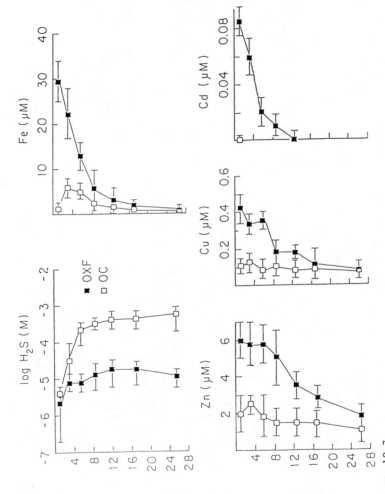

Figure 10.7
The concentration of H_2S, Fe, Zn, Cu, and Cd in pore water from salt-marsh sediment in control areas (OC) and areas that receive a metal-containing, sludge-based fertilizer (OXF). (After A. E. Giblin, 1982, Uptake and remobilization of heavy metals in salt marshes, Ph.D. dissertation, Boston University, Boston, Mass., pp. 123, 125.)

function as efficient "traps" for all metals. By better understanding the biogeochemical processes that alter metal retention, it
may be possible to manipulate wastewater release to maximize metal
removal in natural and artificial wetlands.

Previous studies have shown that some of the assumptions that
have been made in order to construct metal budgets may not be true,
and improved budgets could yield a more accurate picture of what is
occurring. Analytical techniques for measuring metals have improved
tremendously in the last ten years, and complete budgets by input/
output studies are now possible in ecosystems where accurate water
budgets can be constructed. Several areas that require further study
include the following:

1. The role of bed load movement, storm events, and the
microlayer in trace metal transport needs investigation.
2. The hydrology of wetlands, especially subsurface flow,
should be included in studies of fluxes in and out of wetlands.
3. The pore water chemistry of wetlands has been neglected
until recently. Pore water chemistry is a sensitive indicator of
changes that occur in the sediment after deposition and can reveal
more about the postdepositional mobility of metals than can bulk
sediment data.
4. We need to understand the flux of metals through wetlands
well enough to achieve "budget closure." Information on sediment
accumulation of metals should be consistent with the fluxes calculated from input/output data if the system is in equilibrium.
5. We need to better understand the interactive effects of
nutrients and metals. Studies done to date show that nutrient
additions alter primary production, sediment redox, pH, and salinity.
All of these changes can affect metal remobilization in sediments.
Some of these effects were expected, while some would have been quite
difficult to predict. These studies point to the value of conducting
actual field trials to identify interactive factors.
6. Long-term studies are needed if we wish to know the fate of
metals introduced into wetland ecosystems. The capacity of wetlands
to retain metals may change over time, especially since nutrient
additions may alter primary productivity, species composition, and
sediment chemistry.

ACKNOWLEDGMENTS

This is Contribution No. 5208 of the Woods Hole Oceanographic
Institution. Funding for this research was provided by the Andrew W.
Mellon Foundation. I would like to thank Ivan Valiela, John Teal,
John Farrington, Harold Hemond, and Ken Foreman for their comments on
the material in this paper.

REFERENCES

Banus, M., I. Valiela, and J. M. Teal, 1974, Export of lead from salt
marshes, Mar. Pollut. Bull. 5:6–9.
Banus, M., I. Valiela, and J. M. Teal, 1975, Lead, zinc, and cadmium
budgets in experimentally enriched salt marsh ecosystems, Estuar.
Coastal Mar. Sci. 3:421–430.

Bourg, A. C. M., I. Valiela, and J. M. Teal, 1979, Heavy metal budgets in salt marsh ecosystems, Proceedings International Conference on Heavy Metals in the Environment, Commission of the European Communities, London.

Breteler, R. J., J. M. Teal, and I. Valiela,1981a, Retention and fate of experimentally added mercury in a Massachusetts salt marsh treated with sewage sludge, Mar. Environ. Res. 5:211-225.

Breteler, R. J., J. M. Teal, and I. Valiela, 1981b, Bioavailability of mercury in several northeastern U. S. Spartina ecosystems, Estuar. Coastal and Shelf Sci. 12:155-166.

Breteler, R. J., A. E. Giblin, J. M. Teal, and I. Valiela, 1981, Trace enrichments in decomposing litter of Spartina alterniflora, Aquat. Bot. 11:111-120.

Brinson, M. M., 1977, Decomposition and nutrient exchange of litter in an alluvial swamp forest, Ecology 58:601-609.

Chalmers, A. G., 1979, The effects of fertilization on nitrogen distribution in a Spartina alterniflora salt marsh, Estuar. Coastal Mar. Sci. 8:327-337.

Church, T. M., C. J. Lord III, and B. L. K. Somayajulu, 1981, Uranium, thorium, and lead nuclides in a Delaware salt marsh sediment, Estuar. Coastal and Shelf Sci. 13:267-275.

Clymo, R. S., 1978, A model of peat bog growth, in Production ecology of British moors and montane grasslands, O. W. Heal and D. F. Perkins, eds., Springer Verlag, New York, pp. 187-223.

Damman, A. W. H., 1978, Distribution and movement of elements in ombrotrophic peat bogs, Oikos 30:480-495.

Davis, C. B., and A. G. van der Valk, 1978, The decomposition of standing and fallen litter of Typha glauca and Scirpus fluviatilis, Can. J. Bot. 56:662-675.

Day, F. P., Jr., 1982, Litter decomposition rates in the seasonally flooded Great Dismal Swamp, Ecology 63:670-678.

DeLaune, R. D., C. N. Reddy, and W. H. Patrick, Jr., 1981, Accumu-lation of plant nutrients and heavy metals through sedimentation processes and accretion in a Louisiana salt marsh, Estuaries 4:328-334.

Drifmeyer, J. E., and W. E. Odum, 1975, Lead, zinc, and manganese in dredge-spoil pond ecosystems, Environ. Conserv. 2:39-45.

Dunstan, W. M., and H. L. Windom, 1975, The influence of environ-mental changes in heavy metal concentrations on Spartina alterniflora, in Estuarine Research, vol. 2, L. E. Cronin, ed., Academic Press, New York, pp. 393-404.

Dunstan, W. M., H. L. Windom, and G. L. McIntire, 1975, The role of Spartina alterniflora in the flow of lead, cadmium, and copper through the salt marsh ecosystem, in Mineral cycling in south-eastern ecosystems, F. G. Howell, J. B. Gentry, and M. H. Smith, eds., ERDA Symposium Series Conference 740513, ERDA Technical Information Center, pp. 250-267.

Edwards, A. D., and D. E. Davis, 1975, Effects of an organic arsenical herbicide on a salt marsh ecosystem, J. Env. Qual. 4:215-219.

Gaudet, J. J., 1976, Nutrient relationships in the detritus of a tropical swamp, Arch. Hydrobiol. 78:213-239.

Gambrell, R. P., and W. H. Patrick, Jr., 1978, Chemical and microbio-logical properties of anaerobic soils, in Plant life in anaerobic environments, D. D. Hook and R. M. M. Crawford, eds., Ann Arbor Science, Ann Arbor, Mich., pp. 375-424.

Gambrell, R. P., R. A. Khalid, M. G. Veroo, and W. H. Patrick, Jr., 1977, Transformations of heavy metals and plant nutrients in dredged sediments as affected by oxidation reduction potential and pH, contract no. DACW39-74-C-0076, prepared for the Office of Chief of Engineers, U. S. Army Corps of Engineers, Washington, D.C., 124p.

Gardner, L. R., 1973, The effect of hydrologic factors on the pore water chemistry of intertidal marsh sediments, Southeast. Geol. 15:17-28.

Gardner, L. R., 1976, Exchange of nutrients and trace metals between marsh sediments and estuarine waters--a field study, OWRT project B-055-SC, Water Resources Research Institute, Clemson University, Clemson, S. C., 95p.

Giblin, A. E., 1982, Uptake and remobilization of heavy metals in salt marshes, Ph. D. dissertation, Boston University, Boston, Mass.

Giblin, A. E., I. Valiela, and J. M. Teal, 1983, The fate of metals introduced into a New England salt marsh, Water, Air, and Soil Pollut. 20:81-98.

Giblin, A. E., A. C. M. Bourg, I. Valiela, and J. M. Teal, 1980, Uptake and losses of heavy metals in sewage sludge by a New England salt marsh, Am. J. Bot. 67:1059-1068.

Glooschenko, W. A., and J. A. Capoblanco, 1982, Trace element content of Northern Ontario peat, Environ. Sci. and Tech. 16:187-188.

Gorham, E., 1953, Chemical studies on the soils and vegetation of water-logged habitats in the English Lake District, J. Ecol. 41:345-360.

Groet, S. S., 1976, Regional and local variations in heavy metal concentrations of bryophytes in the northeastern United States, Oikos 27:445-456.

Heal, O. W., and R. A. H. Smith, 1978, Introduction and site description, in Production ecology of British moors and montane grasslands, O. W. Heal and D. F. Perkins, eds., Springer Verlag, New York, pp. 3-16.

Hemond, H. F., 1980, Biogeochemistry of Thoreau's Bog, Concord, Massachusetts, Ecol. Monog. 50:507-526.

Howarth, R. W., and A. E. Giblin, 1983, Sulfate reduction in the salt marshes of Sapelo Island, Georgia, Limnol. Oceanogr. 28:70-82.

Howes, B. L., R. W. Howarth, I. Valiela, and J. M. Teal, 1981, Oxidation-reduction potentials in a salt marsh: Spatial patterns and interactions with primary production, Limnol. Oceanogr. 26:350-360.

Klopatek, J. M., 1975, The role of emergent macrophytes in mineral cycling in a freshwater marsh, in Mineral cycling in southeastern ecosystems, F. G. Howell, J. B. Gentry, and M. H. Smith, eds., ERDA Symposium Series Conf. 740513, pp. 367-393.

Kobayashi, J., 1970, Relationship between the "Itai-itai" disease and the pollution of river water by cadmium from a mine, Proceedings of 5th International Water Pollution Research Conference, July-August.

Lindberg, S. E., A. W. Andren, and R. C. Harris, 1975, Geochemistry of mercury in the estuarine environment, in Estuarine Research, vol. I, L. E. Cronin, ed., Academic Press, New York, pp. 64-107.

Lion, L. W. and J. O. Leckie, 1982, Accumulation and transport of Cd, Cu, and Pb in an estuarine salt marsh surface microlayer, Limnol. Oceanogr. 27:111-125.

Lord, C. J. P. III, 1980, The chemistry and cycling of iron,
manganese, and sulfur in salt marsh sediment, Ph. D. dissertation,
University of Delaware, Lewes, Del.

Malmer, N., and H. Sjors, 1955, Some determinations of elementary
constituents in mire plants and peat, Bot. Not. 108:46-80.

McCaffrey, R. J., 1977, A record of the accumulation of sediment and
trace metals in a Connecticut salt marsh, Ph. D. dissertation,
Yale University, New Haven, Conn.

Murray, J. W., V. Grundmanis, and W. M. Smethie, Jr., 1978, Interstitial water chemistry in the sediments of Saanich Inlet,
Geochim. Cosmochim. Acta 42:1011-1026.

Nessel, J. K., 1978, Distribution and dynamics of organic matter and
phosphorus in a sewage-enriched cypress swamp, M. S. thesis,
University of Florida, Gainesville, Fla.

Newell, S. Y., R. E. Hicks, and M. Nicora, 1982, Content of mercury
in leaves of Spartina alterniflora Loisel, Georgia, U. S. A.: An
update, in Estuar. Coastal and Shelf Sci. 14:465-469.

Nicholas, W. L., and M. Thomas, 1978, Biological release and
recycling of toxic metals from lake and river sediment, technical
paper no. 33, Australian Water Resources Council, Australian
Government Publication Service, Canberra, 99p.

Nixon, S. W., 1980, Between coastal marshes and coastal waters: A
review of twenty years of speculation and research on the role of
salt marshes in estuarine productivity and water chemistry, in
Estuarine and wetland processes: With emphasis on modeling,
P. Hamilton and K. B. MacDonald, eds., Plenum Press, New York,
pp. 437-525.

Pellenbarg, R. E., 1978, Spartina alterniflora litter and the aqueous
surface microlayer in the salt marsh, Estuar. Coastal Mar. Sci.
6:187-195.

Pellenbarg, R. E., and T. M. Church, 1979, The estuarine surface
microlayer and trace metal cycling in a salt marsh, Science
203:1010-1012.

Phleger, F. B., and J. S. Bradshaw, 1966, Sedimentary environments in
a marine marsh, Science 154:1551-1553.

Ponnamperuma, F. N., 1972, The chemistry of submerged soils, Adv.
Agron. 22:29-96.

Richardson, C. J., J. A. Kadlec, A. W. Wentz, J. M. Chamie, and R. W.
Kadlec, 1976, Background ecology and the effects of nutrient
additions on a central Michigan wetland, in Proceedings: Third
wetlands conference, M. W. Lefor, W. C. Kennard, and T. B.
Helfgott, eds., report no. 26, Institute of Water Resources,
University of Connecticut, Storrs, pp. 34-72.

Schlesinger, W. H., 1978, Community structure, dynamics, and nutrient
cycling in the Okefenokee cypress swamp-forest, Ecol. Monog.
48:43-65.

Seidel, K., 1976, Macrophytes and water purification, in Biological
control of water pollution, J. Tourbier and R. Pierson, Jr., eds.,
University of Pennsylvania Press, Philadelphia, pp. 121-130.

Settlemyre, J. L., and L. R. Gardner, 1975, A field study of chemical
budgets for a small tidal creek--Charleston Harbor, S. C., in
Marine chemistry of the coastal environment, T. Church, ed.,
ACS symposium series no. 18, American Chemical Society, Washington,
D. C., pp. 152-175.

Sholkovitz, E., 1976, Interstitial water chemistry of the Santa
Barbara Basin sediments, Geochim. Cosmochim. Acta 37:2043-2073.

Siccama, T. G., and E. Porter, 1972, Lead in a Connecticut salt marsh, Biosci. 22:232–234.

Sparling, J. H., 1966, Studies on the relationship between water movement and water chemistry in mires, Can. J. Bot. 44:747–758.

Stumm, W., and J. J. Morgan, 1981, Aquatic chemistry: An introduction emphasizing chemical equilibria in natural waters, 2nd ed., Wiley-Interscience, New York, 780p.

Swanson, V. E., A. H. Love, and I. C. Frost, 1972, Geochemistry and diagenesis of tidal-marsh sediment, Northeastern Gulf of Mexico, Geol. Surv. Bull. 1360, U. S. Government Printing Office, Washington, D. C., 83p.

Valiela, I., 1982, Nitrogen in salt marsh ecosystems, in Nitrogen in the marine environment, E. J. Carpenter and D. G. Capone, eds., Academic Press, New York, pp. 649–678.

Vestergaard, P., 1979, A study of indication of trace metal pollution of marine areas by analysis of salt marsh soils, Mar. Environ. Res. 2:19–31.

Walsh, T., and T. A. Barry, 1958, The chemical composition of some Irish peats, R. Ir. Acad. Proc. 59:305–328.

Williams, R. B., and M. B. Murcoch, 1969, The potential importance of Spartina alterniflora in conveying zinc, manganese, and iron into estuarine food chains, in Proceedings of the second national symposium on radioecology, D. J. Nelson and F. C. Evans, eds., USAEC CONF-67-0503, Springfield, Va., pp. 431–439.

Windom, H. L., 1975, Heavy metal fluxes through salt marsh estuaries, in Estuarine research, vol. I, L. E. Cronin, ed., Academic Press, New York, pp. 137–152.

Windom, H. L., 1977, Ability of salt marshes to remove nutrients and heavy metals from dredged material disposal area effluents, report prepared for the Office of the Chief of Engineers, contract no. DACW21-76-C-0134, U. S. Army Corps of Engineers, Washington, D. C., 578p.

Windom, H. L., W. S. Gardner, W. M. Dunstan, and G. A. Paffenhofer, 1976, Cadmium and mercury transfer in a coastal marine ecosystem, Marine pollutant transfer, H. Windom and R. Duce, eds., D. C. Heath, Lexington, Mass.

Wolfe, S. A., F. A. Cross and C. D. Jennings, 1973, The flux of manganese, iron, and zinc in an estuarine ecosystem, in Radioactive contamination of the marine environment, International Atomic Energy Agency, Vienna, pp. 159–175.

DISCUSSION

Ewel: Is there a hydrological relationship apparent from a comparison of flowing-water versus hill-water wetlands, particularly with reference to the results from the industrial use of peat filters for stripping out heavy metals?

Giblin: Well, I think those results are specific to industrial applications of peat filters, which have very good heavy-metal retention. Most ecosystems don't work like peat filters, though.

Ewel: Some do. If you have a situation where you have a bog . . .

Giblin: Well, for example--yes, a bog--where there is no other place for water to go but through perhaps 40 cm of peat. Then you get almost complete metal retention, in terms of what's adsorbed; but that's not to say that it's not still bio-available.

Kaczynski: I tend to agree in part with both of you, that the retention ability of any wetland is related to the mass of the matrix of the wetland. In fact, in the design considerations of wetlands, people would be well advised to look at the depth of the soils, at least in a highly dynamic situation such as a salt marsh, where you have a lot of transport of sediment out of the system. That is not true, however, in many inland marshes.

Giblin: I don't think it's so much the dynamics of the sediments. I think it's the change in the pH, rate of oxidation, and the hydro-period that leads to a high rate of mobilization.

Kaczynski: I would not disagree with you. But the mass of the whole thing is going to affect the physical adsorption characteristics.

Bastian: But how do you control the sediment losses and the sampling problem of determining what sediments are there when you're analyzing before-and-after tidal action. In other words, do you actually have mass transport exchange or the same sediment material before and after?

Giblin: Most marshes, by definition, are not highly dynamic situations--most marshes are accreting.

Bastian: Relatively dynamic. A bog versus . . .

Giblin: Well, they're all accreting systems for the most part. I did not include studies such as those of eroding streamside marshes. I didn't use those because, obviously, when you start getting into high-erosion sediments, then the game is off; it's a whole different situation.

Brinson: If sulfide is so important in salt marshes, what is the mechanism in freshwater marshes for tying up lead and the other heavy metals?

Giblin: Well, there are sulfides in freshwater systems, as you know, which do contribute. Also, in bogs, for example, you get high levels of bicarbonates and carbonate mineral formations. Most of the research done in Denmark has implicated things like . . .

Brinson: Even at low pH's?

Giblin: Actually, there is a big controversy among the geochemists as to why there is, for example, siderite formation at fairly low pH's. But the other thing that happens in bogs is that, since they are 99% organic matter, you have some organic adsorption.

Bayley: I was under the impression that in bogs there wasn't much retention because of the fact that, if you look above the area of water table fluctuation, you have low retention levels; and in the

area of low water-table fluctuation you also have low retention
levels. So it seems to me that retention is only concentrated in the
zone of water-table fluctuation, and there it can be flushed out
periodically.

Giblin: Wait, that's the difference between retention and mobilization. The bogs that I looked at, which were basically only
Thoreau's Bog and Morhouse Bog where there were complete budgets,
were hydrologically fairly well isolated. So metals are mobile in
the system, but they don't go anywhere because there's low water
flushing. It makes the comparison between the potential retention of
a saltwater and a freshwater wetland a little bit more difficult.
But I think the evidence for some remobilization is there, although
with lead, most of the lead-210 studies have indicated that it's not
extensive. Would you agree with that, Harry?

Hemond: For lead.

Hodson: Since manganese doesn't directly associate in soluble
sulfide in these systems, do you propose a separate mechanism for the
changes you saw in manganese in the fertilized and unfertilized

Giblin: Well, I don't have to propose a mechanism for manganese,
since it is lost under either situation. It's being lost in the
mostly reduced sediments, and it's being lost in less-reduced
sediments, although it is not being completely lost, because there
seems to be precipitation of oxides near the surface.

Hemond: There's another factor that I think may turn out to be
important, although nobody has really looked at it. The high dissolved organic content can be demonstrated to have a very pronounced
chelating effect. With copper, for example, you can put dissolved
copper in bog water and maybe 90% of it will be chelated by the
dissolved organic material. Hence, a greater proportion is mobile
than otherwise would be.

11

The Effect of Natural Hydroperiod Fluctuations on Freshwater Wetlands Receiving Added Nutrients

Suzanne E. Bayley

Freshwater wetlands exist under a wide variety of hydrologic conditions. Such factors as the frequency, timing, and degree of inundation, still versus flowing water, and surface- or groundwater supply have been shown to affect the chemical and biological dynamics of wetlands. For example, Brown (1981) compared the functional components of wetlands with flowing water to those with still water and found higher rates of gross primary production, net biomass production, respiration, litter fall, and organic matter export in flowing-water wetlands. Higher productivity was possibly attributable to greater phosphorus (P) availability and to decreased stress as the anaerobic substrate was continuously flushed.

The potential of a wetland to assimilate added nutrients is dependent on four factors--the hydrologic regime, the oxidation-reduction state of the soil, the initial soil-nutrient status, and the soil organic content--and their interactions. Generally, under flooded conditions the sediments become anaerobic and there is increased solubility of iron (Fe) and P forms, and organic P and organic nitrogen concentrations are higher. Under continuously flooded conditions, there is decreased decomposition, which would decrease the P released from breakdown products, and denitrification rates are higher, with subsequent loss of N_2O and N_2 to the atmosphere. Under aerobic conditions, the wetland sediments generally contain the less-soluble forms of Fe and aluminum (Al), with P retained below the soil/water interface. There is generally increased decomposition, with increased breakdown of organic matter to inorganic P and nitrate.

Alternate wet and dry periods maximize the fixation of inorganic forms of both nitrogen (N) and P. During dry periods, the NO_3 builds up in the soil, which is subsequently rapidly denitrified under anaerobic conditions. There is also a greater fixation of P in seasonally waterlogged soils due to the release of Fe during flooding and its subsequent oxidation and precipitation as the soil dries (Patrick and Mahapatra 1968).

Details of the effects of organic soils on N and P dynamics under alternately wet and dry conditions are sparse. While the general dynamics may follow those described above, lake sediments and organic substrates do not always follow the predicted P, Fe, and N

180

dynamics. There is some indication that organic acids may retard the oxidation of ferrous Fe (Theis and Singer 1974). Submergence of the soil may also lead to increases in the amorphous Fe content, which increases the adsorption sites for P (Kuo and Mikkelsen 1979). Soils low in ferric phosphate have shown a decreased availability of P upon waterlogging (William et al. in Patrick and Mahapatra 1968). In addition, lake studies have shown that, when nitrate concentrations remained above 0.001 mg/l, no net release of P took place under anaerobic conditions. The oxidized N was able to buffer the redox potential of the surface sediment so that the P remained bound to the sediments.

It is apparent that the N and P relationships in wetland systems are not simple, particularly in sediments that are alternately wet and dry. In a study of wastewater application to a wetland in Clermont, Florida, the objectives were to determine the effects of a fluctuating water table on the retention of N and P from treated sewage effluent by the marsh system and its various components (vegetation, soil, and water). These effects were enhanced by the rainfall regime during the study period, with the first year of the study being very dry and the second year wet.

The research marsh was located in the center of a large wetland area adjacent to the town of Clermont's (population 4,800) sewage treatment plant (0.6 MGD). After primary and secondary treatment, the sewage was pumped from oxidation ponds to a percolation pond and thence to experimental plots in the research marsh. The experimental plots were 2000 m^2 of marsh surrounded by a fiberglass dike buried 0.5 m into the peat substrate. Water that was pumped on the plot left through the peat. Treated effluent was applied to plots at three different rates (Plot L, low--1.5 cm/week; Plot M, medium--3.7 cm/week; Plot H, high--9.6 cm/week). A control plot (Plot C) received 4.4 cm/week of water of drinking-water quality. The water was applied each week over a 24-hour period. Fig. 11.1 shows the research site.

The natural marsh vegetation was dominated by <u>Sagittaria</u> <u>lancifolia</u> and <u>Pontederia</u> <u>cordata</u>, with <u>Panicum</u> and <u>Hibiscus</u> species common. The marsh substrate was 88% organic soil, 1.5 m deep, bounded by a sand and then a clay layer. The water table fluctuated from 5 cm below the peat surface to 20 cm above. Further details of the research design and methods are in Dolan et al. (1981). Monthly measurements of total dissolved P, ortho P, total dissolved N, NH_4, NO_3, NO_2, and organic N and P were taken from surface- and ground-water in the plots. Above- and belowground biomass and nutrient content were measured over the growing season (Dolan et al. 1981).

The marsh vegetation was influenced as much by the wet- and dry-year variation as by the treated sewage effluent. Standing crop measurements of the plots showed that there was no difference in biomass between the control, low-, or medium-effluent plots during either wet or dry years (Fig. 11.2). Plot H demonstrated higher biomass than the control plot during the dry year (777 g/m^2 versus 580 g/m^2, respectively) but not during the wet year (853 g/m^2 versus 900 g/m^2, respectively). There was no difference in biomass in Plot H during dry and wet years (777 versus 853 g/m^2). Differences within the effluent plots (adjacent to effluent pipe versus at the plot periphery) were greater than differences between the plots.

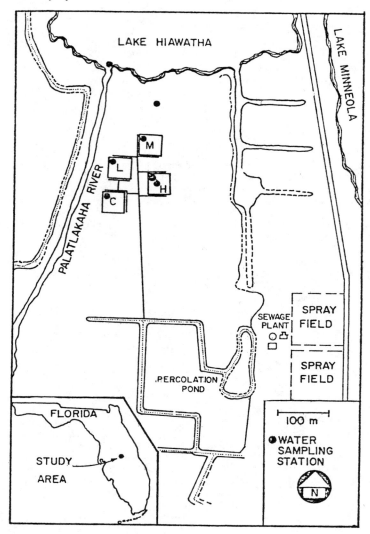

Figure 11.1
Location of the experimental plot in Clermont, Florida.
(From Dolan et al., 1981)

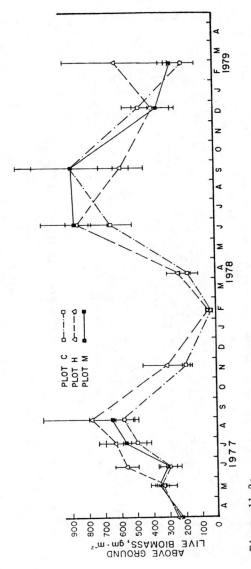

Figure 11.2a
Aboveground live biomass, ±1 SD, in: Plot C (4.4 cc/week freshwater), Plot M (3.7 cc/week treated effluent), Plot H (9.6 cc/week treated effluent).

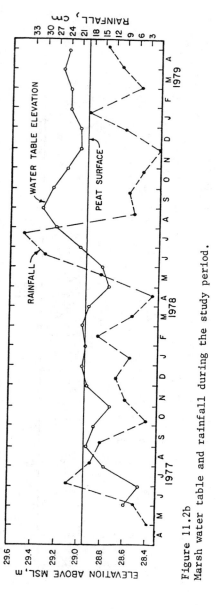

Figure 11.2b
Marsh water table and rainfall during the study period.

184

Comparisons of the daily growth rates in the periphery of the plot with those near the effluent discharge pipe (periphery versus inside) again show the influence of the wet and dry years on the growth rate. Table 11.1 shows the daily net primary production in the control plot and in Plot H during the wet and dry years. Increased growth due to the treated effluent (the inside area) was detected in Plot H during the dry year (9.8 g m^{-2}dy^{-1}). Plot H and Plot C both demonstrated high growth rates during the wet year when additional water was available (8.7 and 7.6 g m^{-2}dy^{-1}). The control plot (drinking water plus the normal seasonal flooding) yielded growth rates almost as high as with 9.6 cm/week of effluent. During the dry year in the area of the periphery (75% of the plots), both Plot H and Plot C had low daily growth rates (1.76 and 2.1 g m^{-2}dy^{-1}). Obviously, neither water nor effluent reached the periphery of the plot. Those areas receiving drinking water during the dry year (4.5 g m^{-2} dy^{-1}) or some rainwater during the wet year (4.0 g m^{-2} dy^{-1}) or some effluent plus rain (5.7 g m^{-2} dy^{-1}) had equivalent daily growth rates.

Not only were growth rates and standing crop highly dependent on the water regime, but the chemical cycling was also strongly influenced by the presence or absence of standing water. Table 11.2 summarizes the differences between plots due to water alone or to treated wastewater. There was no difference between plots C and H in aboveground dead biomass or belowground biomass during either the wet or dry year. Only during the dry year was the aboveground live biomass higher in Plot H. Decomposition rates did not vary between plots during the wet year. P content of the various compartments was the same in both plots during the wet year. During the dry year, Plot H generally had higher tissue concentrations than did Plot C. N concentrations were not as consistent between plots as P content. Aboveground live tissue was higher in Plot H than in Plot C during both years. N content apparently was more influenced by the availability of N in the effluent than by changes in the oxidation state of the substrate. N in the belowground biomass was also higher in Plot H than in Plot C during the wet year. Differences between the plots in N content for other compartments were not detected.

The presence or absence of standing water (or changes in the oxidation state of the substrate) did not have significant effects on the ability of the marsh to remove nutrients from treated wastewater. As shown in Table 11.3, during the two years of the study, most of the P and N was removed by the marsh during both wet and dry years. More N was removed during the wet year, possibly due to denitrification, but there was still a significant N reduction (71%) in the dry year.

To conclude, the marsh removed both N and P during wet and dry years. In fact, only at the highest level of effluent application (9.6 cm/week) could we detect any differences in vegetation or soil chemistry due to the effluent. Plots receiving 1.5 cm/week and 3.7 cm/week of effluent could not be distinguished from the control plot. While the species composition changed with the addition of effluent, it also changed as a result of marsh-water levels. Vegetative growth rates, standing crop, and P tissue content were influenced as much by the presence of standing water as they were by the application of 9.6 cm/week treated effluent. N tissue concentrations were more related to the effluent application than to the water levels.

Table 11.1
Net Productivity of a Freshwater Marsh Comparing the Treated Effluent
Application with Natural Water Level Fluctuation

	Plot C 4.4 cm wk⁻¹ Freshwater Addition			Plot H 9.6 cm wk⁻¹ Nutrient Addition		
	Production ($gm^{-2}dy^{-1}$)		Total P Applied ($gPm^{-2}yr^{-1}$)	Production ($gm^{-2}dy^{-1}$)		Total P Applied ($gPm^{-2}yr^{-1}$)
	Periphery	Inside		Periphery	Inside	
Low-water-level year (1977–78)	2.1	4.5	0.38	1.65	9.8	49.4
Moderate-water-level year (1978–79)	4.0	7.6	0.41	5.7	8.7	42.0

Table 11.2
Summary of the Differences Between Freshwater Plots Due
to Water Alone (Plot C) vs. Treated Wastewater
(Plot H) During Wet and Dry Years

	Dry Year	Wet Year
Biomass		
Aboveground live biomass	+	0
Aboveground dead biomass	0	0
Belowground biomass	0	0
(inner part of plot only)		
Decomposition rate of		
aboveground material	NA	0
by month 3 and 9		
Nitrogen concentrations		
N in aboveground live tissue	+	+
N in aboveground dead tissue	0	0
N in belowground biomass	NA	+
N in belowground soil	NA	0
Phosphorus concentrations		
P in aboveground live tissue	+	0
P in aboveground dead tissue	+	0
P in belowground biomass	+	0
P in belowground soil	$-^a$	0

Note: Symbols are as follows: Plot H higher than Plot C (+);
Plot H lower than Plot C (-); no difference between plots
at $\alpha 0.05$ (0); NA means not sampled.
[a] Except surface 0-25 cm in Plot H is higher (+).

Table 11.3
Annual Retention of Applied Phosphorus and Nitrogen in the
Freshwater Marsh Plot Receiving 9.6 cm wk^{-1} Treated
Wastewater

	Dry Year (%)	Wet Year (%)
Phosphorus		
Well on edge of plot	98.56	99.31
Well below effluent pipe	(97.97)	(93.65)
Nitrogen		
Well on edge of plot	71.19	88.42
Well below effluent pipe	(62.87)	(85.41)

Note: Percent removal based on P and N measured in outflow.

REFERENCES

Brown, S., 1981, A comparison of the structure, primary productivity, and transpiration of cypress ecosystems in Florida, Ecol. Monog. 51:403–427.

Dolan, T. J., S. E. Bayley, J. Zoltek, Jr., and A. J. Hermann, 1981, Phosphorus dynamics of a Florida freshwater marsh receiving treated wastewater, J. Appl. Ecol. 18:205–219.

Kuo, S., and D. S. Mikkelsen, 1979, Distribution of iron and phosphorus in flooded and unflooded soil profiles and their relation to phosphorus adsorption, Soil Sci. 127:18–24.

Patrick, W. H., and I. C. Mahapatra, 1968, Transformation and availability to rice of nitrogen and phosphorus in waterlogged soils, Adv. in Agron. 20:323–359.

Theis, T. L., and P. C. Singer, 1974, Complexation of iron (III) by organic matter and its effect on iron (II) oxygenation, Environ. Sci. and Tech. 8(6):569–573.

DISCUSSION

Giblin: I'm just wondering if the species composition shifts can have anything to do with the differences in phosphorus content?

Bayley: I think the species composition in that area has as much to do with whether the plot is wet or dry. For example, Hibiscus invaded the high–effluent plot, but also invaded other plots when it was dry. I think it could invade under either of those conditions.

Giblin: I mean that the difference in phosphorus retention observed in the dry year may have been masked by a shift in species composition the following year.

Bayley: I don't think so.

Valiela: At the periphery of Plot H, was the phosphate in the water higher than in the control area?

Bayley: Yes, it was. So there was some storage of phosphorus during the wet year. There was a fairly high level of phosphorus in the surface water, but it was higher in the effluent plots than in the control plot.

Valiela: Was this inorganic phosphorus not bound or complexed in the organic matter?

Bayley: No, it was a totally dissolved phosphorus.

Odum: I suppose the Clermont Marsh was already somewhat adapted to wastewaters, since seepage from the main sewage flow had been working its way into that area before the experiment.

Bayley: It might have been possible; it's hard for us to rule that out. We were, of course, bound by where we could put pipes. But we did detect differences between the natural marsh, the control marsh, and the experimental when we averaged out seasonal and short-term fluctuations.

Kappel: What was the method of application in the high rate area? Was it applied in a slug or . . . ?

Bayley: We put it in over a 24-hour period once a week.

Kappel: What was the response in the high-rate area as regards water levels inside the chamber and outside the chamber?

Bayley: We had water-level recorders in both places and they gave the same readings. That is, they receded to the same levels as those outside of the plot within 24 hours.

Kappel: So, basically, the effluent was going in and pushing out what water was there . . .

Bayley: Yes.

Zedler: You suggested that one reason productivity increases is that there is a reduced stress.

Bayley: I was quoting papers by Brinson and Brown.

Zedler: What do you mean by the stress?

Bayley: I think they talked about anaerobic stress resulting from a continuously anaerobic environment. With running waters there would be less stress than with still waters, because there is greater flushing and more oxygenation. Is that what you meant [Mark Brinson]?

Brinson: Yes.

12

Significance of Hydrology to Wetland Nutrient Processing

H. Hemond and W. Nuttle

In many wetland ecosystems, the major nutrient processing occurs at or beneath the surface of the sediments. Accordingly, to understand nutrient processing by a wetland ecosystem, we must understand the manner in which nutrients reach the belowground sites of processing and the manner in which products of nutrient processing may possibly be removed from the sediments and exported. Belowground processes with which we may be concerned in wastewater treatment include uptake by microbes or macrophyte roots, cation exchange, adsorption, dissimilatory reduction by microbes, biologically mediated or chemical oxidation, and a variety of other decomposition processes; in all cases, the processes occur only after the chemicals have been physically carried to the active sites. The major mechanisms of transport are two--namely, active transport by plants and physical transport by water movement. We shall address the latter process.

Water moves through wetland peat according to Darcy's Law:

$$\vec{q} = -\underline{K}\vec{\nabla}\phi$$

where $\vec{q}$ is specific discharge (water flow per unit cross-sectional area), $\underline{K}$ is hydraulic conductivity (we assume for the moment that peat is isotropic), and ϕ is hydraulic head. Peat however, differs from most porous materials in a rather obvious way; it is highly compressible.

Water storage in a porous material results from three mechanisms; (1) change in saturation of the soil pores, (2) change in volume of the soil pores, or (3) compression of water. Mechanism 1 dominates in shallow sandy or gravelly aquifers; mechanism 3 is important in very deep aquifers. Peat is unique in that mechanism 2 may dominate.

The consequences of mechanism 2 are several. A certain amount of water may leave a peat soil without causing desaturation. Water may infiltrate into a fully water saturated peat. Finally, water storage in peat becomes very much conditional upon the surface loading of the soil, and transient mechanical loading of the soil causes nearly instantaneous changes in hydraulic head within the soil. This latter phenomenon presents us with measurement problems. A complete mathematical expression for water flow at a point in an (isotropic) peat soil is as follows:

$$\rho \int_s \underline{K \vec{\nabla} \phi} \cdot \underline{\vec{n} ds} = \rho \underline{V}_s \ (e \ \frac{\partial S}{\partial \psi} + \frac{S \gamma C}{2.3\sigma}) \ \frac{D \psi}{Dt} \qquad (12.1)$$

(See DEFINITIONS for symbol meaning.)

What does this mean to people concerned with the processing of nutrients by wetlands? It means we must understand peat itself--we need to characterize it physically as well as biologically, if we are to predict how water will move into and through it. Our minimum list of important physical peat properties includes: (1) porosity ($\underline{n}$); (2) hydraulic conductivity ($\underline{K}$); (3) compression index ($\underline{C}$) (Fig. 12.1); and (4) the saturation curve (percent saturation as a function of pore pressure (Fig. 12.2).

There is a great paucity of these data for wetland soils. Given these data and given information on the sources and sinks of water in a wetland peat (evaporation, plant uptake, surface water, and ground-water), we can predict the seepage velocity $\vec{\underline{v}}$ of wetland pore water.

Let us take this one step further. Once we have this seepage velocity $\vec{\underline{v}}$, we have a critical quantity that is common to the transport of each and every dissolved nutrient. Let $\underline{c}_i$ be the concentration of a nutrient $\underline{i}$, and $\underline{r}_i$ be a processing rate (a source or sink term) for that nutrient. $\underline{D}$ is a dispersion coefficient related to $\vec{\underline{v}}$, physical peat properties, and the molecular diffusivity of the nutrient molecule. We assume it to be a constant here. Following the one-dimensional advection-dispersion equation in the $\underline{x}$-direction:

$$\frac{\partial c_i}{\partial t} + \underline{v} \frac{\partial c_i}{\partial x} = \underline{D} \frac{\partial^2 c_i}{\partial x^2} + \underline{r}_i \qquad (12.2)$$

Notice that once we have the physical information on the wetland system, plus a nutrient-processing rate, we can compute the nutrient concentrations at any location or time for any set of nutrient inputs to our wetland.

Of course, we're all aware that the most pressing questions of wetland biogeochemistry today relate to that processing term $\underline{r}_i$; herein lies the other reason that an understanding of hydrology is critical. Once we have the physical data and suitable nutrient-concentration data, we can compute the internal nutrient-processing rates for a system ($\underline{r}_i$'s) and learn more about their variation in space, in time, and in relation to the chemical background and the wetland biota.

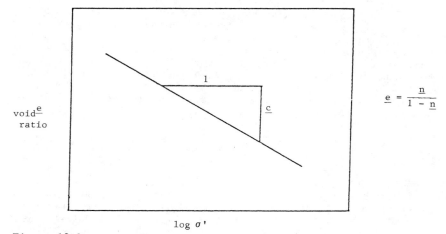

$$e = \frac{n}{1 - n}$$

void $\underline{e}$
ratio

log σ'

Figure 12.1
The definition of the compression index for porous material.
$\underline{C}$ = compression index.

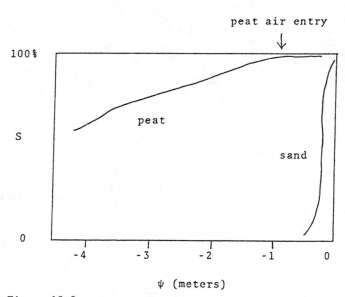

Figure 12.2
The saturation curve for a typical peat soil.
Significant amounts of water under negative pressure
(less than 1 atmosphere) may be stored in peat.

DEFINITIONS

$\vec{q}$ specific discharge = volume of flow per unit area per unit time

$\underline{v}$ seepage velocity = average velocity of pore water

$\underline{n}$ porosity

$\underline{K}$ hydraulic conductivity (defined by Darcy's Law, in text)

ϕ hydraulic potential, piezometric head. Equal to height of water above a reference <u>plus</u> pressure of water divided by its weight density

$\underline{C}$ compression index; see a standard text such as <u>Soil Mechanics</u>, W. T. Lambe and R. V. Whitman, Wiley, New York, 1969, for more information

$\underline{D}$ a dispersion coefficient, which is a measure of a sharp concentration front's tendency to spread out as it travels through a soil; see a text such as <u>Groundwater</u>, R. A. Freeze and J. A. Cherry, Prentice-Hall, Englewood Cliffs, N. J., 1979

ρ mass density of water

γ weight density of water = ρ x acceleration of gravity

$\underline{S}$ percent saturation of soil

$\underline{V}_s$ percent volume of solids in soil

$\underline{s}$ surface of an arbitrary small volume defined for Equation (12.2)

σ effective stress on soil matrix (see a soil mechanics text)

$\underline{e}$ void ratio = $(1-\underline{V}_s)/\underline{V}_s$

ψ pore water pressure

$\underline{c}$ concentration of $\underline{i}$

$\vec{\triangledown}$ gradient operator; see a standard reference text, such as the <u>Handbook of Chemistry and Physics</u>, C. D. Hodgman, ed., Chemical Rubber Co., Cleveland, Oh., 1958

V volume

$\dfrac{D\psi}{Dt}$ total derivative of ψ with respect to time

$\underline{r}_i$ processing rate for nutrient $\underline{i}$ (source or sink term)

DISCUSSION

R. Kadlec: A couple of comments I'd like your reaction to: First, I
don't think you can put that hydraulic conductivity outside that
gradient operator the way you did in Darcy's Law because I am person-
ally convinced that the hydraulic conductivity of peat is, first of
all, variable with depth, which means it ought to be inside that
operator; and second, that peat is probably anisotrophic, meaning
that you probably ought to be dealing with a tensor quantity in peat.
That's number one. Number two, I contend that the diffusivity in
your mass transport equation could be estimated to be insignificant
for flow through peat under almost all circumstances.

Hemond: First of all, those equations are defined at a point. It is
correct that we need to know how the parameters vary--indeed they do
vary--in order to integrate. If we were to put a model together,
typically we would chop our wetland up into slices (a vertical, one-
dimensional model) having different conductivities. We would integrate over all the layers, knowing how they vary with distance. I am
most concerned with our understanding of the hydraulic conductivity
in the system. We can measure how properties such as conductivity
vary with depth in a fairly straightforward fashion; but it is more
difficult to know how the properties vary with pore pressure or with
the void ratio. Also, existing numerical models usually do not
account for changes in (hydraulic) conductivity that occur, for
example, as the peat dries out and shrinks. Another problem is
accounting for changes in Bishop's parameter (which influences pore
water pressure-total stress relationships) as the peat shrinks upon
drying or desaturation. If that is a significant phenomenon, then,
indeed, it has to be considered in modeling. As for isotrophy, it is
the reason that I specified isotropic peat earlier. Once we add two
or three dimensions, we probably do have to deal with anisotrophy.
However, it may not always be a severe problem. To our surprise, the
few points available for Sippewisset Marsh show that there is only a
factor of two to one or so between horizontal and vertical conductivities at the point we sampled.

Valiela: It might be interesting, Harry, if you could give us a few
examples from your own work about what concepts can be developed by
an exercise such as this, which you wouldn't have known without
applying these kinds of models.

Hemond: I should, first of all, point out that much work remains to
be done. Quantitative applications are coming, and some causal
relationships are becoming clearer. Let's talk about Sippewisset.
If we are looking, for example, at the issue of short versus tall
Spartina alterniflora, a flow model incorporating peat characteristics can be helpful--and I should add that we really need a salt
balance here, too, because we have to look at the transport equation
for salt. A flow model can show us that there are at least two
sources of water that are quantitatively important to the rooting
zone. We have fresh water from below and sea water from above; we
could also imagine that we were applying wastewater from above.
Field experiments and model results both tell us the importance of
Spartina transpiration. We can now conceptualize how the soil pore
water status is determined by this balance of hydrologic factors,

including evapotranspiration. In the case of vegetation like
Spartina, we know that its transpiration rate is going to be strongly
influenced by salinity. Once we have a flow model, we can determine
how the hydrology is finally influenced by evapotranspiration.
Transport by the flow system in turn controls the salinity of the
flow water, which in turn influences evapotranspiration. We can see
how the physical and biotic factors in this case are intimately
coupled. This is contrary to what is often assumed, where a hydrologic regime is given, and then we add the plants, and then we add
the chemical transformations. Instead, we can see examples where,
even on short time scales of weeks and months, the two are closely
coupled--the hydrologic regime influences the vegetation, which in
turn directly exerts a major controlling factor on the hydrology.

Bastian: Harry, how much difference would you expect to see [once
this model is developed] between what you describe in the Sippewisset
Marsh and what might occur in a fairly shallow pond full of Potamogeton and a fairly extended sediment? Would you expect to see that
much difference in the water quality?

Hemond: I'm not sure I understand what you are asking.

Bastian: In a shallow pond with static water packed full of pond
weeds, you get a lot of nitrogen transformation and loss. Do you see
any relationships between a wetland area that doesn't have as much
water but, nevertheless, the sediment is still saturated the whole
time?

Hemond: Conceptually, they're very different places, even if the
peat is completely saturated. This shows up when you look at that
mass transport term, in which you have that dispersion coefficient
[D]. That dispersion coefficient includes transport by any kind of
water movement and it also includes molecular diffusion. That D is
very much lower, orders of magnitude lower in a peat system, because
it is impeded by the porous material and it is impeded by the absence
of wind- or thermally driven circulation. The system is transport
limited when you put in peat, whereas in the shallow pond full of
Potamogeton or what-have-you, you still have substantial advective
movement and, hence, there's a lot of dispersion.

SESSION IV

Community Changes

13

Effects of Wastewater on Wetland Animal Communities

Kathleen M. Brennan
(Steven D. Bach and Gregory L. Seegert,
contributing authors)

Little is known of the effects of the discharge of treated
municipal wastewater on the animal communities of freshwater wetland
ecosystems. The addition of nutrients, suspended and dissolved
solids, chlorine, heavy metals, and disease organisms and the alterations of flow periodicity and rate, water level, pH, and alkalinity
constitute changes in environmental conditions that could result in a
wide range of changes in communities of invertebrates, fish, and
other wildlife (amphibians, reptiles, birds, and mammals). Because
of the potential impacts and the lack of direct knowledge of impacts,
wetland use for wastewater treatment must be carefully considered.
Most of the studies performed at wetland treatment sites have focused
on changes in water quality, vegetation, and litter; the only studies
of wildlife at natural wetlands have been performed in Florida,
Michigan, and Wisconsin. Preliminary studies on wildlife have also
been done at two volunteer wetlands in Michigan. Because so few
studies have been done on the primarily terrestrial animals inhabiting wetland ecosystems that receive wastewater, the potential
effects of wastewater application must be identified by inference
from wildlife use of such treatment facilities as effluent lagoons,
stabilization ponds, and old field or forest land irrigated with
wastewater.
On the positive side, the availability of nutrient-bearing water
from a treatment facility constitutes a potential resource that could
be used to restore or maintain a disturbed wetland, create an artificial wetland to provide additional habitat, or maintain a volunteer
wetland (developed as a consequence of the growth of wetland plants
in a flood irrigation field or in a terrestrial plant community
through which a wetland discharge passes), and consequently, to
increase the diversity of wildlife habitat.
The known and anticipated effects of the discharge of treated
municipal wastewater on the animal communities of freshwater wetland
ecosystems are presented in this paper according to three groups of
animals present in wetland ecosystems: (1) invertebrates, (2) fish,
and (3) wildlife (amphibians, reptiles, birds, and mammals). Information in the literature about the effects of wastewater on each of
these groups is summarized and the major types of impacts identified
or postulated are presented in the form of impact factor trains

(see Figs. 13.1, 13.2, and 13.3). In these diagrams, changes in the physical and biological environment are indicated as potential causes of subsequent changes in the animal communities. Effects on animal health, including potential transmission of disease to humans, are mentioned briefly. The use of wastewater for enhancement of wildlife habitat is also discussed in the following paper.

EFFECTS OF WASTEWATER ON ANIMAL COMMUNITIES

INVERTEBRATES

Wetland ecosystems contain abundant and diverse populations of invertebrates, including protozoa, sponges, coelenterates, flatworms, rotifers, annelid worms, nematodes, amphipods, cladocerans, copepods, other crustaceans, mollusks, and various insect groups (Weller 1979). Typical densities of aquatic invertebrates by wetland type are given in Tilton and Schwegler (1978). Wetland systems generally are characterized by much higher diversity and abundance than are nearby unvegetated areas (McKim 1962). The invertebrate populations of wetlands are believed to play a central role in the transfer of energy through detrital, photosynthetic, or predatory portions of the food chain (Weller 1979; Swanson 1978a, 1978b). This is especially true of crustaceans, particularly crayfish. Invertebrates may also be key factors in the transfer of materials between the water and sediment compartments. Gallepp (1979), in laboratory studies, reported that increases in temperature from 10°C to 20°C of lake sediments containing two species of chironomids were followed by a tenfold increase in phosphorus (P) release rates. Without chironomids, temperature change had little effect on release rates from sediments. A Swedish investigator (Graneli 1979) noted that the introduction of chironomid larvae into sediments of eutrophic lakes greatly increased the transport of silica, P, and iron (Fe) from the sediments to the water but did not affect the rate of transfer of inorganic nitrogen (N).

Several studies of the relationships between aquatic invertebrates and waterfowl have been conducted (Swanson 1978a, 1978b; Krull 1970; Kaminski and Prince 1981). Studies on the responses of aquatic macroinvertebrates to prolonged flooding and on the effects of varying nutrient levels on macroinvertebrate production are in progress at the Delta Marsh Research Station in Manitoba, Canada.

Application of wastewater to a wetland ecosystem could cause several impacts on resident invertebrate populations:

Changes in structure and function of resident invertebrate populations, leading in turn to changes at higher trophic levels,

Contamination of the wetland food chain by heavy metals or chlorine and chlorinated hydrocarbons,

Transmission of viral or bacterial diseases to higher food chain organisms such as birds, fish, and mammals (including man).

Changes in invertebrate population structure potentially could be caused by reduction in dissolved oxygen (DO) levels due to elevated biochemical oxygen demand (BOD), introduction of elevated amounts of sediment, introduction of toxic compounds, or changes in vegetation composition due to the introduction of wastewater. These changes could include reductions in abundance and diversity or shifts to more "tolerant" types of benthic organisms. Such changes would alter the pattern of energy flow within the wetland and thus produce significant changes in community structure and function.

Direct bioconcentration of toxic substances within individual organisms and subsequent biomagnification of these substances in wetland food chains presumably could cause similar types of community change. However, Macek, Petrocelli, and Sleight (1979) determined that for many organic chemical residues, biomagnification through food chains was insignificant when compared with bioconcentration of residues directly from the water, and that "the process of biomagnification, as classically defined, probably does not occur within communities of aquatic organisms" (p. 251).

In land application of municipal wastewater effluent, Dindal, Newell, and Moreau (1979) noted that much of the effluent renovation is due to the activities of invertebrates present in the upper layers of the soil. These researchers reported that, in general, populations of mites were suppressed, populations of springtails showed periodic seasonal patterns of increases, and populations of earthworms increased in number, biomass, and community structure. Many organisms with low biomass (mites) were replaced by fewer organisms with greater biomass (earthworms). No gross uptake of heavy metals by earthworms and no heavy-metal toxicity were observed.

Introduction of various diseases also could result in transfer between various portions of the food web, resulting in contamination of higher food chain organisms. The major types of impacts on invertebrate communities are shown in Figure 13.1.

Addition of secondarily treated wastewater potentially could alter the structure and function of resident insect populations. For example, a shift in the composition of vascular plant species could favor some nonaquatic species over others. If the loading rate is sufficient and DO levels are lowered, changes in aquatic insect populations would also be expected. The two most important changes indicative of adverse impact are (1) possible increases in species and numbers of insects that carry diseases that could be transmitted to man or wildlife; and (2) increases in species and numbers of aquatic insects, such as chironomids, that are indicators of significant loss of water quality in the wetland.

However, very little is known about insect populations in natural freshwater wetlands, and even less is known about these populations in wetlands that receive treated wastewater. The only studies of the effects on insects in the latter situation are those of Jetter (1975), Witter and Croson (1976), Davis (1978), McMahan and Davis (1976, 1978), and Fritz and Helle (1978). Jetter (1975) reported that certain species of dipteran flies, including members of the families Psychodidae, Ephydridae, Tipulidae, Dilichopodidae, and Syrphidae, were very abundant (densities of over 3000 individuals per square meter) on mats of floating organic material produced when treated wastewater was added to a cypress dome in Florida. When the volume of wastewater was decreased, the number of individuals of these species declined dramatically. McMahan and Davis (1978) found

Changes in
Environmental Conditions

Possible Impacts on
Benthic Macroinvertebrates

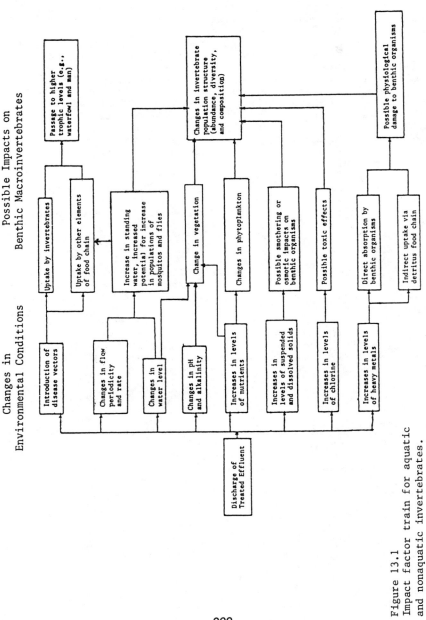

Figure 13.1
Impact factor train for aquatic
and nonaquatic invertebrates.

no evidence of a relationship between cypress dome perturbation and
species diversity. Diversity was retained despite eutrophication due
to the addition of wastewater. Davis (1978) noted that populations
of virus-carrying mosquitoes were lower in a cypress dome that
received sewage effluent than in an untreated control dome. Fritz
and Helle (1978) reported that encephalitis-carrying mosquitoes may
breed within cypress domes receiving treated wastewater, but that
these same species bite "mainly" birds and avoid areas inhabited by
humans. Other varieties of "human pest mosquitoes" were reduced in
numbers in cypress domes receiving treated wastewater because the
water level did not fluctuate, and thus their breeding grounds were
eliminated (Davis 1978).

Because of these potential impacts, monitoring of the status of
wetland invertebrate populations is an important datum to consider in
designing studies of wastewater applications. Witter and Croson
(1976) recommended that indicator species be used to predict the
impact of wastewater application on resident insect populations.
Demgen (1979) summarized the results of studies of aquatic invertebrate populations in a California wetland receiving treated wastewater. (Zooplankton were also included in this study.) She reported
shifts in abundance of major invertebrate species as a result of the
added wastewater but indicated that the new populations appeared to
be stable.

FISH

The major types of impacts on fish communities (Fig. 13.2) are:

Changes in species composition,

Changes in productivity or biomass,

Changes in spawning success,

Toxicity: acute, chronic, or sublethal,

Changes in incidence of disease in fish or in the potential
for fish acting as vectors for mammalian pathogens.

Although it is generally acknowledged that wetlands are available to fish as permanent habitat, as spawning habitat, and for the
production of forage, quantitative data to support this widely held
assumption are lacking (except for the observation that fish do, in
fact, survive and thrive in wetlands generally).

Wetlands are by their nature highly dynamic systems and
frequently are open systems (hydrologically connected to other water
bodies). These two characteristics make quantitative measurements
extremely difficult because of the inherent natural variability of
the wetland. The logistical difficulties of sampling wetlands also
have prevented studies from being undertaken.

Jaworski and Raphael (1978) reported that at least 32 species of
fish commonly use the wetlands in Michigan. Many of these use the
wetlands for spawning. The conditions necessary for successful
spawning are often more rigorous than those necessary to support
other portions of the life cycle. Successful reproduction obviously

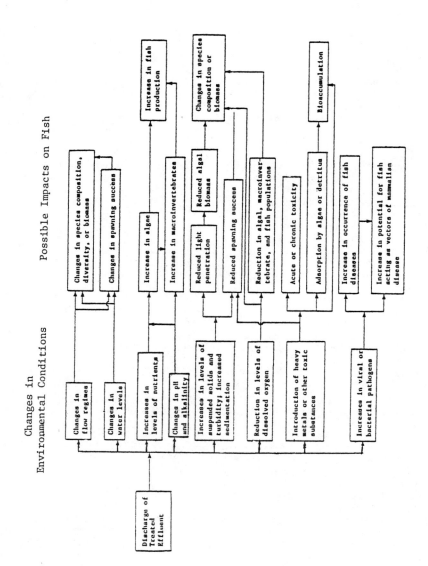

Figure 13.2
Impact factor train for fish.

is necessary for general populations to be self-sustaining. For those species that use wetlands primarily as spawning and nursery areas, successful reproduction in any particular wetland is not necessary each year, provided that alternative spawning areas are available. If alternative areas are not available, then successful spawning during at least some years is crucial to the continued well-being of the species in question.

Literature specifically describing the impacts of wastewater effluents on fishes in wetlands is nearly nonexistent. This is not surprising, because fisheries investigations in wetlands are difficult enough without the addition of a complicating factor such as wastewater effluent. A great deal of information is available regarding the general impacts on fish communities caused by the discharge of wastewater. This information is reviewed in sources such as Tsai (1975), Brungs et al. (1978), and Spehar et al. (1980). The constituents of primary concern are residual chlorine, ammonia, detergents, low DO, pH, and heavy metals. These studies, however, generally deal with discharges into lakes and streams, not wetlands. Thus, their usefulness in predicting impacts on wetland fish communities is uncertain.

In one of the few studies of wetlands that did address fish communities, Jetter and Harris (1976) reported that a cypress dome in Florida receiving sewage effluent was a much poorer habitat for fish than was a natural cypress dome.

Janssen (1970) reviewed the relationship between human disease organisms in polluted waters and fish diseases and concluded that most fish-borne human diseases do not cause disease in fish. Tsai (1975) examined the possibility of fish acting as vectors of mammalian diseases and cited several studies that showed a relationship between highly polluted waters and the presence of certain diseases and tumors in fish. The nature of this relationship was unclear. Tsai (1975) also cited several studies whose authors concluded that fish were not important factors in the transmission of pathogenic bacteria to humans. However, Janssen (1970, p. 284) cited several authors who did establish such links and suggested that "fish may be more important vectors of human infectious disease than generally realized."

OTHER WILDLIFE (AMPHIBIANS, REPTILES, BIRDS, AND MAMMALS)

Wildlife use of a wetland is dependent on the ability of the wetland, often in combination with the wetland edge and adjacent upland areas (Porter 1981), to provide the necessary life requisites of food, shelter, protection from predators, and appropriate areas for breeding, nesting, and other events in the daily, seasonal, and annual activity and population cycles of each species. Thus, the number of species present and the number of individuals of each species present at any one time are dependent on the needs of the species at that time, the availability of the wetland to fulfill those needs, and the overall population status of the species in the region in which the wetland is located (and in other areas as well for migratory species). Severe or favorable weather conditions, heavy predation losses, or population cycles within the species itself may cause changes in the abundance of species from year to year that are not related to the characteristics and condition of the

wetland. The changes in the wildlife community at a particular
wetland that may be due to the introduction of wastewater will result
from changes in the water level, the structure and composition of the
vegetation, the interspersion of vegetation and water, and the
availability of submersed plants, aquatic and terrestrial insects,
and other sources of food. Changes in the number and abundance of
adults and young of prey species such as amphibians, small birds, and
small mammals will result in changes in the number and abundance of
carnivorous species of birds and mammals that constitute the higher
trophic levels.

The changes in water levels associated with the introduction of
treated wastewater also would be expected to result in shifts of the
locations of wildlife breeding and activity sites within the wetland
and between the wetland and adjacent wetland or upland areas. These
water-level changes will also influence the types and amounts of food
organisms present, and the wastewater entering the wetland may
contain pathogenic organisms and heavy metals that can be taken up
directly by individual wildlife or indirectly through the organisms
on which they feed. The general types of changes in wildlife habitat, wildlife use of the wetland, and wildlife health that may occur
subsequent to the introduction of treated wastewater to a wetland are
indicated in Figure 13.3. These include:

Changes in habitat structure and components

Changes in types and distributions of plant communities
(structural diversity, species composition, number and extent
of ecotones)

Changes in natural animal corridors

Change in degree of interspersion between vegetation and open-
water areas

Change in linear extent of edge and open-water areas

Changes in species richness (number of species) and species
density (number of individuals)

Changes in trophic structure and food web

Introduction of or increase in weedy species

Changes in wildlife use patterns between treated and control
areas and between wetland and upland areas

Changes in presence and abundance of indicator species

Changes in incidence of disease, wildlife condition, and
potential for wildlife to act as vectors for human disease

Changes in presence and abundance of **endangered**, threatened, and
rare species.

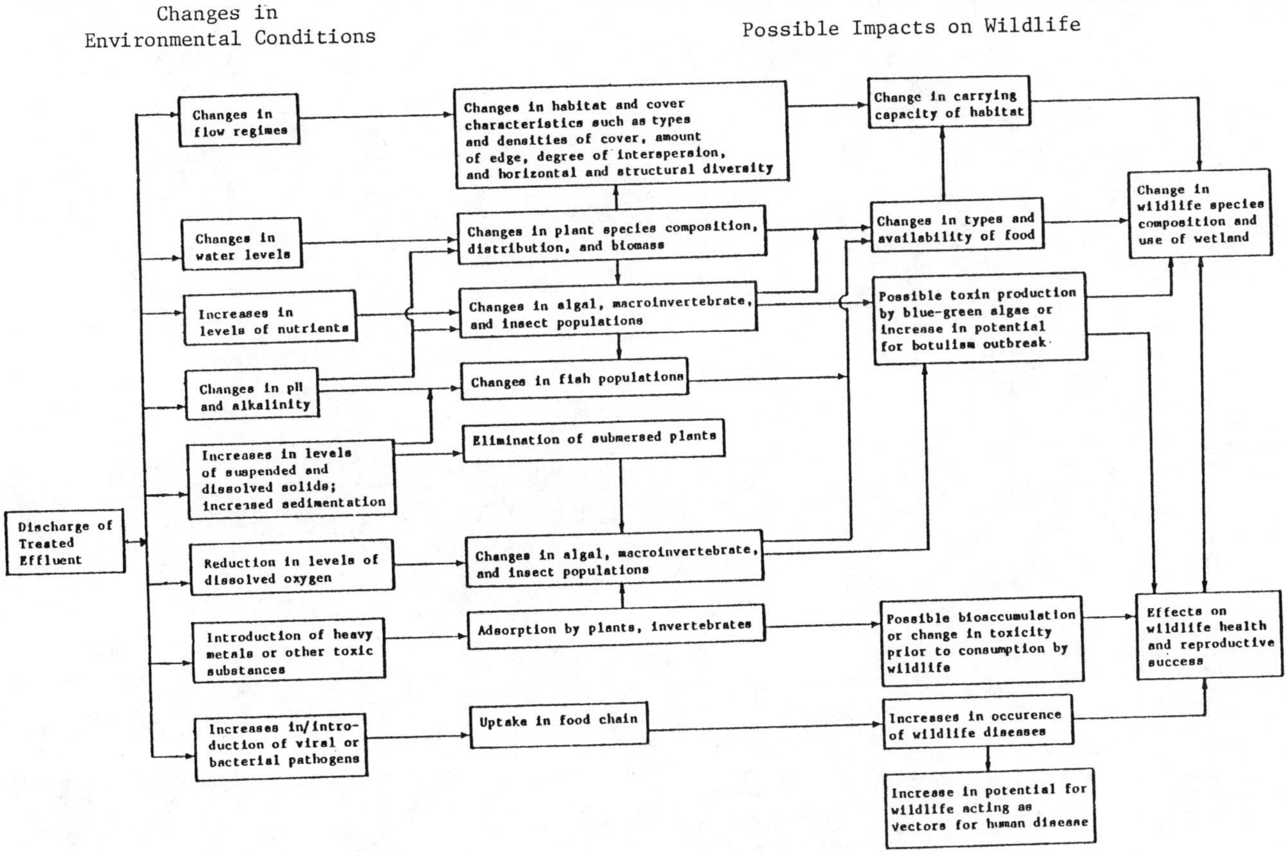

Changes in
Environmental Conditions

Possible Impacts on Wildlife

Figure 13.3
Impact factor train for wildlife.

Wildlife Use of Treatment Facilities

The roles and habitat requirements of many species of wetland wildlife and upland wildlife that use wetlands have been indicated by various authors in Greeson, Clark, and Clark (1979). The open water and undisturbed locations typical of effluent lagoons and stabilization ponds associated with wastewater treatment facilities fulfill some of the requirements of a number of species of wildlife and are particularly attractive to migratory waterfowl. Uhler (1956, 1964) noted that ducks prefer sewage lagoons to natural potholes, which have a more limited food supply. Other species of wildlife associated with wetlands visit these areas periodically to feed on the plants and animals present in or near the water. Thus these parts of the treatment facility must be considered as components of the habitat of these species in a study of the effects of a wastewater discharge to a wetland if the wetland is located near the treatment facility.

The only study known to contain information on the types of bird communities associated with different methods of sewage treatment and disposal was performed in Great Britain as a review of the literature and census data on bird use of treatment facility sites (Fuller and Glue 1980). The authors categorized the species observed into four seasonal groups: breeding, summer-feeding, spring and autumn (migration periods), and winter. The authors concluded that sewage treatment facilities can serve as important feeding sites for passerine (perching) species, primarily songbirds, and that systems with percolating filters and land application types of treatment (surface irrigation) would support the most varied communities of birds. Particular emphasis was placed on wetland species, and the results indicated that high populations of waterfowl and wading birds were present at the "sewage farms" formerly used as treatment sites. (A sewage farm is a farm on which wastewater is used for irrigation of crops or pastureland.) The species composition was determined to be dependent on the quality of the emergent and bankside vegetation on the sides of the irrigation plots and lagoons.

Insect-eating birds fed on the percolating filters and above the effluent lagoons, and birds of prey hunted over these areas. Drained effluent lagoons attracted shorebirds and shallow, filled lagoons, especially if they were bordered by areas of unvegetated mud, attracted large numbers of wading birds. Many passerine species fed on insects on sludge drying beds. Sludge lagoons also served as sources of seeds, as did uncultivated areas and the banks of the lagoons.

Although the variety of birds visiting treatment facilities to obtain food decreased during the winter, the total number of individuals was greater and many waterfowl and wading birds overwintered in the lagoons. The authors concluded that the outdated sewage farms provided a large amount of wetland habitat with both feeding and nesting areas that are not present at the more modern works that have been constructed. However, the development of percolating filters has resulted in the provision of feeding areas for passerine species, and the land application areas provide some habitat that partially compensates for the adverse effects of development on wildlife, particularly in urban areas. The majority of these facilities are small in comparison to the older sewage farms.

In the United States, a number of short-term studies have been conducted on waterfowl use of wastewater lagoons and oxidation ponds (Dornbush and Anderson 1964; Willson 1975; Dodge and Low 1972). Swanson (1977) studied the feeding behavior and food items consumed by adult and immature surface-feeding ducks on stabilization ponds in North Dakota. He suggested that upland nesting cover be improved and that pond shorelines be designed to reduce the likelihood of disease in order to make the ponds safer and more valuable to waterfowl. He also recommended that the potential for heavy-metal uptake and concentration in the food chain and for increased incidence of disease at the ponds be investigated before such habitat enhancement is performed.

Land Treatment Sites

Some estimates of the effects of the application of wastewater to wetlands on terrestrial animals that use such areas may be inferred from the results of studies on the effects of wastewater on terrestrial wildlife and wildlife habitats conducted at the Pennsylvania State University Wastewater Renovation System by Anthony and Wood (1979), Anthony, Bierei, and Kozlowski (1978), Bierei, Wood, and Anthony (1975), Dressler and Wood (1976), Lewis (1977), and Snider and Wood (1975). They noted the following results:

Forage production for deer was lower during the summer, and the levels of nutrients such as protein, P, potassium, and magnesium in the forage species increased; calcium levels were reduced (Anthony and Wood 1979).

Increased soil moisture content enhanced the production of herbaceous plants dramatically, but species diversity declined; many of the species that increased in biomass were unpalatable to wildlife, and the net effect was a reduction in the amount of available forage (Dressler and Wood 1976).

Small-mammal populations increased in autumn, due to the presence of increased food and cover (Bierei, Wood, and Anthony 1975).

Songbird populations increased in the late summer in response to food availability (Snider and Wood 1975); the overall effect appeared to be a less stable community with low diversity and several abundant species (Lewis 1978).

Levels of chromium were significantly higher, and levels of nickel significantly lower, in tissues of rabbits and mice collected from irrigated areas as compared to those in animals collected from control areas. Levels of copper were higher in kidneys of cottontail rabbits from the control area (Bierei, Wood, and Anthony 1975).

These studies were short-term investigations on relatively small sample plots, and Sopper and Kerr (1979) recommended that more detailed studies be conducted over longer periods of time. Sidle and Sopper (1976) noted that different species would accumulate the same heavy metals at different rates due to differences in food habits (species of plants or animals consumed, parts of these organisms consumed, and so forth), and that these rates would also vary at times within a species because of seasonal differences in food availability. These differences were apparently shown in a follow-up study by Anthony and Kozlowski (1982), who reported that concentrations of cadmium and lead were significantly higher in white-footed mice from an irrigated, forested area than in mice from a similar control area, although the concentrations were not at toxic levels. Cadmium/zinc ratios in kidney tissue from small mammals living on the irrigated area were higher than cadmium/zinc ratios for soils and vegetation on these areas, which indicated a potentially hazardous uptake of cadmium by the rodents. However, concentrations of cadmium and zinc were higher in the kidney and liver tissue of meadow voles on control areas of reed canary grass than on irrigated areas with the same vegetation.

Greenwald (1981) reported on the anticipated effects of spray irrigation on songbird populations in an area where the habitats ranged from old fields to mature mixed-oak forest. She noted two major short-term effects: development of a dense ground cover consisting of a few herbaceous species, and damage to woody vegetation in the midstory as a consequence of ice buildup during the winter. She anticipated that the long-term study to be conducted in this area would show the following results:

Short-term increases in diversity and abundance of songbirds, correlated with the increase in mid-story herbaceous vegetation in sprayed forests with more open canopies;

Short-term decrease in diversity and abundance of mature forest songbirds due to the loss of subcanopy woody vegetation that provides nesting, feeding, and perching sites;

Serious long-term changes in diversity and abundance of songbirds in the mature forest because the dense herbaceous growth will compete with hardwood seedlings and forest regeneration will be difficult.

Effects on Wildlife at Wetland Treatment Sites

The majority of the biological studies performed to date at wetland treatment sites in the United States have been focused on changes in the vegetation and litter compartments, particularly as related to treatment effectiveness. The most detailed studies of changes in animal species composition and biomass were performed on cypress domes in Florida by various researchers from the University of Florida. They conducted baseline and experimental studies on the amphibian, avian, and mammalian populations in treated and control domes and correlated the changes in wildlife populations in the treated dome primarily with changes in the water level, water quality, and invertebrate species. The number and diversity of species of birds increased significantly in the treated dome (Ramsay 1978), and aquatic species of mammals were observed that were not present in the control dome (Jetter and Harris 1976). The number of amphibians also increased significantly in the treated dome as animals came there to breed because of the higher water level and greater production of insects that served as food. However, reproduction was not successful due to the poor water-quality conditions, and the treated dome attracted frogs and toads from adjacent forested areas. The authors noted that the treated dome did provide a beneficial function for amphibians during drought periods, when it served as a reservoir for reestablishment of populations in the forested areas and in other domes after fires had destroyed those areas.

An accidental release of sludge into the treated dome resulted in the formation of a mat of organic debris, and the subsequent development of anaerobic conditions under the mat, which were favorable to organisms that could not exist in the presence of oxygen, resulted in a "massive" shift in the species and pathways of the food chains in the dome (Jetter and Harris 1976). The carnivores that constituted the top consumers in the food chains in untreated control domes, primarily wading birds such as herons and egrets, were eliminated from the treated dome because the organisms on which they fed (aquatic insects, crayfish, and fish) could not survive. However, the number of species of insect-eating songbirds increased significantly.

In the upper Midwest, investigations on wildlife populations have been performed at two natural wetlands (Houghton Lake, Michigan, and Drummond, Wisconsin) and two volunteer wetlands (Vermontville, Michigan, and Paw Paw, Michigan). (Volunteer wetlands are land application sites, stabilization ponds, or flood irrigation fields where unplanned wetland vegetation has developed.) The studies at all of these sites are in early stages of development, and few results have been obtained to date. It is expected that a number of the effects on wildlife populations that may occur will be detectable only after changes in the vegetation and other components of the habitat. The length of this lag time is not known, but it is likely to differ at each site.

Kadlec (1979) reported that there were no major shifts in species abundance or species composition of birds at the Houghton Lake, Michigan, treatment site during the period 1975-1977, nor were there any observable effects on mammal populations or mammal use of the area other than a temporary increase in muskrat activity. However, he noted that it is too early in the life of the project, and too few investigations have been done, to draw any conclusions from this information. The long-term effects may not be observed for some years.

Anderson and Kent (1979) described the studies on amphibians, reptiles, birds, and mammals to be conducted at the Drummond, Wisconsin, site. Pretreatment studies have been performed on the bog used as the treatment site, the part of the bog above the discharge point that is used as the control site, a stream flowing from the bog, the lake into which the stream flows, and the surrounding upland areas. Post-treatment studies were performed for four months after the initiation of the discharge. The raw data on the species collected were presented in Anderson and Kent (1979), but no analyses accompanied this information. These data are too few and too preliminary for identification of any significant effects due to the introduction of wastewater to the site.

Only records of observations of wildlife were made by Bevis (1979) at the Vermontville, Michigan, and Paw Paw, Michigan, sites. Additional studies need to be performed at these sites also before any conclusions can be drawn on the beneficial and adverse effects on wildlife populations and habitats at volunteer wetland treatment sites.

The benefits to waterfowl and other species of wildlife from the use of wastewater for habitat enhancement in California marshes were reported by Cederquist and Roche (1979) and Cederquist (1980a, 1980b) for a wastewater discharge to a natural wetland and by Demgen (1979) and Demgen and Nute (1979) for a discharge to an artificial wetland.

EFFECTS ON ANIMAL HEALTH

The major concerns for effects on wildlife health revolve around the effects on the condition and health of individual organisms and populations as a whole, rather than on the direct transmission of disease due to the operation of the treatment system (see Friend, this volume). The majority of the organisms that cause disease in wildlife are already present in wetland environments; the issues of concern are the potential for the creation of physical and/or biological conditions that would increase the populations of these organisms dramatically, that would increase the susceptibility of wildlife to these pathogens, or that would reduce viability of individual wildlife and thus increase morbidity and mortality due to other factors or pathogens.

Sanderson et al. (1980) noted that toxic substances and other contaminants in the environment can also cause alterations in basic biochemical functions in wildlife that may result in subtle changes in behavior. These changes, in turn, may affect the productivity and survival of these individuals. They cautioned that although an environment that has been polluted, disturbed, and simplified may evidence high primary and secondary productivity, the productive species typically are not those characteristic of the natural condition of that environment.

POTENTIAL FOR ENHANCEMENT OF WILDLIFE HABITAT

The availability of a continuous source of water and nutrients associated with a discharge to a wetland from a treatment plant or with a wetland treatment system constitutes a potential resource for the maintenance, enhancement, and creation of wetland habitats. Artificial wetlands in particular appear to be promising. If properly constructed and managed, such areas should be able to provide both treatment of wastewater and a source of the environmental requirements of wildlife during different seasons and life stages. They also could serve as habitats for rare species or as valuable educational resources. The use of wastewater to restore wetlands that have been drained or otherwise altered has also been proposed. This technique is particularly attractive for use in agricultural areas of the Midwest.

On the basis of the preliminary results obtained to date at a few study sites, it appears that the introduction of wastewater to natural wetlands is followed by:

A decrease in species richness and diversity;

A decline or disappearance of the more "sensitive" species that have relatively narrow tolerances for changes in physical and chemical parameters;

An increase in the number and abundance of "weedy" opportunistic species and more common species that can adjust to a wider range of conditions and habitats.

Some of the researchers who have conducted ecological investigations at these sites have recommended, on the basis of their preliminary results, that wastewater not be discharged to natural wetlands or that it be discharged only on a carefully controlled experimental basis until more knowledge is obtained on these ecosystem simplification effects (Stearns 1978; Sloey, Spangler, and Fetter 1978; and others). Thus the potential for the use of treated wastewater to enhance pre-existing wetlands may be less than the potential for the creation of new wetlands or the restoration of former wetlands, because of the likelihood of adverse effects on the wildlife and fish communities presently occupying or using such areas. However, in dry years or under other adverse conditions, the temporary introduction of wastewater may help to maintain the natural vegetation and wildlife communities in these wetlands, with a lesser likelihood of long-term adverse effects as long as no toxic substances are present in the wastewater.

NATURAL WETLANDS

The introduction of wastewater into the natural wetlands presently used as treatment sites has altered the environment sufficiently to change the species composition of the predischarge plant and animal communities (Jetter and Harris 1976; Bevis 1979). However, the overall characteristics of a wetland remain, even though the characteristics of this disturbed habitat no longer fulfill the requirements of some species and the alteration of physical, chemical, and biological conditions has resulted in the introduction or increase of other species not previously present or not present in abundance because some of their requirements were not met. This phenomenon has been well documented by Jetter and Harris (1976) for cypress domes in Florida that received treated wastewater contaminated with sludge.

Wastewater has been used on a seasonal basis for several years for habitat enhancement at several duck clubs located in the Suisun Marsh in California without noticeable adverse effects on wildlife populations (Cederquist and Roche 1979; Cederquist 1980a, 1980b). One club received only wastewater, two received different combinations of wastewater and slough water, and a fourth, used as a control, received only slough water. No problems were noted except for algal blooms during the spring and decreased oxygen levels associated with pond draining in the spring and pond flooding in the

autumn. The wastewater is used for irrigation of surrounding agricultural lands during the dry season and for marsh management during
waterfowl migration periods. The use of the treated wastewater
reduces the amount of river water required to prevent the intrusion
of seawater into this coastal marsh.

Blumer (1978) proposed measures for recreation and restoration
of wetlands in Florida that would increase the natural nutrient-
assimilation capacity of these areas and also improve their value as
wildlife habitat. These measures were based on experience gained
through the operation of the artificial wetland treatment systems at
Brookhaven National Laboratory in New York (Small 1978).

The state of the art of management of freshwater marshes for
wildlife was summarized by Weller (1978). Detailed summaries of
techniques for wetland habitat improvement and manipulation practices
were prepared by Burger (1973) and Yoakum et al. (1980). Such information is also contained in many articles in the wildlife management
literature and in various federal and state government reports.
Management for a particular species alone was not recommended by
Weller (1978) because of the relatively high cost and difficulty of
such techniques and the limited applicability of these techniques to
other wetland areas. He noted that management for an entire community of organisms, based on natural patterns of succession, would be
more cost-effective and provide more benefits to the public in the
long term (both from the point of view of harvest and recreational
opportunities and of preservation of natural values and functions)
and would be more consistent with the natural variability of wetland
ecosystems and species population cycles.

Weller (1978) also presented an outline of recommended procedures for management of marsh vegetation through regulation of water
levels, based on work by Weller and Fredrickson (1974). He noted
that marshes generally do not remain productive if flooded continually, and that periodic drawdowns are necessary for germination of
emergent plants. The timing and extent of such drawdowns should be
correlated with the reproductive cycles and seasonal needs of wildlife for optimum enhancement of wildlife populations. Chabreck
(1976) included similar topics in his discussion of management of
wetlands for wildlife habitat enhancement. His recommendations were
designed for application in coastal wetlands of the southern and
southeastern states. Guidelines for the restoration and subsequent
management of various types of wetlands in Wisconsin were developed
by Bedford, Zimmerman, and Zimmerman (1974). Information on the
characteristic plant and animal components of wetland ecosystems in
the Midwest and the requirements of particular species of wildlife is
available in publications such as Wheeler and March (1979) and
Greeson, Clark, and Clark (1979). Sloey, Spangler, and Fetter (1978)
reviewed the various techniques for management of a wetland for
nutrient assimilation and the seasonal aspects of nutrient storage in
each type of wetland. Any research plans developed to investigate
the potential for wildlife habitat enhancement at various wetland
treatment sites also should be correlated with the nutrient uptake
and storage capabilities of such sites. Wetland restoration
conferences and workshops have been held in the southern United
States (such as Cole 1979). Information on restoration of midwestern
wetlands is available in Richardson (1981).

ARTIFICIAL WETLANDS

The problems associated with the disposal of dredged spoil have
provided the stimulus for a considerable amount of research in the
area of wetland habitat development, primarily in coastal areas. The
most comprehensive work has been conducted as part of the Dredged
Material Research Program, sponsored by the U. S. Army Corps of
Engineers Waterways Experiment Station in Vicksburg, Mississippi
(Soots and Landin 1978). They coordinated laboratory and field
studies on the creation of wetlands at nine coastal or riverine sites
during the mid-1970s and also conducted assessments of the environmental impacts of such activities. The results of these studies were
published during 1978 in a series of technical reports (DS-78-15
through DS-78-79) and impact studies (Technical Report DS-77-23 and
Appendices).

Demgen (1979) and Demgen and Nute (1979) reported benefits to
wildlife populations from the construction of a small wetland in
California. In 1978, the EBC Company began a pilot project that
combines marsh and forest habitats for improvement of effluent
quality, habitat development, and timber production. This pilot
project was initiated in 1974 to demonstrate the feasibility of using
secondary effluent for creation of wildlife habitat and to test
various techniques for the improvement of both water quality and
wildlife habitat. A series of five interconnected plots, totalling
8.2 ha (20.3 acres), is located in a tidal area. Wastewater passes
by gravity flow through the system and discharges to a slough that
connects to Suisan Bay. Ecofloats, water-level control devices,
paddle wheel-type aerators, and an underground irrigation system with
an infiltration unit under each redwood tree have been used as components of this "marsh-forest system" (EBC Company 1979). The last
cell in the fourth plot contains habitat-enhancement devices called
Ecofloats, which are manufactured by the EBC Company. These consist
of a raft-like flotation structure that provides resting and breeding
areas for waterfowl, below which are suspended sacs filled with
decorticated redwood bark that provide substrates for aquatic invertebrates. Another plot is planted with seed-bearing grasses and
bulrushes to provide food for migratory waterfowl, and a third is
maintained as an open-water area with four vegetated islands. The
combination of habitat types and the availability of food, cover, and
protected nesting sites has resulted in intensive use of the site by
waterfowl. Mouse, muskrat, amphibian, reptile, and fish populations
have become established at the site. Mosquitofish have been
harvested for use in mosquito control efforts by the local mosquito
abatement district, and the crayfish and other invertebrates could be
marketed for bait and as food for tropical fish, respectively. The
educational and recreational benefits provided by the site are also
being utilized. The system is operating properly and appears to be
meeting the desired objectives (Demgen 1979). Small (1976) reported
that the small artificial wetlands constructed by researchers at
Brookhaven National Laboratory also attracted amphibians, waterfowl,
and shorebirds.

Weller (1978) listed the data needs that could be reduced
through investigations in an artificial marsh situation where a
number of similar units of habitat would be managed experimentally.
Such a facility was recently constructed at the Delta Waterfowl

Research Station in Manitoba, Canada, and the results of investigations at that site will serve as the basis for future developments in management of inland freshwater marshes. However, the results of the first long-term studies to be performed at that site may not be available for several years.

Garbisch (1978) conducted a survey of marsh establishment/ restoration projects and reported that only 18 of 105 completed or ongoing projects were located in or near freshwater locations. He noted that the technology of marsh rehabilitation is of recent origin and has been developed primarily in and for application to brackish and saltwater areas. He also cautioned that general guidelines may be helpful for a project at a particular site, but that the program developed must be tailored to the unique characteristics of the site. Garbisch also reported that the maximum level of functioning of these created areas is attained within one to three years and that these functions (vegetative and wildlife productivity, water purification, and so forth) appear to be comparable to those of natural wetlands. Information on various techniques for the construction of artificial wetlands also is available from state wildlife agencies in the form of research reports and brochures for landowners. Additional information is also provided by Wile, Miller, and Black (this volume).

VOLUNTEER WETLANDS

The only work known to have been done in volunteer wetlands is that of Bevis (1979), who studied the volunteer wetlands that developed at treatment sites in Vermontville, Michigan, and Paw Paw, Michigan. He noted that such areas could serve as refuges for, and possibly as reservoirs of, rare species of plants and animals. Some species, particularly plants, would need to be transplanted or stocked at those locations. The access limitations and present use of the sites would limit their use by the general public for recreational or harvest purposes, although scientific studies would be possible.

Tusack-Gilmour (1980) described a desert wash in Nevada in which marsh and riparian vegetation became established as a consequence of the introduction of wastewater. An avian community that is uncommon in desert areas has developed, and the Las Vegas Wash has become a valuable habitat for migratory and resident shorebirds and waterfowl. Amphibians and mammals not normally present in desert environments have also been observed in the wetlands. The author noted that the increasing urbanization of the adjacent area probably was the major stimulus for this influx, and that the value of the wash as a refuge from such pressures would increase as more development occurred.

The potential for the development of volunteer wetland vegetation as a consequence of wastewater discharge in the Midwest appears to be high, given the general climatic and soil conditions and the previous history of wetlands in many parts of the region. It is likely that a number of these sites presently exist in association with wastewater treatment facilities.

CONCLUSION

The studies performed on wetlands that receive wastewater have been concerned primarily with changes in hydrology, changes in water quality, nutrient uptake and retention by soils, sediments, and plants, and changes in plant life forms and plant communities. Too little information is available on the animal communities of the affected areas to determine the types and extent of any changes that may have occurred. Information on predischarge animal communities is nonexistent at almost all of the sites. Because of (1) the mobility of animals, (2) their capability to adjust their activities if weather conditions and other environmental factors are not favorable, and (3) the fact that they may be present only at certain times of the day or seasons of the year, the composition of the wildlife community of a wetland at any one time is less reliable as an indication of the value to and the use of the wetland by wildlife than are the structure and composition of the plant community and the presence and abundance of food sources such as benthic macroinvertebrates and insects. Additional studies need to be done to identify potential adverse effects on animal communities, their duration, and their implications for changes in the structure and function of wetland animal communities.

An inventory of known discharges of wastewater to wetlands in Illinois, Indiana, Michigan, Minnesota, Ohio, and Wisconsin (Brennan and Garra 1981) was performed in association with this literature review. The results of this inventory show that the use of natural wetlands for the discharge of treated wastewater is relatively common. However, the intentional inclusion of wetlands as part of the treatment process is rare. Both types of situations may become more attractive due to economic factors. Although the short-term benefits of the use of natural wetlands for the disposal or treatment of wastewater (cost-effectiveness, treatment efficiency, and convenience) appear promising, the long-term ability of these areas to treat wastewater is questionable. Research at wetland treatment sites in the Midwest and in Florida has indicated that the introduction of wastewater may cause permanent changes in the components of wetland communities and thus result in the simplification of these ecosystems and the elimination of some of their characteristic species, functions, and values. The long-term effects of the discharge of treated wastewater at current study sites will not be evident for many years. More information on the potential effects and long-term utility of this technology, particularly the effects on animal communities, is needed before knowledgeable regulatory decisions on wetlands proposed as discharge or treatment sites can be made.

The construction of artificial wetlands for the treatment of wastewater would avoid any detrimental effects that might result from the use of natural wetlands and also could provide supplementary habitats for wetland wildlife and possibly reservoirs for rare species. Few animal-related studies have been performed at the small number of artificial wetland sites presently in existence, and, thus, the information base is too small and too short-term for any conclusions to be drawn. More data are expected to be obtained on the benefits and drawbacks of this technology, particularly for animal communities, as more such facilities are developed.

ACKNOWLEDGMENTS

This paper is adapted from a draft technical report entitled "The Effects of Wastewater Treatment Facilities on Wetlands in the Midwest," which was prepared in 1981 by WAPORA, Inc., for the U. S. Environmental Protection Agency, Region 5, under Contract No. 68-01-5989. The section on invertebrates was written by Steven D. Bach; the section on fish was written by Gregory L. Seegert.

REFERENCES

Anderson, R. K., and D. Kent, 1979, Progress report: Drummond, Wisconsin, tertiary treatment demonstration project study, report to the U. S. Fish and Wildlife Service, University of Wisconsin, Stevens Point, 22p.

Anthony, R. G., and G. W. Wood, 1979, Effects of municipal wastewater irrigation on wildlife and wildlife habitat, in Utilization of municipal sewage effluent and sludge on forest and disturbed land, W. E. Sopper and S. N. Kerr, eds., Pennsylvania State University Press, University Park, pp. 213-223.

Anthony, R. G., and R. Kozlowski, 1982, Heavy metals in tissues of small mammals inhabiting wastewater-irrigated habitats, J. Environ. Qual. 11(1):20-22.

Anthony, R. G., G. R. Bierei, and R. Kozlowski, 1978, Effects of municipal wastewater irrigation on select species of mammals, State of knowledge in land treatment of wastewater, vol. 2, H. L. McKim, coordinator, proceedings of an international symposium, August 20-25, 1978, Hanover, N. H., U. S. Army Corps of Engineers, pp. 281-287.

Bedford, B. L., E. H. Zimmerman, and J. H. Zimmerman, 1974, The wetlands of Dane County, Wisconsin, Dane County Regional Planning Commission, in cooperation with the Wisconsin Department of Natural Resources, Madison, 581p.

Bevis, F. B., 1979, Ecological considerations in the management of wastewater-engendered volunteer wetlands, abstracts of the conference on freshwater wetlands and sanitary waste disposal, The Michigan Wetlands Conference, July 10-12, 1979, Higgins Lake, Mich., 19p.

Bierei, G. R., G. W. Wood, and R. G. Anthony, 1975, Population response and heavy metals concentrations in cottontail rabbits and small mammals in wastewater irrigated habitat, Faunal response to spray irrigation of chlorinated sewage effluent, G. W. Wood et al., eds., Publication 87, Institute for Research on Land and Water Resources Research, Pennsylvania State University Press, University Park, pp. 1-9.

Blumer, K., 1978, The use of wetlands for treating wastes--wisdom in diversity, Environmental quality through wetlands utilization, M. A. Drew, ed., proceedings of a symposium, February 28-March 2, 1978, Tallahassee, Florida, Coordinating Council on the Restoration of the Kissimmee River Valley and Taylor Creek-Nubbin Slough Basin, Tallahassee, Fla., pp. 182-201.

Brennan, K. M., and C. G. Garra, 1981, Wastewater discharges to wetlands in six midwestern states, in Selected proceedings of the Midwest conference on wetland values and management, B. Richardson, ed., June 17-19, 1981, St. Paul, Minn., Minnesota Water Planning Board, St. Paul, pp. 285-293.

Brungs, W. A., R. W. Carlson, W. B. Horning II, J. H. McCormick, R. L. Spehar, and J. D. Yount, 1978, Effects of pollution on fresh-water fish, Water Pollut. Control Fed. J. 50(6):1582-1637.

Burger, G. V., 1973, Practical wildlife management, Winchester Press, New York, 218p.

Cederquist, N. W., 1980a, Suisun Marsh management study, progress report on the feasibility of using wastewater for duck club management, July 1980, U. S. Dept. of the Interior, Water and Power Resources Service, Sacramento, Calif., 45p.

Cederquist, N. W., 1980b, Suisun Marsh management study, 1979-1980 progress report on the feasibility of using wastewater for duck club management, September 1980, U. S. Dept. of the Interior, Water and Power Resources Service, Sacramento, Calif., 61p.

Cederquist, N. W., and W. M. Roche, 1979, Reclamation and reuse of wastewater in the Suisun Marsh of California, in proceedings of the water reuse symposium, vol. 1, March 25-30, 1979, Washington, D. C., American Water Works Association Research Foundation, Denver Col., pp. 685-702.

Chabreck, R. H., 1976, Management of wetlands for wildlife habitat improvement, in Estuarine processes, vol. I. Uses, stresses, and adaptation to the estuary, M. Wiley, ed., Academic Press, New York, pp. 227-233.

Cole, D. P., ed., 1979, Proceedings of the sixth annual conference on wetlands restoration and creation, Hillsborough Community College, Tampa, Florida, sponsored by Tampa Port Authority and the Environmental Studies Center at Cockroach Bay, Fla., Hillsborough Community College, Tampa, Fla., 357p.

Davis, H. G., 1978, Effects of the treated effluent on mosquito populations and arbovirus activity, in Cypress wetlands for water management, recycling and conservation, H. T. Odum, K. C. Ewel, J. W. Ordway, and M. K. Johnston, eds., fourth annual report to the National Science Foundation and the Rockefeller Foundation. Center for Wetlands, University of Florida, Gainesville, pp. 361-428.

Demgen, F. C., 1979, Wetlands creation for habitat and treatment at Mt. View Sanitary District, California, in Aquaculture systems for wastewater treatment: Seminar proceedings and engineering assessment, R. K. Bastian and S. C. Reed, project officers, no. 430/9-80-006, U. S. EPA, Office of Water Program Operations, Municipal Construction Division, Washington, D. C., pp. 61-73.

Demgen, F. C., and J. W. Nute, 1979, Wetlands creation using secondary treated wastewater, American Water Works Association Research Foundation Water Reuse Symposium, March 25-30, 1979, Washington, D. C., vol. I, AWWARF, pp. 727-739.

Dindal, D. L., L. T. Newell, and J. Moreau, 1979, Municipal waste-water irrigation: Effects on community ecology of soil inverte-brates, in Utilization of municipal sewage effluent and sludge on forest and disturbed land, W. E. Sopper and S. N. Kerr, eds., Pennsylvania State University Press, University Park, pp. 197-205.

Dodge, D. E., and J. B. Low, 1972, Logan lagoons good for ducks, Utah Sci. 33(2):55-57.

Dornbush, J. N., and J. R. Anderson, 1964, Ducks on the wastewater pond, Water and Sewage Works 3(6):271-276.

Dressler, R. L., and G. W. Wood, 1976, Deer habitat response to irrigation with municipal wastewater, J. Wildl. Manage. 40(4):639-644.

EBC Company, 1979, The marsh-forest system: A pleasant and positive answer for water reclamation, 9p.

Fritz, W. R., and S. C. Helle, 1978, Cypress wetlands as a natural treatment method for secondary effluents, in Environmental quality through wetlands utilization, M. A. Drew, ed., proceedings of a symposium, February 28-March 2, 1978, Tallahassee, Florida, Coordinating Council on the Restoration of the Kissimmee River Valley and the Taylor Creek-Nubbin Slough Basin, Tallahassee, Fla., pp. 69-81.

Fuller, R. J., and D. E. Glue, 1980, Sewage works as bird habitats in Britain, Biol. Conserv. 17(3):165-182.

Gallepp, G. W., 1979, Chironomid influence on phosphorus release in sediment-water microcosms, Ecology 60:547-556.

Garbisch, E. W., Jr., 1978, Wetland rehabilitation, in Proceedings of the national wetland protection symposium, J. H. Montanari and J. A. Kusler, co-chairmen, June 6-8, 1977, Reston, Virginia, FWS/OBS-78/97, U. S. Dept. of the Interior, Fish and Wildlife Service, Office of Biological Services, Washington, D. C., pp. 217-219.

Graneli, W., 1979, The influence of Chironomus plumosus larvae on the exchange of dissolved substances between sediment and water, Hydrobiologia 66:149.

Greenwald, M., 1981, Prediction of songbird responses to habitat alteration resulting from wastewater irrigation, M. S. thesis, Pennsylvania State University, University Station.

Greeson, P. E., J. R. Clark, and J. E. Clark, eds., 1979, Wetland functions and values: the state of our understanding, proceedings of the national symposium on wetlands, American Water Resources Association, Minneapolis, Minn., 674p.

Janssen, W. A., 1970, Fish as potential vectors of human bacterial diseases, in A symposium on diseases of fishes and shellfishes, S. F. Snicszko, ed., special publication no. 5, American Fisheries Society, Washington, D. C., pp. 284-290.

Jaworski, E., and C. N. Raphael, 1978, Fish, wildlife, and recreational values of Michigan's coastal wetlands: Phase I of coastal wetlands value study in Michigan, report prepared for Great Lakes Shorelands Section, Division of Land Resource Programs, Michigan Dept. of Natural Resources, U. S. Fish and Wildlife Service, Twin Cities, Minn., 209p.

Jetter, W., 1975, Effects of treated sewage on the structure and function of cypress dome consumer communities, in Cypress wetlands for water management, recycling and conservation, H. T. Odum, K. C. Ewel, J. W. Ordway, and M. K. Johnston, eds., second annual report to the National Science Foundation and the Rockefeller Foundation, Center for Wetlands, University of Florida, Gainesville, pp. 588-610.

Jetter, W., and L. D. Harris, 1976, The effects of perturbation on cypress dome animal communities, in Cypress wetlands for water management, recycling and conservation, H. T. Odum, K. C. Ewel, J. W. Ordway, and M. K. Johnston, eds., third annual report to the National Science Foundation and the Rockefeller Foundation, Center for Wetlands, University of Florida, Gainesville, pp. 577-653.

Kadlec, R. H., 1979, Wetland tertiary treatment at Houghton Lake, Michigan, in Aquaculture systems for wastewater treatment: Seminar proceedings and engineering assessment, R. K. Bastian and S. C. Reed, project officers, 430/9-80-006, U. S. EPA, Office of Water Program Operations, Municipal Construction Division, Washington, D. C., pp. 101-139.

Kaminski, R. M., and H. H. Prince, 1981, Dabbling duck and aquatic macroinvertebrate responses to manipulated wetland habitat, J. Wildl. Manage. 45(1):1-15.

Krull, J. N., 1970, Aquatic plant macroinvertebrate associations and waterfowl, J. Wildl. Manage. 34(4):707-718.

Lewis, S. J., 1977, Avian communities and habitats on natural and wastewater irrigated vegetation, M. S. thesis, Pennsylvania State University, University Park.

Macek, K. J., S. R. Petrocelli, and B. H. Sleight III, 1979, Considerations in assessing the potential for, and significance of, biomagnification in aquatic food chains, in Aquatic toxicology, L. L. Marking and R. A. Kimerle, eds., American Society for Testing Materials, Philadelphia, Penn., pp. 251-268.

McKim, J., 1962, The inshore benthos of Michigan waters of southeastern Michigan, M. S. thesis, University of Michigan, Ann Arbor.

McMahan, E. A., and L. R. Davis, Jr., 1976, Effects of waste and mosquito control treatments on insect density and diversity, in Cypress wetlands for water management, recycling and conservation, H. T. Odum, K. C. Ewel, J. W. Ordway, and M. K. Johnston, eds., third annual report to the National Science Foundation and the Rockefeller Foundation, Center for Wetlands, University of Florida, Gainesville, pp. 671-677.

McMahan, E. A., and L. R. Davis, Jr., 1978, Density and diversity of micro-arthropods in wastewater treated and untreated cypress domes, in Cypress wetlands for waste management, recycling and conservation, H. T. Odum, K. C. Ewel, J. W. Ordway, and M. K. Johnston, eds., fourth annual report to the National Science Foundation and the Rockefeller Foundation, Center for Wetlands, University of Florida, Gainesville, pp. 429-461.

Porter, B. W., 1981, The wetland edge as a community and its value to wildlife, in Selected proceedings of the Midwest conference on wetland values and management, B. Richardson, ed., Minnesota Water Planning Board, St. Paul, Minn., pp. 15-25.

Ramsay, A., 1978, The effect of the addition of sewage effluent on cypress dome bird communities, M. S. thesis, School of Forest Resources and Conservation, University of Florida, Gainesville,

Richardson, B., ed., 1981, Selected proceedings of the Midwest conference on wetland values and management, Minnesota Water Planning Board, St. Paul, Minn., 660p.

Sanderson, G. C., E. D. Ables, R. D. Sparrowe, J. R. Grieb, L. D. Harris, and A. N. Moen, 1980, Research needs in wildlife, Transactions of the 44th North American Wildlife and Natural Resources Conference, pp. 166-175.

Sidle, R. C., and W. E. Sopper, 1976, Cadmium distribution in forest ecosystems irrigated with treated municipal wastewater and sludge, J. Environ. Qual. 5(4):419-422.

Sloey, W. E., F. L. Spangler, and C. W. Fetter, Jr., 1978, Management of freshwater wetlands for nutrient assimilation, in Freshwater wetlands: Ecological processes and management potential, R. E. Good, D. F. Whigham, and R. L. Simpson, eds., Academic Press, New York, pp. 321-340.

Small, M. M., 1976, Data report, marsh/pond system, preliminary report no. 50600, U. S. Energy Research and Development Administration, Brookhaven National Laboratory, Upton, N. Y., 28p.

Small, M. M., 1978, Artificial wetlands as non-point source wastewater treatment systems, Environmental quality through wetlands utilization, M. A. Drew, ed., proceedings of a symposium, February 28-March 2, 1978, Tallahassee, Florida, Coordinating Council on the Restoration of the Kissimmee River Valley and Taylor Creek-Nubbin Slough Basin, Tallahassee, Fla., pp. 171-181.

Snider, J. R., and G. W. Wood, 1975, The effects of wastewater irrigation on the activities and movements of songbirds, in Faunal response to spray irrigation of chlorinated sewage effluents, G. W. Wood et al., eds., publication no. 87, Institute for Research on Land and Water Resources Research, Pennsylvania State University, University Park, pp. 20-49.

Soots, R. F., Jr., and M. C. Landin, 1978, Development and management of avian habitat on dredged material islands, technical report DS-78-18, U. S. Army Corps of Engineers, Waterways Experiment Station, Vicksburg, Miss., 96p.

Sopper, W. E., and S. N. Kerr, 1979, Renovation of municipal wastewater in eastern forest ecosystems, in Utilization of municipal sewage effluent and sludge on forest and disturbed land, W. L. Sopper and S. N. Kerr, eds., Pennsylvania State University Press, University Park, pp. 61-76.

Spehar, R. L., R. W. Carlson, A. E. Lemke, D. I. Mount, Q. H. Pickering, and V. M. Snarski, 1980, Effects of pollution on freshwater fish, Water Pollut. Control Fed. J. 52(6):1703-1768.

Stearns, F., 1978, Management potential: Summary and recommendations, in Freshwater wetlands: Ecological processes and management potential, R. E. Good, D. F. Whigham, and R. L. Simpson, eds., Freshwater wetlands: ecological processes and management potential. New York, NY: Academic Press, Inc.; pp. 357-363.

Swanson, G. A., 1977, Diel food selection by Anatinae on a waste-stabilization system, J. Wildl. Manage. 41(2):226-231.

Swanson, G. A., 1978a, A simple lightweight core sampler for invertebrates in shallow wetlands, J. Wildl. Manage. 42(2):426-428.

Swanson, G. A., 1978b, A water column sampler for quantitating waterfowl foods, J. Wildl. Manage. 42(3):670-672.

Tilton, D. L., and B. R. Schwegler, 1978, The values of wetland habitat in the Great Lakes Basin, in Wetland functions and values: The state of our understanding, P. E. Greeson, J. R. Clark, and J. E. Clark, eds., American Water Resources Association, Minneapolis, Minn., pp. 267-277.

Tsai, C., 1975, Effects of sewage treatment plant effluents on fish: A review of the literature, Chesapeake Research Consortium, Inc., Baltimore, Md., 229p.

Tusack-Gilmour, D., 1980, Effluent creates an oasis in Nevada, Water and Wastes Eng. 17(9):22-24,56.

Uhler, F. M., 1956, New habitats for waterfowl, North Am. Wildl. Conf. Trans. 21:453-469.

Uhler, F. M., 1964, Bonus from waste places, in Waterfowl tomorrow, J. P. Lindusky, ed., U. S. GPO, pp.643-653.

Weller, M. W., 1978, Management of freshwater marshes for wildlife, in Freshwater wetlands: Ecological processes and management potential, R. E. Good, D. F. Whigham, and R. L. Simpson, eds., Academic Press, New York, pp. 267-284.

Weller, M. W., 1979, Birds of some Iowa wetlands in relation to concepts of faunal preservation, Iowa Acad. Sci. Proc. 86(3):81-88.

Weller, M. W., and L. H. Fredrickson, 1974, Avian ecology of a managed glacial marsh, The Living Bird 12:269-271.

Wheeler, W. E., and J. R. March, 1979, Characteristics of scattered wetlands in relation to duck production in southeastern Wisconsin, technical bulletin no. 116, Wisconsin Dept. of Natural Resources, Madison, Wisc., 61p.

Willson, M. F., 1975, Wastewater lagoons attract migrating waterfowl, Nat. Wildl. Fed. Cons. News 40(23):11-13.

Witter, F. A., and S. Croson, 1976, Insects and wetlands, in Freshwater wetlands and sewage effluent disposal, proceedings of a national symposium, D. L. Tilton, R. H. Kadlec, and C. J. Richardson, eds., May 10-11, 1976, University of Michigan, Ann Arbor, pp. 271-295.

Yoakum, J., W. P. Dasmann, H. R. Sanderson, C. M. Nixon, and H. S. Crawford, 1980, Habitat improvement techniques, in Wildlife management techniques manual, S. D. Schemnitz, ed., The Wildlife Society, Washington, D. C., pp. 329-403.

14

Terrestrial Communities: From Mesic to Hydric

William B. Jackson

Changes in animal and plant communities when the environment is converted from xeric or mesic to near-hydric should be instructive for this wetlands discussion. The early forest and grassland effluent disposal systems fit such a model (C. W. Thornthwaite Associates 1969).

SEABROOK FARMS

Seabrook Farms, a producer of frozen foods in southern New Jersey, when faced with an effluent-disposal crisis more than three decades ago, opted for a land disposal system using an oak woodland. This operation, shifting the environment from a relatively xeric to near-hydric condition, is described more fully by Jackson, Bastian, and Marks (1972) and Mather and Parmelee (1963).

The 140-acre woodland of predominately white oak (Quercus alba) received 10,000,000 gal of effluent daily from the food processing operations. Rotary nozzles, each covering about an acre, distributed the effluent. Some areas received in excess of 1,000 in of supplementary water annually. Though significant changes occurred, the biological communities constituted functioning bio-filters during their more than two decades of operation.

The soils generally had high (79%) sand content, and water filtration was rapid. This situation was less true in some depressed areas and where clay lenses existed.

Some trees were killed by the mechanical damage resulting from the force of the spray jet; others died from oxygen starvation when saturation of the root zone occurred. When the canopy opened, increased light coupled with abundant moisture stimulated rank weed growth (Sambucus, Phragmites, Erigeron, Phytolacca, Chenopodium, Polygonum, and others). As a result, marsh communities existed on hilltops and throughout the former woodland. Changes were not uniform but were influenced by the degree of spray impact.

Examination of tree rings indicated that growth of many of those trees not in the mechanical impact zone of the sprayers was stimulated, with annual rings doubling in width. However, after two decades of operation, the total basal area of woody species had

Table 14.1
Selected Trapping Data for White-footed Mouse (<u>Peromyscus</u>
<u>leucopus</u>) as Related to Wastewater Spray Disposal at
Seabrook Farms (Jackson et al. 1972)

	Corrected Trap Success (%)	
Year	Unsprayed Area	Sprayed Area
1950	5	3
1952	6	12
1957	5	12
1967	3	12
1970	6	13

<u>Source</u>: Data from Jackson et al. 1972.

decreased more than 50%, while a tripling had occurred in unsprayed
reference areas. Overall, the community had changed from one dominated by woody species to one of herbaceous species.

Our specific study focused on the small-mammal populations. The
white-footed mouse (<u>Peromyscus leucopus</u>) was the only species present
in sufficient numbers to permit analysis (Table 14.1). Initially
(1950–1951) this rodent existed at low levels (<4% trap success).
With the spectacular increase in herbaceous vegetation (1952), trap
success increased to 12%. In subsequent years, it remained above 7%,
while in sampling of the unsprayed reference area populations, trap
success was at or less than 6%.

The mice were not deterred by the water, some being caught
adjacent to spray nozzles. Essentially the mice were inhabiting or
utilizing a massive "edge-effect" area created within the woods.
Live-trapping data indicated that they moved readily between the oak
and weed communities. Only the <u>Phragmites</u> community was little
utilized by the mice.

Other small mammals were present, but in small numbers. The
jumping mouse (<u>Zapus hudsonius</u>) and red-backed vole (<u>Clethrionomys</u>

gapperi), typical of humi'd environments, did not increase substantially. Shrews (Blarina brevicauda and Sorex cinereus) responded
similarly. The limited reproductive data we were able to collect did
not indicate any changes in fecundity in wildlife at the site.

Bird and large-mammal populations were not followed, but local
hunters suggested that an increase of bobwhite quail had occurred;
more white-tail deer were observed.

Soil filtration/biodegradation effectively removed phosphates
and nitrates, and the whole operation had slight impact on peripheral
ground- or surface-water quality. While chloride levels were raised
in off-site water, they were still below "federal permissible
criterion" levels; biochemical oxygen demand (BOD) levels were not
elevated. Invertebrates in receiving streams were not impacted.

Within the spray woods, the soil levels of calcium (Ca),
magnesium (Mg), soluble nitrogen (N), phosphorus (P), and potassium
(K) increased severalfold (especially at 6-in depth). Organic matter
increased, but soil porosity was not destroyed. The soil pH rose
from 4.1 to 5.7.

Mosquito and tick populations increased dramatically.
Disease/vector studies were not possible, however.

A conventional assessment would term this south Jersey woodlot
destroyed. The relatively open woods floor with scattered blueberry,
mountain laurel, and other shrubs had been destroyed and replaced by
a creeping, unpicturesque mass of weeds. Though the biological
processes were usually aerobic, there were times and sites where this
was not the case. During beet processing by the company, the zone of
immediate impact was a bit more visible. Yet, despite all the
travail of an experiment, Seabrook Farms was proud of their operational and functional disposal system, and the downstream residents,
despite their fears, were little affected.

PENN STATE STUDY

The Penn State project also pioneered in woodland effluent
disposal (Sopper and Kardos 1973). Disposal operation had less
drastic environmental impacts than at Seabrook, but application rates
were lower. The conifer and mixed oak forests were not altered in
species and form, though growth rates generally were stimulated.
Fewer trees survived in treated areas.

Cottontail rabbit (Sylvilagus floridanius) and mice (Peromyscus
spp.) populations were greater in the fall, but not in the spring for
the treated environments, compared to control areas (Bierei, Wood,
and Anthony 1975). White-tail deer were not favored by the environmental changes, since the bulk of increased forage production was
with species not important in the deer diet, but deer were not
deterred from the site (Dressler and Wood 1975).

Studies of summer bird populations indicated only slight
response to treatments, and that response was primarily related to
increased shrub cover (and food supply) utilization by only a few
species (Savidge and Davis 1971; Snider and Wood 1975).

Companion arbovirus studies were undertaken, but no increased
epidemiological hazard was detected. Unfortunately, none of the data
has been published.

MICHIGAN STUDIES

The Muskegon (Michigan) project, serving approximately 160,000 persons, took sewage from the total community (including industry), passed it through a series of aerated lagoons, and distributed the clarified effluent for crop irrigation (Baur and Matsche 1973). This project, massive and far-reaching as it was, paid scant attention to bird and mammal populations in croplands and adjacent woodlands. An interesting component has been the blackbird depredations on the maturing corn growing in the irrigated croplands.

At Middleville, Michigan, effluent from facultative ponds is used to irrigate tree plantations (Urie, Cooley, and Harris 1978; Cooley 1979). Irrigated Populus plantings were girdled by mice and virtually eliminated by the third year, while in the nonirrigated plantings, where grass cover was sparse, much less mouse damage occurred. Similarly, in irrigated scotch pine (Pinus sylvestris) plantings, mice girdled 26% to 45% of the four- and five-year-old trees; no trees were girdled in the nonirrigated plots. Rodent damage was related to the protective cover (grass and weeds), the girdling rate being a direct correlate of herbage weight (Cooley 1980). The animals involved are presumed to have been meadow voles (Microtus Pennsylvanicus) (Colvin and Jackson 1982).

The Houghton Lake (Michigan) project represents another approach, that of utilization of a peatland community. The impact of the sewage effluent (less than 100 million gal annually on 300 acres of wetland) appears to have been slight. No changes in the plant and animal communities that cannot be attributed to normal variations have been detected, but animal data at this time consist of species lists obtained from various patterns of transect sampling (Kadlec et al. 1978; Kadlec, Tilton, and Schwegler 1979; Rabe 1979).

CONCLUSIONS

Responses of terrestrial vertebrates are rarely addressed in sufficient detail in project investigations. Certainly it is true that herbivores will shift in relation to the food supply. This fact would argue that floristic studies are of greater value. Yet the birds and mammals, especially the carnivores, are food chain monitors. For example, while concern has often been voiced about heavy-metal accumulation, few studies of effluent projects have studied the resident mammals. An exception is Anthony, Bierei, and Kozlowski (1978), who point to accumulations of lead and cadmium in white-footed mice in the Penn State study.

The potential for biomagnification through the animal community remains a concern of unknown dimensions (e.g., Larkin et al. 1978; Lennette and Spath 1978). Increasing water levels may facilitate breeding of mosquitos (including tree-hole species) and flies. The moist habitat and luxuriant vegetation can elevate tick populations. Such vectors could increase transmission rates of arboviruses both within and adjacent to treatment sites.

Both birds and small mammals are known reservoirs for several of the viral encephalitides. Yet we apparently have no data to document enhancement or lack of such diseases in treatment systems. Dog heartworm, a mosquito-vectored parasite, is of increasing concern to rural and suburban residents.

Enteric organisms are of obvious concern. However, in the Houghton Lake project, the peatland itself contained moderate levels of nonfood coliforms, and it was decided that no chlorination of the effluent was necessary (Kadlec, Richardson, and Kadlec 1975; Kadlec et al. 1978). While echovirus was present in the aeration ponds, it was not detectable further in the system (Kadlec, Tilton, and Schwegler 1979). Such concerns need further resolution.

The question of "normal variation" is vexing. Usually we do not have the opportunity to obtain five- or ten-year samples, and operationally induced variations become intertwined with "normal" variation. In a decade-long study of one small woodlot (Jackson et al. 1980) we found that soil moisture levels in the spring and late summer were especially critical for seed germination and seedling survival, respectively. Interactions with soil type, soil chemistry, and topography also occurred. Consequently, the composition of the herbaceous community and the availability of woody seedlings could shift dramatically, and a single series of quadrat data could result in a nonrepresentative assessment of the community. A determination of effect/no effect operations might readily be erroneous.

Enrichment and enhancement of these environments may have other impacts. Will the carrying capacity of wetlands for red-winged blackbirds be increased? Certainly the corn farmers in Ohio and elsewhere might be less than enthusiastic about such a program! Will skunk or raccoon populations be stimulated? What impact will that have on wildlife rabies? On depredations of waterfowl or upland game bird nests?

Usually we are focusing efforts on "damaged" ecosystems, on subclimax or disclimax communities, on cut-over or scrub environments. Even so, community structure is present, and environmental inputs will influence those communities and ecosystems. Even though we may affect a "degraded" environment, the objective can be enhancement rather than further degradation or increasing environmental hazard.

Where we are talking about utilizing only a few hundred acres scattered here and there, the impact is likely to be self-limited. But as the feasibility (and necessity) of wetlands utilization increases, community impact may be much more significant and of more concern to related human populations.

REFERENCES

Anthony, R. G., G. R. Bierei, and R. Kozlowski, 1978, Effects of municipal wastewater irrigation on select species of mammals, State of knowledge in land treatment of wastewater, Vol. 2, U. S. Army Corps of Engineers, Hanover, N. H., pp. 281-287.

Baur, W. J. and D. E. Matsche, 1973, Large wastewater irrigation systems: Muskegon County, Michigan and Chicago metropolitan region, in Recycling treated municipal wastewater and sludge through forest and cropland, W. E. Sopper and L. T. Kardos, eds., Pennsylvania State University Press, University Park, pp. 345-363.

Bierei, G. R., G. W. Wood, and R. G. Anthony, 1975, Population response and heavy metal concentrations in cottontail rabbits and small mammals in wastewater irrigated habitat, research publication 87, Institute for Research on Land and Water Resources, Pennsylvania State University, University Park, pp. 1-9.

Cooley, J. H., 1979, Effects of irrigation with oxidation pond effluent on tree establishment and growth on sand soils, in Utilization of municipal sewage effluent and sludge on forest and disturbed land, W. E. Sopper and S. M. Kerr, eds., Pennsylvania State University Press, College Park, pp. 145–153.

Cooley, J. H., 1980, Christmas trees enhanced by sewage effluent, Compost Sci./Land Util. 21:28–30.

Colvin, B. A. and W. B. Jackson, 1982, Christmas trees, rodent damage, and habitat manipulation (abstract), Ohio J. Sci. 82(2):6.

Dressler, R. L. and G. W. Wood, 1975, Deer habitat response to irrigation with municipal wastewater, research publication no. 87, Institute for Research on Land and Water Resources, Pennsylvania State University, University Park, pp. 10–19.

Jackson, W. B., R. K. Bastian, and J. R. Marks, 1972, Effluent disposal in an oak woods during two decades, Publ. in Climatol. 25:20–36.

Jackson, W. B., E. S. Hamilton, A. Limbird, and G. Frey, 1980, Annual report, Davis-Besse terrestrial monitoring contract, Toledo Edison Co., Toledo, Ohio, 42p.

Kadlec, R. H., D. E. Hammer, D. L. Tilton, L. Rosman, and B. Yardley, 1978, Houghton Lake wetland treatment project, first annual operations report, Wetland Ecosystem Research Group, University of Michigan, Ann Arbor, 88p.

Kadlec, R. H., C. J. Richardson, and J. A. Kadlec, 1975, The effects of sewage effluent on wetlands ecosystems, Semi-annual report no. 4 to NSF-RANN, University of Michigan, Ann Arbor, 199p.

Kadlec, R. H., D. L. Tilton, and B. R. Schwegler, 1979, Wetlands for tertiary treatment, a three-year summary of pilot scale operations at Houghton Lake, report to NSF-RANN, University of Michigan, Ann Arbor, 96p.

Larkin, E. P., et al, 1978, Land application of sewage wastes: Potential for contamination of foodstuffs and agricultural soils by viruses, bacterial pathogens, and parasites, state of knowledge in land treatment of wastewater, Vol. 2, U. S. Army Corps of Engineers, Hanover, N. H., pp. 215–232.

Lennette, E. H. and D. P. Spath, 1978, Comparison of health considerations for land treatment of wastewater, state of knowledge in land treatment of wastewater, Vol. 1, U. S. Army Corps of Engineers, Hanover, N. H., pp. 27–34.

Mather, J. R., and D. M. Parmelee, 1963, Water purification by natural filtration, Publ. in Climatol. 16:481–510.

Rabe, M. L., 1979, Impact of wastewater discharge upon a northern Michigan wetland wildlife community, unpublished report to Houghton Lake Sewer Authority and Michigan Dept. of Natural Resources, 41p.

Savidge, I. R., and D. E. Davis, 1971, Bird populations in an irrigated woodlot, 1963–1967, Bird-banding 42:249–263.

Snider, J. R., and G. W. Wood, 1975, The effects of wastewater irrigation on the activities and movements of songbirds, research publication 87, Institute for Research on Land and Water Resources, Pennsylvania State University, University Park, pp. 20–39.

Sopper, W. E., and L. T. Kardos, eds., 1973, Recycling treated municipal wastewater and sludge through forest and cropland, Pennsylvania State University Press, University Park, 479p.

C. W. Thornthwaite Associates, 1969, An evaluation of cannery waste disposal by overland flow spray irrigation, Campbell Soup Company, Paris Plant, Publ. in Climatol. 22:1-73.

Urie, D. H., J. H. Cooley, and A. R. Harris, 1978, Irrigation of forest plantations with sewage lagoon effluents, state of knowledge in land treatment of wastewater, Vol. 2, U. S. Army Corps of Engineers, Hanover, N. H., pp. 207-213.

DISCUSSION

Larson: I really have a comment rather than a question. Your white-footed mice might be responding not only to edge but also to life forms. They are fairly arboreal.

Jackson: They're very much arboreal, and they're very much at home in the trees. But they respond in a variety of ways. They are ubiquitous, adaptable things.

Odum: You referred to this project in the past tense. What is happening to that land now? What is coming back?

Jackson: Well, Seabrook Farms went out of business as a frozen food and agricultural operation about a decade ago. The whole operation was shut down. This study was essentially unfunded to begin with, and we have not been able to get back. In fact, we can't get the legal clearances and/or money to get back in and repeat some of these things. It would be a fascinating thing to go back, find those old quadrats and make some evaluations. I have been back to the site and looked at it a little bit. It's beginning to regenerate the oaks, and there are more seedlings.

Whigham: Are those open areas still covered with herbaceous vegetation or have they gone to woody species?

Jackson: A lot of them are still covered with herbaceous stuff. Elderberry began taking over in massive quantities. Pokeweed, lambs quarters, Virginia creeper, and the like came in first, then the elderberry began to override that.

15

Vegetation in Wetlands Receiving Sewage Effluent: The Importance of the Seed Bank

Dennis F. Whigham

Van der Valk (1981,1982) has proposed a model for studying changes in wetland vegetation that is based on three life-history attributes: (1) life span, (2) propagule longevity, and (3) propagule establishment requirements. The model assumes that knowledge of these attributes for species in a wetland would enable one to predict vegetation composition under various hydrologic (and other environmental) conditions. Although the model needs to be modified or expanded to permit quantification of population properties (e.g., biomass, aerial coverage, and so forth), it provides a suitable framework to evaluate the types of changes that might occur when wetlands are used for wastewater management.

The purpose of this paper is to discuss the effects that wastewater application might have on wetland vegetation. Because two of the three attributes used in van der Valk's model relate to the seed bank, the paper will focus primarily on the effects that altered hydrologic and nutrient patterns have on recruitment from the seed bank.

SEED BANKS IN WETLANDS

Seed banks have been shown to be important in a number of terrestrial ecosystems (see discussion in Harper 1977), but there have been only a few studies of seed banks in wetlands even though water-level manipulations have been used for many years to regulate wetland vegetation (Keddy and Reznicek 1982; Meeks 1969). Seed banks have been shown to be very important in the vegetation dynamics of midwestern prairie pothole wetlands that are subject to periodic droughts (van der Valk and Davis 1976, 1978; Millar 1969). Seed banks have also been shown to be important in littoral wetlands where long-term patterns of species diversity are maintained because of periodic natural drawdowns (Keddy and Reznicek 1982; Dykyjová and Květ 1978). Seed banks have also been studied in freshwater wetlands that are subjected to daily tidal activity (Leck and Graveline 1979; Ristich, Fredrick, and Buckley 1976), and Junk (1970) has suggested that buried seeds are important in Amazonian riverine and lacustrine wetlands that are subject to dramatic annual changes in water levels.

I have found no studies of the relationship between vegetation composition and seed banks for permanently flooded herbaceous wetlands in which the period of drawdown is very brief. Neither have I found any studies of seed banks in forested wetlands, although Curtis Richardson (personal communication) suggests that recruitment from the buried seed pool is not very important in northern bogs. In contrast, Christensen et al. (1981) have suggested that the seed bank may be important in forested southeastern Pocosin wetlands where the peat substrate is subjected to fire during periodic droughts. In tundra wetlands, recruitment from the seed pool seems to be unimportant for herbs and minimally important for shrubs (Callaghan and Collins 1981).

Although the data base on wetland seed banks is meager, it appears that they are most important in wetlands that are subject to daily, seasonal, annual, or less frequent periods of drawdown.

Under what conditions would the seed bank be least important? Any attempt to answer this question requires an understanding of the autecology of wetland plants, and a brief review of van der Valk's model will demonstrate the importance of autecological data. The model (Table 15.1) includes three life spans: annuals (A), perennials that do not form large clones and/or spread slowly (P), and perennials that form large clones by vegetative growth (V); and two types of propagule (primarily seed) longevity: species with long-lived propagules that are called seed bank species (S) and species with short-lived propagules, called dispersal species (D). Finally, the model includes two seedling-establishment scenarios: plants that require a period of drawdown for germination and establishment are called Type I species, and plants that can become established under flooded conditions are called Type II species. When all possible combinations of the three categories are considered, there are 12 types of species (Table 15.1).

Water-level fluctuations are critical to the model. Figure 15.1 shows one example of how species are eliminated when a wetland normally exposed to fluctuation in the water levels is permanently flooded. In permanently flooded wetlands, all Type I species would be eliminated and their propagules eliminated from the seed bank unless they were annually replenished by dispersal from other areas. The highest diversity of life-history types would be expected to occur in wetlands where water levels fluctuate, particularly with a period of drawdown occurring when Type I species would germinate.

A second example is presented in Table 15.2, which shows that the diversity of annuals (A)--species that are dependent on the seed bank--declines in a freshwater tidal wetland along a drawdown gradient. Maximum diversity occurs on the high marsh, where the substrate is exposed twice daily. Lowest diversity occurs in permanently inundated ponds. There is evidence that seeds of most species in freshwater tidal wetlands are distributed throughout the wetland (Whigham, Simpson, and Leck 1979; Whigham and Simpson 1982) so that the existing vegetation in permanently flooded areas does not mirror the species composition of seeds in the seed bank. The seed bank in permanently flooded habitats, therefore, has less influence on vegetation composition compared to high marsh and stream bank areas where the vegetation closely mirrors the composition of the seed bank (Simpson et al. 1984).

Table 15.1
Classification of Wetland Plants Used in van der Valk's
Succession Model

Type of Plant	Propagule Longevity	Establishment Requirement	Symbol
Annual (A)	Short-lived (D)	Drawdown (I)	AD-I
		Flooded (II)	AD-II
	Long-lived (S)	I	AS-I
		II	AS-II
Perennial (P)	S	I	PS-I
		II	PS-II
	D	I	PD-I
		II	PD-II
Vegetative perennial (V)	S	I	VS-I
		II	VS-II
	D	I	VD-I
		II	VD-II

Source: Data from A. G. van der Valk, 1981.

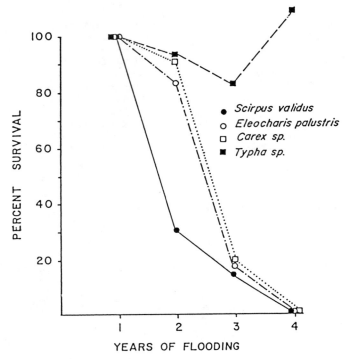

Figure 15.1
Survival of dominant species in a wetland following
establishment of permanently flooded conditions.
(After S. W. Harris and W. H. Marshall, 1963)

EFFECTS OF WASTEWATER ADDITIONS

    Wastewater application would most likely increase the depth of
flooding, duration of flooding, and/or frequency of flooding. In
addition, there would be significant increases in nutrient loading.
Nutrient additions have been shown to cause increased biomass production in both coastal salt marshes (Valiela, Teal, and Sass 1975;
Valiela, Teal, and Persson 1976) and the following inland freshwater
wetlands: herbaceous wetlands (Zoltek et al. 1979) and cypress domes
in Florida (Odum and Ewel 1978), bogs in Michigan (Kadlec 1980;
Tilton and Kadlec 1979), and artificial wetlands in New York (Small
1976; Woodwell 1977; Woodwell et al. 1974). Production did not
increase in a freshwater tidal wetland receiving chlorinated,
secondarily treated wastewater (Whigham, Simpson, and Lee 1980).
    Increased production could have an indirect effect on the seed
bank due to the competitive elimination of some species by aggressive
Type V species that form monocultures. Examples would be Typha spp.,
surface mats of floating vegetation such as Lemna (Odum and Ewel

Table 15.2
Importance Values of Annual Species in Three Habitats
Within a Freshwater Tidal Wetland

Species	High marsh	Pondlike	Pond
Bidens laevis	98.9	–	–
Polygonum punctatum	39.5	44.6	49.6
Impatiens capensis	32.6	–	–
Polygonum arifolium	20.2	–	–
Zizania aquatica	9.9	32.6	–

Source: Data from Whigham 1974.

1978), and Phragmites. Expansion of aggressive Type V species
following wastewater addition has been documented by Tilton and
Kadlec (1979).

In other instances, the seed bank may be influenced by loss of
some species following wastewater addition. Species losses have been
found for bogs (Curtis Richardson, personal communication) and freshwater tidal wetlands (Whigham, Simpson, and Lee 1980). In the latter
instance, recovery was very rapid following cessation of wastewater
application (Robert Simpson, personal communication).

The seed bank would be expected to become less important in
wetlands where the hydrology is altered and permanent standing water
conditions are created. This situation is shown in Figure 15.2. By
maintaining permanent standing water, all species that require a
drawdown for establishment (AS-I, PS-I, VS-I, AD-I, PD-I, and VD-I)
could be eliminated. If VD-I and VD-II species (e.g., Typha)
ultimately dominate the site, AS-II, AD-II, short-lived PD-II, and
short-lived PS-II might also be eliminated.

There would probably be no change in the importance of the seed
bank in wetlands where seeds are of minor importance (e.g., salt
marshes, northern bogs, and wetlands that are permanently flooded or
are already dominated by Type V species). One would also predict
that there would be no change in the importance of the seed bank in
wetlands that have regular drainage (freshwater tidal wetlands),
periodic drawdown (Playa wetlands, herbaceous wetlands in Florida),
or undergo climate extremes (Prairie glacial wetlands).

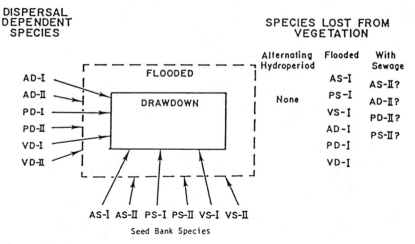

Figure 15.2
Diagrammatic representation of van der Valk's (1982) model and the
changes that would result from water level changes and nutrient
additions. (Refer to Table 15.1 for description of symbols.)

MANAGEMENT IMPLICATIONS

Much of the previous discussion is, unfortunately, speculative.
Van der Valk (1981, 1982) has shown that the model can be used to
predict vegetation changes associated with water-level manipulations.
There are no data for wetlands that have received wastewater for
extended periods of time to test the model. The model does, however,
provide insight into management strategies to be used to minimize
impacts due to wastewater irrigation.

As has been suggested in the previous section, the addition of
wastewater can influence vegetation composition by enabling some
species to become more important and by eliminating others. Changes
can occur because of increased nutrient loading alone or in concert
with changing hydrologic patterns. Deleterious effects can be
avoided by minimizing changes in the hydrologic regime. In particular, increasing the depth of flooding, frequency of flooding, and
duration of flooding should be avoided because those conditions have
a negative impact on the seed pool. In addition, it has been shown
that primary production and, consequently, nutrient retention are
less under flooded conditions. Wastewater addition would be expected
to have minimal effects in systems that have frequent drawdown, since
diversity appears to be greatest in those types of wetlands, and the
seed pool would be minimally affected under such conditions.

REFERENCES

Callaghan, T. V., and N. J. Collins, 1981, Life cycles, population dynamics, and the growth of tundra plants, in Tundra ecosystems: A comparative analysis, L. C. Bliss, D. W. Heal, and J. J. Moore, eds., Cambridge University Press, London, England, pp. 257–284.

Christensen, N., R. Burchell, A. Liggett, and E. Simms, 1981, The structure and development of Pocosin vegetation, in Pocosin wetlands, C. J. Richardson, ed., Hutchinson Ross Publishing Co., Stroudsburg, Penn., pp. 43–61.

Dykyjová, D., and J. Kvét, eds., 1978, Pond littoral ecosystems, Springer-Verlag, New York, 264p.

Harper, J. L., 1977, Population biology of plants, Academic Press, New York, 892p.

Harris, S. W., and W. H. Marshall, 1963, Ecology of water level manipulations of a northern marsh, Ecology 44:331–343.

Junk, W., 1970, Investigations on the ecology and production-biology of the flooding meadows (Paspalo-Echinochloetum) on the middle Amazon, Amazoniana 2:449–495.

Kadlec, R. A., 1980, Wetlands for tertiary treatment, in Wetland function and values: The state of our understanding, P. E. Greeson, J. R. Clark, and J. E. Clark, eds., American Water Resources Association, Minneapolis, Minn., pp. 490–504.

Keddy, P. A., and A. A. Reznicek, 1982, The role of seed banks in the persistence of Ontario's coastal plain flora, Am. J. Bot. 69:13–23.

Leck, M. A., and K. J. Graveline, 1979, The seed bank of a freshwater tidal marsh, Amer. J. Bot. 66:1006–1015.

Meeks, R. L., 1969, The effect of drawdown date on wetland plant succession, J. Wildl. Manage. 33:817–821.

Millar, J. B., 1969, Observations on the ecology of wetland vegetation, in Saskatchewan wetlands seminar, Canadian Wildlife Service Report Series, no. 6, Canadian Wildlife Service, Ottawa, Can., pp. 48–69.

Odum, H. T., and K. C. Ewel, eds., 1978, Cypress wetlands for water management, recycling, and conservation, fourth annual report, Center for Wetlands, University of Florida, Gainesville, 879p.

Ristich, S. S., S. W. Fredrick, and E. H. Buckley, 1976, Transplantation of Typha and the distribution of vegetation and algae in a reclaimed estuarine marsh, Torrey Bot. Club Bull. 103:157–164.

Simpson, R. L., R. E. Good, M. A. Leck, and D. F. Whigham, 1984, The ecology of freshwater tidal wetlands, Bioscience, 33:255–259.

Small, M. H., 1976, Marsh/pond sewage treatment plants, in Freshwater wetlands and sewage effluent disposal, D. L. Tilton, R. H. Kadlec, and C. J. Richardson, eds., School of Natural Resources, College of Engineering, University of Michigan, pp. 197–214.

Tilton, D. L., and R. A. Kadlec, 1979, The utilization of a freshwater wetland for nutrient removal from secondarily treated wastewater effluent, J. Environ. Qual. 8:328–334.

Valiela, I., J. M. Teal, and N. Y. Persson, 1976, Production and dynamics of experimentally enriched salt marsh vegetation: Belowground biomass, Limnol. Oceanogr. 21:245–252.

Valiela, I., J. M. Teal, and W. J. Sass, 1975, Production and dynamics of salt marsh vegetation and the effect of sewage contamination: Biomass, production, and species composition, J. Appl. Ecol. 12:873–882.

van der Valk, A. G., 1981, Succession in wetlands: A Gleasonian
    approach, Ecology 62:688-696.
van der Valk, A. G., 1982, Succession in temperate North American
    wetlands, in Wetlands: Ecology and management, B. Gopal, R. E.
    Turner, R. G. Wetzel, and D. F. Whigham, eds., International
    Scientific Publications, Jaipur, India, pp. 169-179.
van der Valk, A. G., and C. B. Davis, 1976, The seed banks of prairie
    glacial marshes, Can. J. Bot. 54:1832-1838.
van der Valk, A. G., and C. B. Davis, 1978, The role of seed banks in
    the vegetation dynamics of prairie glacial marshes, Ecology
    59:322-335.
Whigham, D. F., 1974, Preliminary ecological studies of a freshwater
    tidal marsh along the Delaware River, Biology Department, Rider
    College, Lawrenceville, N. J., 66p.
Whigham, D. F., and R. L. Simpson, 1982, Germination and dormancy
    studies of Pontederia cordata, Torrey Bot. Club Bull. 103:524-528.
Whigham, D. F., R. L. Simpson, and M. A. Leck, 1979, The distribution
    of seeds, seedlings, and established plants of Arrow arum
    (Peltandra virginica (L.) Kunth) in a freshwater tidal wetland,
    Torrey Bot. Club Bull. 106:193-199.
Whigham, D. F., R. L. Simpson, and K. Lee, 1980, The effects of
    sewage effluent on the structure and function of a freshwater
    tidal marsh ecosystem, Water Resources Research Institute, Rutgers
    University, New Brunswick, N. J., 106p.
Woodwell, G. M., 1977, Recycling sewage through plant communities,
    Am. Sci. 65:556-562.
Woodwell, G. M., J. Ballard, M. Small, E. V. Pecan, J. Clintin, R.
    Wetzler, F. German, and J. Hennessy, 1974, Experimental eutrophi-
    cation of terrestrial and aquatic ecosystems, first annual report,
    BNL 50420, Brookhaven National Laboratory, Upton, N. Y., 28p.
Zoltek, J., Jr., S. E. Bayley, A. J. Hermann, L. R. Tortora, T. J.
    Dolan, D. A. Graetz, and N. L. Erickson, 1979, Removal of
    nutrients from treated municipal wastewater by freshwater marshes,
    final report to the City of Clermont, Florida, Center for
    Wetlands, University of Florida, Gainesville, 325p.

DISCUSSION

Larson: I'd like to make a couple of comments. First, in selecting
papers for this session, we did not select from the rather well
developed literature on manipulation of water levels for wildlife
purposes. There is quite a lot to be drawn out of that literature
without reinventing the wheel. The other point is that some years
ago the National Wetlands Technical Council, under NSF sponsorship,
held a symposium in Athens, Georgia, and came out with a moderately
well distributed report. It contained the very important conclusion
that a great deal more work needed to be done on hydrology. And this
afternoon we've had a specific reference to hydrology. I must
confess that at least as regards my examination of the literature,
not enough funding from the important federal sources, nor enough
effort to secure funding from private sources, has gone into hydrology, whether surficial or subsurface. A great deal of what we've
talked about this afternoon and an immense amount of what we'll talk
about at this whole workshop is going to remain wide open because we
have not engaged enough hydrologists to look at wetlands. To the

extent we don't do that, we will continue to have many unanswered questions. Are there any questions for Dennis?

Zedler: Dennis Whigham's and Barbara Bedford's papers both suggest that what is causing all the species composition change is the hydrology. It seems to have nothing to do with nutrients. To me, that's a bit surprising. You didn't say it was all due to the water level, but that certainly seems like the most important impact of wastewater treatment.

Whigham: I think that in the case of Typha, for example, it will increase its presence in the wetland just because of the additional nutrients.

Zedler: It becomes more productive, but it won't shift to a new type of wetland just because of nutrients.

Guntenspergen: Joy, there is some work I've been doing for my Ph.D. thesis in artificially fertilizing freshwater marshes. In some cases, I have seen species compositional changes due entirely to the nutrient additions and having nothing at all to do with hydrology.

Valiela: John Teal and I have a system in which we irrigate one hectare of salt marsh with a solution of simulated sewage effluent. We also have adjoining plots where we add only freshwater, and we can measure no change at all due to the addition of freshwater. All the change that we measure is associated with nutrient additions and, also, the elevation of the marsh surface in relation to tidal heights.

Larson: Are you maintaining the depth and replicating the perio- dicity of the normal inundation?

Valiela: No. We irrigate during low tide in order to get the maximum impact possible. We still don't get an impact. In fact, salinity of the pore water hardly changes, in spite of the fact that we're adding a couple of inches of rain per week.

Richardson: A few years ago, Al Wentz, in his dissertation on the phosphorus content in the Houghton Lake study, compared some plots where we had added simulated sewage effluent to some where we added just plain water in the same amount for the control. We lost several species of early aster on the nutrient plots, but not on the control plots.

Bedford: I wasn't implying that the only changes observed were due to water. What I was saying was that we don't have enough experi- mental evidence to suggest what proportion of the change is due to nutrients and what is due to water. In most cases you're going to get both of them at once. So there are some things you can infer from the water-level changes. For the nutrients, I tried to suggest some basis from which we might infer what's happening from the magnitude of change. If you have a system that's already nutrient- saturated, not nutrient-limited, you're not likely to get the magni- tude of change that you would in a nutrient-poor system.

Kaczynski: I'd like to make a practical observation from the point of view of someone designing these systems. If you're getting serious species changes, it sounds to me as if you have not done your homework in terms of looking at the application. If you had under-sized the wetland relative to the amount of effluent you're putting in (in other words, you're overapplying the effluent), you could use agricultural application techniques to determine the acreage of application relative to the water portion of the effluent. I'd suggest you go back and look at that.

Bedford: We need a basis for comparison with the experimental work that's been done on natural systems. For example, most of the existing work has been done in the South, but we don't have a basis for comparing response to the different loading rates in the North. Changes are undoubtedly related to loading rates, but we don't have enough data. I believe you're right. We might obtain better initial estimates by using irrigation application techniques.

Ewel: Victor [Kaczynski], you're making an assumption that a substantial species change is serious. But I would challenge you that in some cases substantial species changes are not necessarily serious. You can get the same change in production with a major species change in one kind of system and no species change in another. Who's to say that one is serious and one is not? So, I think we have to find a better measure than species composition before we can ask people to apply sophisticated design criteria.

Kaczynski: I would agree, if you're interested in wetland function.

Ewel: It would be nice if we all had gas analyzers or something to clamp over these species and measure productivity. But it's not that easy.

SESSION V

Environmental Health

16

Public Health Implications of Sewage Applications on Wetlands: Microbiological Aspects

Michael P. Shiaris

With an ever-expanding population, the demand for water for urban, agricultural, and industrial use is growing. At the same time, the volume of biological and chemical wastes that enter our waters is also increasing. The impending dilemma requires a switch to increased and more efficient water recycling. The potential use of wetlands as wastewater treatment sites is a promising approach (U. S. EPA 1979; Kadlec and Tilton 1979); hence, the subject of this workshop.

Any future application of sewage to wetlands must be free of unreasonable risks to public health. As a rule, untreated municipal wastewater contains many components that pose a threat to the well-being of wildlife and humans. Pathogens from human and animal feces, organic toxins, industrial waste, heavy metals, and pesticides are present in varying amounts in all municipal sewage. A variety of specific treatments to attain a satisfactory water effluent are available. The projected use of the recycled wastewater and the potential for human contact will dictate which treatment and the degree of treatment necessary to reduce the health risk (Shuval 1982).

Microorganisms are of paramount importance in any public health consideration of wastewater. Clearly, the direct hazards of micro-organisms as agents of disease in wastewater are well recognized, if not fully understood. The role of microorganisms as agents of bio-transformations, however, is not as well recognized, and certainly not as well understood relative to public health. In a wetlands treatment site, the microbial component may actively detoxify hazardous chemicals and, conversely, may transform or "activate" some relatively nontoxic chemicals to more toxic forms.

The present purpose is to highlight briefly the potential health risks of microorganisms and microbial activities in sewage waste-water. Relatively little information is available on the public health hazards that may be encountered by applying wastewater to wetlands. On the other hand, there is considerable information on land-application health aspects that have been reviewed elsewhere (Lance and Gerba 1978; Kowal, in press; Pahren et al. 1979). Regulatory authorities postulate that the health risks for wetland systems will not be greater than the hazards encountered in

conventional wastewater treatment, assuming that materials will not
be harvested for human consumption (U. S. EPA 1979). Therefore,
while much of the work discussed herein was not conducted in wetland
systems, it will be assumed that the health risks discussed are
applicable to wetlands as well.

PATHOGENS

The hazards to public health due to pathogens in wastewater are
fairly well understood with the exception of viruses. Pathogens may
be transmitted to humans by direct contact with wastewater or by
indirect means, such as consumption of contaminated plants and
animals, aerosols, and other contaminated objects. The degree of
danger to human health depends on the ability of the pathogens to
persist, either by surviving treatment conditions or by proliferating
in nonhuman hosts indigenous to the wetland. In marine environments,
pathogens may be concentrated by shellfish. Numerous worldwide
outbreaks of typhoid fever and infectious hepatitis have been linked
to the consumption of raw seafood (Shuval 1978). The risk of
infection by recreational use of contaminated water is not certain.
However, at least one study (Cabelli et al. 1976) provided evidence
for significantly higher incidence of gastrointestinal disorders
among bathers compared to nonbathers in sewage-contaminated waters.
No correlation was present among bathers and nonbathers in nearby
uncontaminated waters. In Israel, Katzenelson et al. (1976) reported
a two- to fourfold increase in the incidence of communicable diseases
(shigellosis, salmonellosis, typhoid fever, and hepatitis) in kibbutz
communities practicing wastewater irrigation. While the evidence for
occupational exposure risk to disease is inconclusive, it seems a
reasonable precaution to limit contact between treatment workers and
the wastewater.
Pathogens present in sewage are traditionally divided into three
or four groups: bacteria, viruses, and parasites (protozoa and
helminths). A comprehensive review of public health aspects of land
treatment has been prepared by Kowal (in press), and only some major
aspects of bacteria and viruses will be discussed here. Further
treatment of the subject of parasites can be found in a review by
Pahren et al. (1979).

BACTERIA

Of all the pathogens, bacterial pathogens are the most susceptible to wastewater treatment. Bacteria of major concern to public
health are listed in Table 16.1.
The most numerous pathogen in municipal wastewater is Salmonella
spp. Typically, sewage sludge contains between 1.4×10^2 to 1.4×10^5 Salmonella spp. per 100 g of dry weight (Langeland 1982).
Consequently, Salmonella spp. is the bacterium of major concern to
public health. In 1979 alone, over 30,000 cases of salmonellosis
were reported to the Center for Disease Control (CDC) in Atlanta,
Georgia (MMWR 1980). Over 600 cases of typhoid fever, a more severe
disease than salmonellosis, were reported. Approximately 2% to 3% of
treated typhoid fever cases are fatal, with the mortality increasing
to 10% in untreated cases. Salmonella spp. may survive relatively

Table 16.1
Pathogenic Bacteria of Major Concern in Wastewater

Organism	Disease	Nonhuman Reservoir
Campylobacter fetus	Acute gastroenteritis	Cattle, dogs, cats, and poultry
Escherichia coli (pathogenic strains)	Acute diarrhea	—
Leptospira spp.	Leptospirosis	Domestic and wild animals, rats
Salmonella paratyphi	Paratyphoid fever	—
Salmonella typhi	Typhoid fever	—
Salmonella spp.	Salmonellosis	Domestic and wild animals, birds and turtles
Shigella spp.	Shigellosis, bacillary dysentary	—
Vibrio cholera	Cholera	—
Yersinia enterocolitica Yersinia pseudotuberculosis	Yersiniosis	Wild and domestic birds and mammals

long periods in water (Yoshpe-Purer and Shuval 1972), although high infective doses (10^5 to 10^8 organisms) are required for infection (Kowal, in press).

Only recently has it become apparent that Campylobacter fetus subspecies jejuni may be a major agent of acute gastroenteritis with diarrhea. Between 4% and 8% of all patients with diarrhea have C. fetus present in their stool (MMWR 1979), and C. fetus may be as common as Salmonella spp. in wastewater.

A third pathogenic bacterium of major concern in the United States is Shigella spp. Over 15,000 cases of shigellosis were reported to the CDC in 1979. Four shigellae species--S. sonnei, S. flexneri, S. boydii, and S. dysenteriae--are responsible for the disease state. Absolute numbers of shigellae in wastewater are much lower than salmonella; however, the infective dose may be as low as 10 to 100 organisms. Therefore, the mere presence of Shigella spp. in contaminated waters is cause for concern.

Leptospirosis (causative agent, Leptospira spp.) is not as prevalent as salmonellosis and shigellosis in the United States. Less than 500 cases were reported between 1974 and 1978 (Martone and Kaufmann 1980). Unlike most waterborne bacterial pathogens, Leptospira spp. enters municipal wastewater primarily from the urine of domestic and wild animals (Kowal, in press).

The remaining pathogens may be present in sewage, but in much lower abundance. Because the main mode of entry of bacterial pathogens is by the feces of infected individuals, their abundance in wastewaters depends on the frequency of disease in the immediate population. The numbers of infectious agents present in the stools of infected individuals are presented in Table 16.2.

Occasionally, the occurrence of a potential pathogen in water is not reflective of either the incidence of disease or the amount of fecal pollution to which the water is subjected. For example, Vibrio cholera is commonly found in estuarine waters (Kaper et al. 1979), yet the incidence of cholera in the United States is rare. Only 12 cases have been reported in the past 70 years, 1 in 1973 in Texas and 11 in 1978 in Louisiana (Blake et al. 1980). V. cholera is thought to be a normal resident of estuarine waters (Hood and Ness 1982).

Bacterial pathogens of minor concern to public health in wastewaters, due either to their extremely low frequency of occurrence or their opportunistic mode of disease, include the following: Brucella spp., Citrobacter spp., Clostridium spp., Coxiella burnetti, Enterobacter spp., Erysipelothrix rhusiopathiae, Francisella tularensis, Klebsiella spp., Legionella pneumophila, Listeria monocytogenes, Mycobacterium tuberculosis, Proteus spp., Pseudomonas aeruginosa, Serratia spp., Staphylococcus aureus, and Streptococcus spp.

Fecal coliforms and fecal streptococci have been enumerated in the environment as traditional indicators of fecal pollution. Therefore, their abundance in waters reflects the risks, under most conditions, to public health by bacterial pathogens. Health officials have relied on the indicator bacteria to undergo the same fate as bacterial pathogens under identical environmental conditions. Table 16.3 summarizes removal rates of fecal coliforms, fecal streptococci, and Salmonella spp. by various wastewater treatments, as reported by Kowal (1982). Similar reduction might be expected for wetland applications. In wetland treatment sites, removal of fecal coliforms depends on residence time at the site, the amount of contact with the soil, and flow rates (Kadlec and Tilton 1979).

Table 16.2
Pathogenic Bacteria in Human Feces of Infected Persons

Organism	Number/g Net Weight
Campylobacter fetus	?
Escherichia coli (enteropathogenic strains)	10^8
Salmonella paratyphi (A,B,C)	10^6
Salmonella typhi	10^6
Salmonella spp.	10^6
Shigella spp.	10^6
Vibrio cholerae	10^6
Yersina entercolitica, Y. pseudotuberculosis	10^5

Source: From N. E. Kowal, Health effects of land treatment: Microbiological, EPA research report, EPA-600-1-82-007, U. S. EPA, Cincinnati, Oh., 78p.

Table 16.3
Bacterial Removal by Various Wastewater Treatment

Treatment	Retention Time (days)	Removal (%)	
Single anaerobic pond	3.5–5	46–85	E. coli
Single facultative or aerobic pond	10–37	80–99	E. coli
Series of three or more ponds	10–37	99.99	E. coli
One to two aerobic ponds	3.5–5	90	Salmonella spp.
One to two aerobic ponds	30–40	Complete removal of pathogenic bacteria	
Aerated lagoons	-	60–99.99	Total coliforms
		99	Fecal coliforms
		99	Salmonella

Source: From N. E. Kowal, Health effects of land treatment: Microbiological, EPA research report, EPA-600-1-82-007, U. S. EPA, Cincinnati, Oh., 78p.

The conventional approach to removal of pathogenic bacteria from effluents has been chlorination. However, major disadvantages include cost, if biochemical oxygen demand (BOD) levels are high as in raw sewage, and the production of carcinogenic trihalomethanes and other potentially hazardous chlorinated organic compounds. Alternative choices for pre- and post-treatment of the wastewater will have to be site specific, based on regular monitoring for pathogenic bacteria.

VIRUSES

Viral agents of human disease that may cause asymptomatic infectious to permanently debilitating diseases are present in municipal wastewater (Table 16.4). Accounting for 40,000 to 50,000 reported cases of hepatitis in the United States annually, hepatitis A virus may be the most serious public health hazard (Pahren et al. 1979). Like many of the enteroviruses, hepatitis A virus commonly infects young children, causing an asymptomatic disease state. In an adult, however, hepatitis may cause permanent liver damage.

The enteroviruses are ubiquitously distributed in waters and wastewaters. Epidemics due to enteroviruses are particularly prevalent in the summer months. Indeed, gastroenteritis of suspected viral etiology (probably enteroviral) is the major waterborne illness in the United States (Haley et al. 1980).

Rotavirus is also of important concern to public health. Severe diarrhea with accompanying dehydration caused by rotavirus may be fatal in infants (Flewett and Woode 1978). Rotavirus infection in adults, however, is self-contained.

Viruses are more resistant to removal treatments than bacteria and tend to persist longer in aquatic environments. They also interact differently with other environmental components. For example, many viruses readily adsorb to naturally occurring silts and clays (Goyal and Gerba 1979). As a result, fecal coliforms and other bacterial indicators of fecal pollution are not useful indicators for waterborne viral pathogens (Grabow 1968). Goyal and Gerba (1979) reported that shellfish standards based on coliform counts do not necessarily correlate to enterovirus contamination. Enterovirus may also be detected in seawater with very low coliform counts (Katzenelson 1978). Differences in the persistence and environmental response of individual viral types are further causes for consternation in the search for a universal viral indicator. It has become apparent that no one virus can serve as an adequate indicator of viral pollution (Grabow 1968).

The detection of any pathogenic viruses in waters designated for human use constitutes a health hazard (Berg 1967). In contrast with bacterial pathogens, only a few pathogenic viruses (<10 particles) may cause infection. In addition, viral numbers in the stool of infected individuals are typically large. For example, 10^{10} to 10^{11} rotavirus particles per wet gram are present in the diarrheic stool of infected patients (Flewett and Woode 1978). Rotavirus can also survive long periods in both fresh and estuarine waters (Hurst and Gerba 1980).

More practical and efficient detection methods are required in order to assess the health significance of pathogenic viruses. Commonly employed techniques may detect only 1/10 to 1/100 of the

Table 16.4
Human Enteric Viruses in Wastewater

Virus	Disease
Enteroviruses	
Poliovirus	Poliomyelitis
Coxsackievirus A, B	Aseptic meningitis, herpangia, epidemic myalgia pericarditis, pneumonia, rashes, common colds, fever, congenital heart anomalies, hepatitis
Echovirus	Aseptic meningitis, paralysis, encephalitis, fever, rashes, common colds, epidemic myalgia, pericarditis, myocarditis, diarrhea
New enteroviruses	Pneumonia, bronchiolitis, aseptic meningitis, encephalitis, hand-foot-and-mouth disease, epidemic myalgia
Hepatitis A virus	Infectious hepatitis
Rotavirus	Acute gastroenteritis
Norwalk-like agents	Epidemic gastroenteritis
Adenovirus	Sporadic gastroenteritis
Reovirus	?
Papovavirus	Progressive multifocal leukoencephalopathy?
Astrovirus	Gastroenteritis?
Calicivirus	Gastroenteritis?
Coronavirus-like particles	Gastroenteritis?

viruses present in the water. Presently, efforts are being made to improve viral concentration steps and detection techniques. For example, Smith and Gerba (1982) have employed immunofluorescence techniques to detect rotavirus as low as ten infectious units per liter of sewage.

Removal of virus by conventional treatment is variable and may depend on the properties of individual viruses (Christie 1967; Shuval 1970; Sheladia, Ellender, and Johnson 1982). Even chlorinated effluents may contain virus if the residual chlorine concentration in the effluent is too low (Kott 1973). The factors responsible for viral survival and inactivation are not adequately understood.

MICROBIAL TRANSFORMATION OF POLLUTANTS

Public health aspects of toxic pollutants in municipal waste-waters are greatly influenced by the active microorganisms in the system. Microbial metabolic processes may transform pollutants to nontoxic forms or activate pollutants to more toxic compounds. Frequently the waste products of microbial metabolism themselves may be hazardous to human health.

Municipal wastewater contains thousands of compounds that may serve as growth substrates for microbial growth. All too frequently, industrial wastes are released into municipal sewage systems, thereby increasing the load of toxic compounds (Woodwell 1977). Unfortunately, little is known of the quantities and types of toxins found in sewage. Quantities and types would, of course, be determined in large part by the nature of the surrounding industry. No systematic survey of toxins in sewage has been conducted to date (Pahren et al. 1979).

The total decomposition of organic compounds to inorganic compounds, or mineralization, in soils and waters is primarily due to microbial activity. There are few abiotic mechanisms in the environment that can mineralize organics (Alexander 1980). Quite often, however, toxic compounds of public health concern are resistant to microbial mineralization. Aside from their toxic properties, it is their recalcitrant properties, their ability to persist, and their potential to accumulate in the environment that render them hazardous. Well-known examples are polychlorinated biphenyls (PCBs), polycyclic aromatic hydrocarbons (PAHs), DDT, and kepone. It is becoming increasingly apparent that compounds may also be transformed by microbial mechanisms other than mineralization. The ability of microorganisms to transform compounds that do not support their growth has been termed cometabolism or cooxidation (Leadbetter and Foster 1959).

Cooxidative processes will likely occur in wetland treatment sites. Perry (1979) has reviewed some of the compounds that are cooxidized by microorganisms. Direct evidence for the cooxidation of four herbicides in sewage water was shown by Jacobson, O'Mara, and Alexander (1980). Cooney and Shiaris (1982) demonstrated that estuarine sediments contain both bacteria that utilize phenanthrene (a PAH) as a sole carbon source and bacteria that cooxidize phenanthrene (Table 16.5). Many aspects of microbial cooxidation in waters important to public health are not known: (1) Which potential toxins are cooxidized? (2) What is the extent of cooxidation? (3) Do the products of cooxidation accumulate? (4) If so, are they toxic? and

Table 16.5
Enumeration of Phenanthrene Utilizers and Cooxidizers in
Four Chesapeake Bay Sediments

Site	Total Number PHE Degraders Enumerated	Number of PHE Utilizers (% of total)	Number of PHE Cooxidizers (% of total)
2	46	5 (12.2)	36 (87.8)
8	49	35 (73.5)	13 (26.5)
9	237	132 (55.7)	105 (44.3)
9	28	17 (60.7)	11 (39.3)
All sites	355	190 (53.5)	165 (46.5)

Source: Data from Cooney and Shiaris 1982.

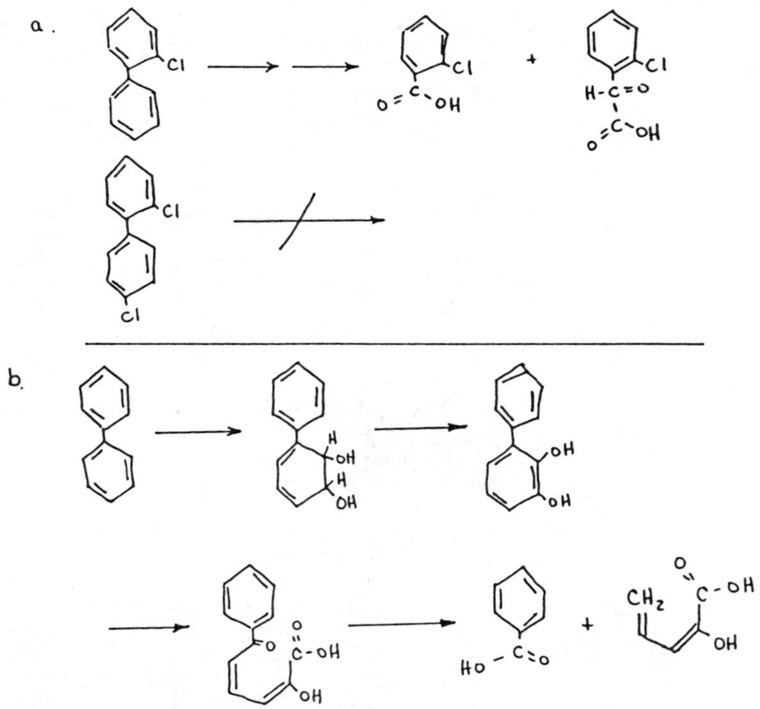

Figure 16.1
Microbial transformation of chlorinated biphenyls.
(a) Stable byproducts formed in cultures of natural
populations of freshwater bacteria. (b) Proposed
pathway of biphenyl biodegradation.

(5) What environmental factors affect cooxidation?

Another microbial process of potential concern to public health is the "activation" of a chemical to a more toxic form. For example, Dean-Raymond and Alexander (1976) showed that carcinogenic nitrosamines may form from secondary or tertiary amines plus inorganic nitrogen. Once more, these carcinogens can move into food crops and groundwater. Microorganisms in sediments are capable of transferring methyl groups to inorganic mercury to form highly toxic methylmercury (Wood, Kennedy, and Rosen 1968). Selenium, tellurium, arsenic, and tin are also subject to methylation in sediments (Jernelov and Martin 1975; Hallas, Means, and Cooney 1982). The extent and public health significance of these phenomena in aquatic ecosystems are not well understood.

Predicting microbial transformations in wetland environments is at best difficult. Little is known of the toxic and organic composition of municipal wastewater. An estimated 2000 new chemicals are synthesized each year and introduced into the environment (Kurzel and Cetrulo 1981). The influences of physical, chemical, and other biological factors are not well recognized. Knowledge of the fate of toxic chemicals in pure culture studies is inadequate, and even simulated environments (i.e., microcosms) may not be sufficient (Pritchard et al. 1978).

A recent study of PCB biodegradation in freshwater reservoirs (Shiaris and Sayler 1982) demonstrates some of the problems with predictability. Natural populations of freshwater microorganisms were shown to degrade 2-chlorobiphenyl but not 2,4'-dichlorobiphenyl. Two biotransformation products were found to accumulate: chlorobenzoic acid and chlorobenzoylformic acid. The accumulation of chlorobenzoic acid was expected from pure culture studies and others. However, the accumulation of chlorobenzoylformic acid was a novel observation, a product not expected from knowledge of known pathways of biphenyl metabolism (Catelani et al. 1973).

There is no reason to expect the fate of chlorinated compounds to be similar to their unchlorinated analogues (Wood 1982) or to fates determined in the laboratory. The public health hazards presented by potential biotransformation products are unknown. This relatively unexplored area deserves more attention, particularly in view of the prediction that environmental pollutants may be responsible for at least 5% to 10% and possibly more of all human birth defects (Kurzel 1980).

Finally, the interaction of pollutants with microbial processes in aquatic environments may raise public health concerns. The effect of PCB on bacterial nitrification in two freshwater reservoirs (Sayler et al. 1982) is an example (Fig. 16.1). In a relatively pristine reservoir, Center Hill, PCB concentrations as low as 10 ppb inhibited nitrification. At the lowest inhibitory concentration (10 ppb), nitrate oxidation was inhibited to a greater extent than was ammonia oxidation, resulting in the accumulation of nitrite, a potential health hazard. Nitrification was not inhibited in the eutrophic reservoir, Fort Loudon Reservoir, even at 1000 ppb PCB concentrations. Hence, knowledge of the ambient PCB concentrations alone is not sufficient to predict toxic effects. In this case, other environmental factors, perhaps the organic carbon load, influenced the PCB effects on nitrification.

Microorganisms may also be a key component in the first step of toxic-compound accumulation in food chains. Bacteria can concentrate

chlorinated hydrocarbon insecticides from aquatic systems (Grimes and
Morrison 1978). Detritus-feeding organisms in turn may potentially
transfer the toxins to higher levels of the food chain.

SUMMARY

Microorganisms occupy a central role in wastewater relations to
public health. An awareness of the microbial role in sewage application to wetlands is necessary to assure that no unreasonable risks
are present to human and wildlife health. The following highlights
summarize the state-of-knowledge of microbiological health implications for wetland treatment systems:

1. Bacterial pathogens are well understood in wastewater
environments. Organisms of major concern to human health are
Salmonella spp., Shigella spp., Campylobacter fetus, and Leptospira
spp. Of all the pathogens, bacteria are the most susceptible to
conventional treatments.
2. Viruses of major concern to public health include hepatitis
A virus, rotavirus, and enteroviruses, which are the major agents of
waterborne disease in the United States. Traditional indicators of
fecal pollution are not reliable indicators of viral pollution.
Viruses tend to be more persistent in aquatic environments than
bacteria, and no single virus may serve as a universal indicator of
viral pollution. Improved techniques are required for the detection
and enumeration of viral pathogens. A better understanding of the
fate of viruses in the aquatic environment is necessary for public
health assessment.
3. Microorganisms in wastewaters play an important role as
detoxifying agents. However, the extent of pollutant detoxification
in aquatic systems is not well understood.
4. Microorganisms also have the potential to transform
relatively nontoxic compounds to toxic forms. The extent of this
phenomenon in wastewater is not known.
5. Indirectly, microorganisms may contribute to detrimental
public health conditions in aquatic ecosystems. Hazardous metabolic
wastes, such as nitrite, may potentially accumulate in the environment, or the microbial biomass may act as a source for the
bioaccumulation of toxins in the wetlands food chain.

REFERENCES

Alexander, M., 1980, Biodegradation of chemicals of environmental
    concern, Science 211:132-138.
Blake, P. A., D. T. Allegra, J. D. Snyder, T. J. Barret, L.
    MacFarland, C. T. Caraway, J. C. Feeley, J. P. Craig, J. V. Lee,
    N. D. Puhr, and R. A. Feldman, 1980, Cholera--a possible focus in
    the United States, New Eng. J. Med. 302:305-309.
Berg, G., 1967, Transmission of viruses by the water route, Wiley
    Interscience Publishers, New York, 149p.
Cabelli, V. J., A. P. Dufour, M. A. Levin, and P. W. Haberman, 1976,
    The impact of pollution on the marine bathing beaches: An epide-
    miological study, in Middle Atlantic continental shelf and the New
    York bight: Proceedings of the symposium, vol. 2, M. G. Gross,

ed., American Society of Limnology and Oceanography, Lawrence, Kansas, pp. 424–432.

Catelani, D., A. Colombi, C. Sorlini, and V. Trecanni, 1973, 2-Hydroxy-6-oxo-6-phenylhexa-2,4-dienoate: The meta-cleavage product from 2,3-dihydroxybiphenyl by Pseudomonas putida, J. Biochem. 134:1063–1066.

Christie, A. E., 1967, Virus reduction in the oxidation lagoon, Water Pollut. Control Fed. J. 105:50–54.

Cooney, J. J., and M. P. Shiaris, 1982, Utilization and co-oxidation of aromatic hydrocarbons by estuarine microorganisms, Dev. Indust. Microbiol. 23:177–185.

Dean-Raymond, D., and M. Alexander, 1976, Plant uptake and leaching of dimethylnitrosamine, Nature 262:394–396.

Flewett, T. H., and G. N. Woode, 1978, The rotaviruses, Arch. Virol. 57:1–23.

Goyal, S. M., and C. P. Gerba, 1979, Comparative adsorption of human enteroviruses, simian rotavirus, and selected bacteriophages to soils, Appl. Environ. Microbiol. 38:241–247.

Grabow, W. O. K., 1968, The virology of wastewater treatment—a review paper, Water Res. 2:675–701.

Grimes, D. J., and S. M. Morrison, 1978, Bacterial bioconcentration of chlorinated hydrocarbon insecticides from aqueous systems, Microbial Ecol. 2:43–59.

Haley, C. E., R. A. Gunn, J. M. Hughes, E. C. Lippy, and G. F. Craun, 1980, Outbreaks of waterborne disease in the United States, 1978, J. Infect. Dis. 141:794–799.

Hallas, L. E., J. C. Means, and J. J. Cooney, 1982, Methylation of tin by estuarine microorganisms, Science 215:1505–1506.

Hood, M. A., and G. E. Ness, 1982, Survival of Vibrio cholerae and Escheria coli in estuarine waters and sediments, Appl. Environ. Microbiol. 43:578–584.

Hurst, C. J., and C. P. Gerba, 1980, Stability of simian rotavirus in fresh and estuarine water, Appl. Environ. Microbiol. 43:1–5.

Jacobson, S. N., N. L. O'Mara, and M. Alexander, 1980, Evidence for cometabolism in sewage, Appl. Environ. Microbiol. 40:917–921.

Jernelov, A., and A. -L. Martin, 1975, Ecological implications of metal metabolism by microorganisms, Ann. Rev. Microbiol. 29:61–77.

Kadlec, R. H., and D. L. Tilton, 1979, The use of freshwater wetlands as a tertiary wastewater treatment alternative, CRC Crit. Rev. Environ. Control 9:185–212.

Kaper, J., H. Lockman, R. R. Colwell, and S. W. Joseph, 1979, Ecology, serology, and enterotoxin production of Vibrio cholera in Chesapeake Bay, Appl. Environ. Microbiol. 37:91–103.

Katzenelson, E., 1978, Concentration and identification of viruses in seawater, Rev. Intl. Oceangr. Med. 48:9–16.

Katzenelson, E., I. Buium, and H. I. Shuval, 1976, Risk of communicable disease infection associated with wastewater irrigation in agricultural settlements, Science 194:944–946.

Kott, Y., 1973, Hazards associated with the use of chlorinated oxidation pond effluent for irrigation, Water Res. 7:853–862.

Kowal, N. E., Health effects of land treatment: Microbiological, EPA research report, EPA-600-1-82-007, U. S. EPA, Cincinnati, Oh., 78p.

Kurzel, R. B., and C. L. Cetrulo, 1981, The effect of environmental pollutants on human reproduction, including birth defects, Environ. Sci. Technol. 15:626–640.

Lance, J. C., and C. P. Gerba, 1978, Pretreatment requirements before land application of municipal wastewater, in State of knowledge in land treatment of wastewater, vol. 1, H. L. McKim, ed., CRREL, U. S. Army Corps of Engineers, Hanover, N. H., pp. 293-304.

Langeland, G., 1982, Salmonella spp. in the working environment of sewage treatment plants in Oslo, Norway, Appl. Environ. Microbiol. 43:1111-1115.

Leadbetter, E. R., and J. W. Foster, 1959, Oxidation products formed from gaseous alkanes by the bacterium Pseudomonas methanica, Arch. Biochem. Biophys. 82:491-492.

Martone, W. J., and A. F. Kaufman, 1980, Leptospirosis in humans in the United States, 1974-1978, J. Infect. Disease 140:1020-1022.

Mortality and Morbidity Weekly Report, 1979, Campylobacter enteritis in a household--Colorado, MMWR 28:273-274.

Mortality and Morbidity Weekly Report, 1980, Human Salmonella isolates--United States, 1979, MMWR 29:189-191.

Pahren, H. R., J. B. Lucas, J. A. Ryan, and G. K. Dotson, 1979, Health risks associated with land application of municipal sludge, Water Pollut. Control Fed. J. 51:2588-2601.

Perry, J. J., 1979, Microbial cooxidations involving hydrocarbons, Microbiol. Rev. 43:59-72.

Pritchard, P. H., A. W. Bourquin, H. L. Frederickson, and T. Manziarz, 1978, in Proceedings of the workshop: Microbial degradation of pollutants in marine environments, A. W. Bourquin and P. H. Pritchard, eds., EPA-600/9-79-012, U. S. EPA, Gulf Breeze, Fla., pp. 251-272.

Sayler, G. S., M. P. Shiaris, W. Beck, and S. Held, 1982, Effects of polychlorinated biphenyls and environmental biotransformation products on aquatic nitrification, Appl. Environ. Microbiol. 43:949-952.

Sheladia, V. L., R. D. Ellender, and R. A. Johnson, 1982, Isolation of enterovirus from oxidation pond waters, Appl. Environ. Microbiol. 43:971-974.

Shiaris, M. P., and G. S. Sayler, 1982, Biotransformation of PCB by natural assemblages of freshwater microorganisms, Environ. Sci. Technol. 16:367-369.

Shuval, H. I., 1970, Detection and control of enteroviruses in the water environment, in Developments in water quality research, H. I. Shuval, ed., Ann Arbor-Humphrey Science Publishers, Ann Arbor, Mich., pp. 47-71.

Shuval, H. I., 1978, Studies on bacterial and viral contamination of the marine environment, Rev. Intl. Oceangr. Med. 50:43-50.

Shuval, H. I., 1982, Health risks associated with water recycling, Water Sci. Technol. 14:6E1-6E10.

Smith, E. M., and C. P. Gerba, 1982, Development of a method for detection of human rotavirus in water and sewage, Appl. Environ. Microbiol. 43:1440-1450.

U. S. Environmental Protection Agency, 1979, Aquaculture systems for wastewater treatment: Seminar proceedings and engineering assessment, EPA 430/9-80-006, U. S. EPA, Washington, D. C.

Wood, J. M., 1982, Chlorinated hydrocarbons: Oxidation in the biosphere, Environ. Sci. Technol. 16:291A-297A.

Wood, M. M., F. S. Kennedy, and C. G. Rosen, 1968, Synthesis of methyl-mercury compounds by extracts of a methanogenic bacterium, Nature 220:173-174.

Woodwell, G. M., 1977, Recycling sewage through plant communities, Am. Sci. 65:556-562.

Yoshpe-Purer, Y., and H. I. Shuval, 1972, Salmonellae and bacterial indicator organisms in polluted coastal water and their hygienic significance, in Marine pollution and sea life, M. Ruivo, ed., Fishing News Ltd., pp. 574-580.

DISCUSSION

Miller: Mike, there is one issue you raised earlier when you talked about bacteria--we talked about this before and I thought we should bring it out--and that is this question of retention time. We are saying bacterial treatment is related to retention time. I think a very important point we are finding in our artificial wetlands is that the actual hydrologic retention time may be different from the theoretical one. For our channel marshes or our open marshes that I was talking about yesterday, the theoretical retention time can be more or less the same, but when they are tested with dye, the leading edge of the dye slug gets to the outlet at drastically different times. We have also been looking at tracer $E.$ $coli$ during the week prior to my coming here, and our microbiologist was amazed to find out that $E.$ $coli$ was actually moving through our systems faster than the dye. In fact, in one test when the dye slug had moved halfway through the marsh in 57 hours, an $E.$ $coli$ tracer introduced halfway down the system was picked up at the outlet in 24 hours. Now, we do not understand this whole thing yet, but the theoretical retention time may be a misleading factor.

Shiaris: Well, that is something that you probably would not monitor normally anyway. My suggestion to you is that perhaps the $E.$ $coli$ was concentrated in some of the upper layers [the hydrophobic film of the surface] and moved faster than the subsurface dye. But I think many people have shown that the retention time is of most importance; the longer the residence time, the greater chance there is of reducing the coliform count. Is that your main point?

Miller: Well, the point is that our theoretical hydrological retention time doesn't necessarily relate to the retention time of the fastest-moving pathogens. For our channel systems, with retention time of, say, seven days, I would hazard a guess that, in terms of dye anyway, the first dye will not get there until five days. But, in the open marsh with the same theoretical retention time, the first dye can get there in two days.

Shiaris: What you are saying is that if you go by theoretical values and then calculate the necessary flow rate to get a 99% reduction of coliforms, it will be incorrect. I think that what you have to do is continually monitor the influent and the effluent rather than rely on a calculated retention time.

Jackson: Would you comment a little more on your concern for putting what is usually a relatively small amount of effluent, possibly containing organisms such as coliforms, salmonella, leptospira, and so forth, into a relatively large environment in which these organisms already exist.

Shiaris: You are saying the environment is already contaminated. That is not necessarily so.

Jackson: What?

Shiaris: Are you saying that the environment is already contaminated with these organisms?

Jackson: Well, those organisms are already there as a result of the natural conditions that exist. I do not know whether you want to call the environment contaminated or not. The organisms are there because of wildlife and a variety of other factors. All you are doing is adding some more, perhaps in smaller quantity than had already existed.

Shiaris: I do not think that is the case. I think there are much higher quantities in municipal wastewater. Furthermore, the specific types of organisms from human waste may be of greater public health concern to humans than the indigenous flora.

Jackson: In terms of E. coli at Houghton Lake, that was not the situation.

Shiaris: E. coli is not a pathogen, it is an indicator of fecal pollution and potential human pathogens. This includes animal fecal pollution as well.

Jackson: But a lot of people are using E. coli as a magic sort of thing.

Shiaris: Right. And that is not necessarily the case. I hope I pointed that out, anyway. At least for viral pollution, E. coli is not a very good indicator. The use of fecal streptococci may aid in differentiating between human and animal sources. I might have some help here, I can see that Milt Friend is interested in talking.

Friend: I think the concept that has been brought up is one that is not necessarily substantiated. We tend to talk about pathogens as being widely distributed in nature as if anywhere one went and took a sample, one would be able to isolate that pathogen. That situation is far from reality, and what you were talking about, essentially, in the presentation here is point source of introduction of a finite number of organisms.

Shiaris: I do not think it is a matter of putting a few pathogens into a system that already has them. I think it is a matter of potentially polluting a system with pathogens that were not present before.

Sutherland: Some other aspects of the fecal coliform: Probably, most stabilization pond systems are adequate for removing them. That is, the systems take them down to a level acceptable for total body-contact recreation in receiving waters, although this method is under review by the EPA. So, if the effluent ponds were not put into wetlands, it would still be okay from the people point of view. When you get out in the wetlands and measure the fecals, you are picking

up fecals from other sources. So, there is a good reason to review
fecals as an indicator.

Shiaris: Agreed. I hope I made the point that there is a need to
monitor viruses--for example, pathogenic viruses--because fecal
coliforms just simply are not good indicators.

Bastian: But one problem with conventional treatment plants is that
they do not monitor viruses that they discharge into drinking water.

Shiaris: True, but they are chlorinating those effluents.

Bastian: There is no requirement any more for chlorination on the
federal basis, because of their concern over chlorine. It is down to
a case-by-case, city-by-city situation. I did not hear much discussion on concern for things other than pathogens. For example,
interactive-type concerns: What's in wastewater and what do you see
for human health linkages as regards exposure, by-products, that type
of thing?

Shiaris: For toxins?

Bastian: Sure. The other thing you did not mention at all is
dealing with the extensive literature on worker exposure within
sewage treatment plants--that is, what that really means as regards
exposure of the general public to a wetlands project.

Shiaris: That work is inconclusive.

Bastian: We are talking about the difference between millions of
dollars being spent on treatment plants--for example, concerning
whether to cover aeration basins where we get much more exposure--as
measured by counts of organisms affecting people who live within a
mile of the treatment plant.

Shiaris: It seems to be inconclusive. Some people have concluded
that there is no problem in occupational risk for sewage treatment
work. And, as I mentioned, in some areas of land disposal, there
obviously is a problem.

Bastian: It also makes a lot of difference whether you are applying
raw wastewater or whether you have something that has been relatively
well stabilized.

Shiaris: Absolutely.

Mikulak: You have dealt with the problem of the importance of
residence time in the removal of bacteria. Do you have any feel for
the importance of residence time in the removal of viruses?

Shiaris: I think the same situation exists. I am not sure about
viruses in wetlands. As I said, they are adsorbed to clay particles
and such, and it may turn out that they are rapidly inactivated or at
least adsorbed to the soil if there is a lot of soil contact in the
wetland. So, it may not be a problem. But the longer the viruses
are in the system, the better removal you get.

Mikulak: The same general time frame that you spoke of for bacteria?

Shiaris: I think a longer time frame, except that we are talking about particle/water interaction. So, you might get quick removal of viruses. This situation should be investigated.

Hodson: I am not sure that viruses would be removed or inactivated by the same mechanism. It is my understanding that some groundwater hydrologists are now using polio viruses as a tracer simply because they are there and they seem to drop off only with dilution--not by inactivation of the virus.

Shiaris: It depends very much upon the physical properties of the soil. Certain clays are very effective in adsorbing viruses and other materials are not. I spoke with someone here yesterday who was following a virus and told me it was rapidly inactivated.

Brennan: There are some areas that have not been covered by your talk. Is anyone else going to touch on them? For instance, protozoa and helminths.

Shiaris: I don't think so. I did reference some review articles that do cover those areas in more detail. There are many aspects I did not mention, as I said. I hope Dr. Friend will cover the area of animal reservoirs.

Grimes: I'd like to talk briefly about three points that have been made here. First of all, I do not think that fecal coliforms are probably the way to go as an indicator in wetlands. In my talk, I shall be talking about coliforms as though they are the standard, and I'm not going to address the problems with them, but I do feel there are some very well substantiated problems with using them as indicators. Secondly, I would like to comment on the remark by Bill Jackson. I think he implied that release of a finite number of pathogens into a habitat with larger numbers of pathogens would perhaps not be a problem. A problem that does exist is that pathogens of a given species are not the same. There are drastically different strains within species, and I think we are looking at a situation where a highly virulent strain of a given species could be released into a habitat. So, I think this is an area of concern. The third thing I would like to mention is percent reduction. I think Table 16.3 had percent reduction. I think we have to be very cautious about looking at percent reduction in water: Are we really looking at die-off, sedimentation, or attachment of the organisms to interfaces? Hence, do they become nonmeasurable because we are not looking at the interfaces (be those interfaces particles, leaf surfaces, invertebrate surfaces, or whatever)? It has been shown that bacteria are very quickly partitioned to interfaces and hence disappear, so to speak, out of the water. And, finally, these bacteria can also become nonculturable. They are still viable and potentially infectious. But, because they become nonculturable, we consider that a percent reduction.

Shiaris: There is increasing evidence that many allochthonous
bacteria (at least the indicator organisms) can accumulate in
sediments. Therefore, sediments may act as a reservoir for coliforms
and potentially for pathogens.

Berger: Conventional chlorination is very often employed for
disinfection. Is it possible that a wetland receiving sewage offers
no such opportunity?

Shiaris: No. In many cases, if a wetland is going to be used and
the effluent has too many pathogens or coliforms, I am sure it will
have to be chlorinated or ozonated.

Berger: Physically, how can it be?

Shiaris: If it comes out as a point source? Obviously, it would
have to. That is a consideration that has to be made.

Mikulak: Some of the discharges in California are chlorinated and
dechlorinated before the wastewater discharge is allowed through a
wetland.

Shiaris: Because of bacterial contamination?

Mikulak: I think it's state law.

17

Wildlife Health Implications of Sewage Disposal in Wetlands

Milton Friend

Wildlife health implications of sewage disposal in wetlands are more complex than may be readily apparent. Consideration of this subject must go beyond simple cause-and-effect relationships. Although uncomplicated disease transmission involving direct exposure of wildlife to pathogens present in sewage effluent can occur, effluent-induced changes in the microenvironment of wetlands may be of far greater importance in the development and maintenance of disease problems among wildlife dependent upon those wetlands. This presentation highlights some of the direct and indirect wildlife health considerations.

DISEASE CONSIDERATIONS

DIRECT INTRODUCTION OF DISEASE

Direct introduction of disease into wildlife populations requires suitable means of transmitting threshold levels of pathogens into susceptible hosts. The three major components of this event are transmission, threshold levels, and susceptible hosts. Contaminated water is an efficient vehicle for transmission of many diseases; therefore, sewage disposal in wetlands represents a suitable system for disease transmission. However, environmental maintenance of levels of pathogens required for initiation of disease outbreaks is a variable that greatly influences the potential for disease problems to occur. Pathogens have finite life spans that vary with the agent and other factors. Therefore, the amount, timing, and frequency of pathogen discharge into wetlands is of considerable importance. In addition, pathogens have variable host spectrums. Some pathogens, such as duck plague virus, may be group specific (Leibovitz 1971), and others, such as Salmonella spp., are capable of infecting a wide range of hosts (Steele and Galton 1971). As a consequence, the origin of materials ultimately being deposited in wetlands is worthy of consideration. Wetland discharges containing large amounts of agricultural waste from poultry processing plants, feedlots, and other types of animal industry are potentially the most likely discharges to contain pathogens transmissible to wildlife, while those discharges involving domestic wastes are least likely to contain pathogens of concern to wildlife.

INDIRECT DEVELOPMENT OF DISEASE

Environmental factors are important in many disease problems of humans, domestic animals, and wildlife. Although there are many considerations associated with this concept, two deserve special attention in evaluating wildlife health implications of sewage disposal. The first consideration involves changes induced in the microenvironment of wetlands, and the second involves interactions between environmental pollutants and microbial pathogens.

Persistence of pathogenic agents outside living host systems is related to the chemical and physical properties of the micro-environment that these agents have entered. Environmental conditions that favor the persistence of these agents increase the probability for disease problems to develop. Our laboratory (the National Wildlife Health Laboratory) has been investigating this type of relationship in the epizootiology of avian cholera in Nebraska's Rainwater Basin.

Avian cholera was first recognized as a disease of waterfowl using this important spring staging area in 1975 (Zinkl et al. 1977). Since then, avian cholera epizootics have occurred annually, with the heaviest mortality always confined to western portions of the basin. In addition, specific wetlands have been focal points for recurring avian cholera losses. Preliminary field and laboratory results indicate that differences in physical and chemical characteristics of the surface waters occur between problem and nonproblem wetlands. Further, these differences appear to be related to the survival of Pasteurella multocida, the causative agent of avian cholera. Similar differences in the enzootic distribution of avian cholera in California and Texas and field observations of environmental conditions suggest the same type of relationship is probable in those areas.

In addition to infectious disease, such toxic processes as avian botulism and blue-green algal poisoning require specific environmental conditions for their development. An apparent but undefined relationship between avian botulism and sewage disposal exists. Outbreaks of Clostridium botulinum type C frequently occur in sewage ponds in California and occasionally elsewhere in the United States (NWHL records). Additional botulism outbreaks have been associated with sewage outfalls and discharges in North America and other parts of the world (Martinovich et al. 1972; Moulton, Jensen, and Low 1976; Muller 1967; Wilson and Locke 1982).

Interactions between environmental pollutants and microbial pathogens are real-world events (Durham 1967). Therefore, the impact of chemical contributions of sewage effluent reaching wetlands is as much a wildlife health concern as are pathogen contributions. Suppression of the immune response and other alterations in host body defense mechanisms represent a reasonable wildlife health concern since depression of this essential biological activity can result in increased susceptibility to disease and the activation of latent infections (Fenner et al. 1974; Gabliks and Friedman 1969; Hough and Robinson 1975; Wassermann et al. 1969). The potential for synergistic responses between chemical agents and between chemical and microbial agents complicates the establishment of biologically safe levels of these components in sewage effluents (Dieter and Ludke 1978; Friend and Trainer 1970a, 1970b, 1974; Hayes 1975; Ludke 1977).

MIGRATORY BIRD CONSIDERATIONS

Waterfowl and shorebirds are the migratory bird species groups at greatest risk from sewage effluent discharges in wetlands. These species are dependent upon wetlands for many of their essential needs, are attracted to specific sites of sewage discharge, and have highly gregarious social behavior. These characteristics enhance the potential for exposure to environmental pollutants and pathogenic agents present in sewage effluent. In addition, transmission of infectious disease is facilitated by the dense concentrations of waterfowl and shorebirds that frequently occur.

This situation is compounded by the fact that wetland acreage has been declining at an alarming rate. About 40% of the total wetlands available to waterfowl at the time of first settlement of the United States (excluding Alaska and Hawaii) no longer exist; the current rate of decline in existing wetlands is estimated to be 300,000 acres per year (Ladd 1978). Despite this loss of essential habitat, waterfowl management efforts are focused on maintaining and even increasing current waterfowl populations. This goal can be achieved only by greater bird use of available habitat--that is, producing and sustaining more birds on less acreage. The quality of the habitat has a direct relationship to disease problems (Friend 1981a).

Increased public sensitivity to disease has resulted in losses from this cause becoming less acceptable than during earlier times of greater waterfowl abundance and less-intensive utilization of this resource. The significance of disease has also been increased by the emergence of infectious diseases, such as avian cholera and duck plague. These types of diseases are of greater concern than noninfectious diseases because an outbreak at one location can potentially serve as a focal point for spread to other locations (Friend 1981b). Also, survivors of outbreaks may become carriers capable of initiating future outbreaks of the same disease. Therefore, outbreaks of infectious disease in migratory birds are not necessarily isolated events and should not be considered in that perspective.

The mobility of wildlife and the role of disease carriers must both be accommodated in any consideration of wildlife health implications of sewage disposal in wetlands. The wildlife component of those wetlands interfaces with the surrounding environment in a manner that facilitates the transmission of pathogens to other wildlife populations, domestic animals, humans, and even agricultural crops. Special consideration must also be given to wetlands used by major segments of discrete populations of migratory birds and to wetlands utilized by endangered species. Disease outbreaks on these wetlands place major portions of wildlife populations at risk; therefore, creation of environmental conditions conducive to the development of disease problems on these areas is not likely to be acceptable.

GUIDELINES

Wildlife health implications of sewage effluent discharges into wetlands have not been adequately addressed. Until this is done, only general disease-management recommendations can be developed. The following are offered as interim guidelines that might be applied in the evaluation of <u>specific</u> wetlands for sewage wastewater treatment. However, at best, these guidelines represent only minimum considerations for risk assessment and should be reevaluated and upgraded as appropriate research findings become available.

<u>Unsuitable wetlands</u>: There are some wetlands with wildlife values so great that their present use for disposal of sewage effluent seems unwise given the many unknowns regarding impacts on wildlife health. Wetlands frequently utilized by endangered wildlife species fall in this category. The consequences of disease problems developing in these wetlands represent an anticipated unacceptable cost-benefit ratio. Wetlands occasionally utilized by endangered species also deserve special attention. When migratory birds are the species of concern, it would be prudent to avoid effluent discharges into these wetlands from 60 days before the earliest known date (day and month) of recorded endangered species use to 30 days after the latest known date of use.
Wetlands being used as primary habitats for discrete wildlife populations, for major segments of regional wildlife populations (25% or more of the total population), or as major staging areas and migration concentration points for migratory birds are also poor risks for sewage effluent disposal. If circumstances dictate that wetlands within these categories must be utilized for effluent disposal, serious consideration should be given to the avoidance of discharges for a period comparable to that suggested above for endangered species.

<u>Concentration of wildlife</u>: Discharges of sewage effluent into wetlands should be done in such a manner that no major alterations in established migratory patterns of wildlife utilizing these wetlands occur. Discharge methods and rates should not concentrate wildlife in small areas as a result of enhanced food sources or create open water in northern climates during periods of the year when those wetlands would normally be ice covered.

<u>Type of discharges</u>: To the extent possible, discharges of sewage effluent into wetlands should be limited to materials from urban sources. The makeup of urban sewage should be evaluated prior to each specific decision regarding the acceptability of this material for wetland discharge. Agricultural and industrial sources represent higher-risk situations for wildlife because of the types of pathogens and chemicals likely to be present in wastewater from those sources. Only sewage effluent that has received at least secondary treatment should be considered for discharge into wetlands.

<u>System development</u>: Adequate control structures should be developed so that alternative disposal sites can be utilized in the event a wildlife health problem develops on a wetland being used for effluent discharge. The ability to drain a wetland area rapidly or the ability to dilute surface waters by the rapid addition of water

(noneffluent source) are also highly desirable and should be accommodated to the extent possible. In areas of high rates of evaporation, effluent discharge rates should be adjusted to maintain stable water levels during hot summer months.

Monitoring: Active wildlife-health monitoring programs are as important as the water-quality and other routine monitoring programs currently being utilized to assess the biological effects of effluent discharges in wetlands. The time is long overdue to establish such programs. The wildlife species involved, type of wetland, and other circumstances will dictate the type of monitoring best suited for a specific wetland. General guidance for wildlife-health monitoring programs should be developed as a result of consultation with individuals knowledgeable in diseases of wildlife.

CONCLUSIONS

Wildlife health concerns associated with disposal of sewage effluent in wetlands are of three primary types: (1) introduction of pathogens, (2) introduction of pollutants that adversely impact on host body defense mechanisms, and (3) changes in the physical and chemical properties of wetlands that favor the development and maintenance of disease problems.

Unlike the situation with human health concerns, introduction of pathogens is not the major concern regarding wildlife health. Instead, the focus of attention needs to be directed at environmental changes likely to take place as a result of effluent discharges into different types of wetlands. Unless these changes are adequately addressed from a disease perspective, marshes utilized for sewage disposal could become disease incubators and wildlife death traps. This result would be unfortunate because the backlash would likely negate the potentially beneficial aspects of the use of sewage wastewater for the creation of new wetlands and have a severe impact on progress being made towards evaluation of the compatibility of wildlife and sewage effluents.

REFERENCES

Dieter, M. P., and J. L. Ludke, 1978, Studies on combined effects of organophosphates or carbamates and morsodren in birds. II. Plasma and cholinesterase in quail fed morsodren and orally dosed with parathion or carbofuran, Bull. Environ. Contam. Toxicol. 19:389-395.

Durham, W. F., 1967, The interaction of pesticides with other factors, Residue Rev. 18:22-103.

Fenner, F., B. R. McAuslan, C. A. Mims, J. Sambrook, and D. O. White, 1974, The biology of animal viruses, student ed., Academic Press, New York, p. 420.

Friend, M., 1981a, Waterfowl diseases--changing perspectives for the future, in Fourth international waterfowl symposium, Ducks Unlimited, Chicago, Ill., pp. 189-196.

Friend, M., 1981b, Waterfowl management and waterfowl disease: Independent or cause and effect relationships? N. Am. Wildl. Conf. 46:94-103.

Friend, M., and D. O. Trainer, 1970a, Some effects of sublethal
  levels of insecticides on vertebrates, J. Wildl. Dis. 6:335–343.
Friend, M., and D. O. Trainer, 1970b, Polychlorinated biphenyl:
  Interaction with DHV, Science 170:1314–1316.
Friend, M., and D. O. Trainer, 1974, Experimental dieldrin-duck
  hepatitis virus interaction studies, J. Wildl. Manage. 38:896–902.
Gablicks, J., and L. Friedman, 1969, Effects of insecticides on
  mammalian cells and virus infections, in Biological effects of
  pesticides in mammalian systems, H. F. Kraybill, P. D. Albertson,
  and M. Krauss, eds., N. Y. Acad. Sci. Ann. 160:254–271.
Hayes, W. J., Jr., 1975, Toxicity of pesticides, Williams and Wilkins
  Company, Baltimore, pp.182–224.
Hough, V., and T. W. E. Robinson, 1975, Exacerbation and reactivation
  of herpesvirus hominis infection in mice by cyclophophamide, Arch.
  Virol. 48:75–83.
Ladd, W. N., Jr., 1978, Continental habitat status and long-range
  trends, in Third international waterfowl symposium, Ducks
  Unlimited, January 27–29, 1978, New Orleans La., pp. 14–19.
Leibovitz, L., 1971, Duck plague, in Infectious and parasitic
  diseases of wild birds, J. W. Davis, R. C. Anderson, L. Karstad,
  and D. O. Trainer, eds., Iowa State University Press, Ames, pp.
  22–33.
Ludke, J. L., 1977, DDE increases the toxicity of parathion to
  coturnix quail, Pest. Bioch. Physiol. 7:28–33.
Martinovich, D., E. Carter, D. A. Woodhouse, and I. P. McCausland,
  1972, An outbreak of botulism in wild waterfowl in New Zealand,
  N. Z. Vet. J. 20:61–65.
Moulton, D. W., W. I. Jensen, and J. B. Low, 1976, Avian botulism
  epizootiology on sewage oxidation ponds in Utah, J. Wildl. Manage.
  40:735–742.
Muller, J., 1967, [Botulism in wild ducks diagnosed for the first
  time in Denmark], J. Medlemsbl. Dan. Dyrlaegeforen. 50:887–890.
Steele, J. H., and M. M. Galton, 1971, Salmonellosis, in Infectious
  and parasitic diseases of wild birds, J. W. Davis, R. C. Anderson,
  L. Karstad, and D. O. Trainer, eds., Iowa State University Press,
  Ames, pp. 51–58.
Wassermann, M., D. Wasserman, Z. Gershon, and L. Zeller-Mayer, 1969,
  Effects of organochlorine insecticides on body defense systems, in
  Biological effects of pesticides in mammalian systems, H. F.
  Kraybill, P. D. Albertson, and M. Krauss, eds., N. Y. Acad. Sci.
  Ann. 160:393–401.
Wilson, S. S., and L. N. Locke, 1982, Bibliography of references to
  avian botulism: Update, special scientific report--wildlife 243,
  U. S. Fish and Wildlife Service, Washington, D. C., pp. 1–10.
Zinkl, J. G., N. Dey, J. H. Hyland, J. J. Hurt, and K. L. Heddleston,
  1977, An epornitic of avian cholera in waterfowl and common crows
  in Phelps County, Nebraska, in the spring, 1975, J. Wildl. Dis.
  13:194–198.

DISCUSSION

Brennan: Can you comment on the potential relationship between changes in water level and avian botulism?

Friend: Fluctuating water levels are commonly associated with botulism problems. Botulism tends to ignite when areas that have not been recently flooded are inundated and other environmental conditions, including presence of the bacteria, are suitable. I was present at a die-off in Mexico's central highlands that was so severe that I could not step on the ground without stepping on a carcass. That area had just experienced the highest water levels in many years. Windrows of dead birds marked the edges of the receding water levels. This year, at the Bear River marshes in Utah, we have been experiencing very high water and are anticipating a substantial botulism die-off. In California, botulism losses in one area are already up to 6,000 birds, which is way ahead of what we would anticipate. Again, the die-off followed a period of high water in that area. In general, if you want to avoid botulism, stabilize water levels to the degree you can.

Zedler: What about cholera? What are the features of wetlands that bring on cholera?

Friend: We have been studying environmental conditions present in problem and nonproblem wetlands by measuring differences in chemical and physical parameters. These differences are being studied initially using a simple laboratory system to evaluate the effects of the various factors detected on the persistence of Pasteurella multocida. What we find is that there are measurable differences between problem and nonproblem wetlands. There are also detectable visual differences in the quality of these wetlands. So, in essence, it appears that problem areas are capable of maintaining a high threshold level of the causative bacteria. The incubation period for avian cholera can be as short as 6 to 8 to 10 hours after exposure, at which time the bird is dead. Millions of virulent bacteria are released into the environment after carcasses have been disassembled by scavengers or decomposition. If you enhance the potential for survival of these bacteria at these high levels by having suitable environmental conditions, you perpetuate the outbreak. However, sunlight penetration within the water column and its ultraviolet destruction of the bacteria and other environmental factors probably play an important role in the lack of bird mortality in nearby wetlands. Similar conditions appear to exist in California and Texas. The lines of demarcation appear to be quite pronounced between problem and nonproblem wetlands.

Zedler: But, basically, it is some physical/chemical composition of the water?

Friend: Yes. I would say hardness has a role, conductivity to a degree, but the chemicals--phosphates, things of this nature--appear to be very different between problem and nonproblem areas.

R. Kadlec: We heard yesterday that in the six-state Region 5, there are, I believe, 96 wetlands receiving wastewater, most for ten years or more. In your experience, or in anything you have heard, have any documented outbreaks occurred that have been related to one of those wetlands?

Friend: I do not know the specific region you are talking about, but I offer a general comment relative to the detection of wildlife disease. Nonhunting mortality results in an annual loss of about 20 million waterfowl, about equal to hunting mortality. However, little of this nonhunting mortality is documented by body counts, particularly such dramatic ones as described earlier. Disease tends to be of a chronic nature, and the efficient elimination of carcasses by scavengers and predators can give one a false sense of security that disease is not present. One more point: I do have a fair amount of documentation of avian cholera and avian botulism die-offs in areas receiving sewage effluent. However, I am not willing at this time to attribute the occurrence of these diseases [avian cholera and avian botulism] to sewage effluent. Further study of the relationship between sewage effluent discharge and avian cholera and avian botulism outbreaks is needed.

18

Microbiological Studies of Municipal Waste Release to Aquatic Environments

D. Jay Grimes

Release of wastes to aquatic environments has, in past years, brought about serious public health repercussions. A recent but classic example occurred in Dubuque, Iowa, in 1974. Forty-five persons who swam in the Mississippi River contracted bacillary dysentery caused by Shigella sonnei. A retrospective study by Rosenberg et al. (1976) linked the cases with swimming exposure to the river, demonstrated large numbers (17,500/100 ml) of fecal coliform bacteria in the swimming area, and isolated S. sonnei from Dubuque sewage effluent and from the downstream swimming area that was of the same biotype as that which caused the cases.

That human pathogens can survive wastewater treatment has been unequivocally documented (Shuval 1982). While their density usually decreases during treatment, pathogens are, nevertheless, released in varying numbers with the final effluent. During treatment, the virulence of the pathogens can actually be enhanced (Grabow, Middendorff, and Prozesky 1973; Grabow and Prozesky 1973), and a recent study by Mach and Grimes (1982) demonstrated that antibiotic resistance can be genetically transferred in situ from resistant to sensitive pathogens. One implication of this work is that other traits known to be transmissible--for example, enterotoxin and colonizing factors--can be propagated within the confines of a sewage plant for later release to the environment.

Upon release of municipal wastes to aquatic environments, many of the more sensitive pathogens die, or at least disappear from the water column. I emphasize the word "disappear" because recent studies have demonstrated two very interesting phenomena that have bearing on this conference. First of all, both pathogens and fecal indicator bacteria are capable of surviving in water for long periods of time (McFeters et al. 1974), albeit often in a nonculturable state (Xu et al. 1982). Figure 18.1 illustrates this observation for Vibrio cholerae contained in sterile Chesapeake Bay water at 4°C. Note that the acridine orange direct count and fluorescent antibody count remained virtually constant throughout the test period, whereas the number of culturable cells decreased by as much as five orders of magnitude (Xu et al. 1982). Similar experiments using Escherichia coli were conducted, and the results were virtually the same. Total numbers of E. coli remained constant for several weeks, while the

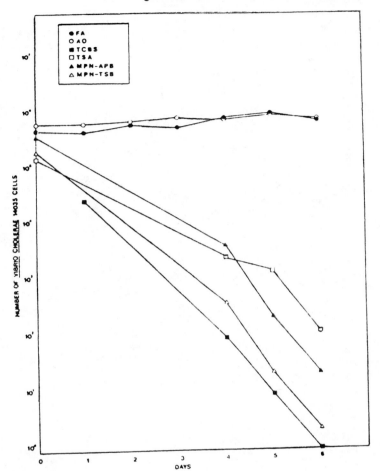

Figure 18.1
Survival of <u>Vibrio cholerae</u> ATCC 14035 in filter-
sterilized (0.2 μm) Chesapeake Bay water, 11°/oo
salinity pH 7.4, at 4°C (redrawn from Xu et al.
1982).

culturable component of the population rapidly diminished (Xu et al.
1982). In the coming weeks, these experiments will be extended to
other waterborne pathogens, both in the laboratory as well as under
field conditions, and in both the presence and absence of selected
toxic chemical water pollutants. Second, it has become increasingly
apparent that many bacteria are capable of surviving, and even
growing, in sediment. In situ experiments, using dialysis culture
techniques, were performed by LaLiberte and Grimes (1982). As
illustrated in Figure 18.2, fecal coliform densities remained fairly
constant for over 100 hours in unsterile, inoculated sediment and

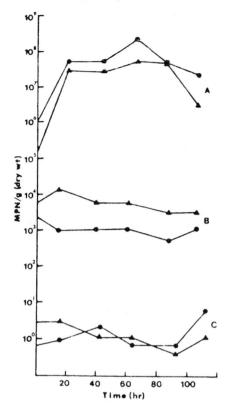

Figure 18.2
FC densities in autoclaved,
inoculated sediment (A): in
unsterile, inoculated sediment
(B): and in unsterile, uninoculated
sediment (C). Symbols ●, mud: ▲,
sand (from LaLiberte and Grimes,
1982).

increased by 2 orders of magnitude in sterile, inoculated sediment.
The conclusion to be drawn from these studies is that once released
to the water column, pathogens and fecal indicators do not always
die.   Instead they may become nonculturable and they may settle into
bottom sediments; in either case, they would be undetectable by the
conventional culturing techniques relied upon to assess the safety of
water.
      Sedimented bacteria are subject to resuspension by a variety of
physical forces, including wave action, animal disturbance, swimming,
powerboat passage, and dredging. I conducted four different studies
on the bacteriological water-quality effects of dredging in the
Mississippi River and here report on the salient features of one
project (Grimes 1980). The dredge site (Fig. 18.3) was 10 miles
below the Minneapolis-St. Paul treatment plant, a 200 million gallons
per day (MGD) plant operating at 75% treatment efficiency and having
a history of frequent raw sewage bypasses (Grimes 1980). Fecal
coliform (FC) counts upstream to the dredging activity averaged 700
membrane colony-forming units per 100 ml of sample (Fig. 18.4). The
dredge effluent averaged 12,000 FCs/100 ml, and downstream FC counts

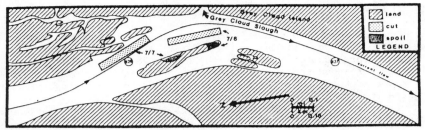

Figure 18.3
Map showing the navigation channel of the Upper Mississippi River in the
vicinity of Grey Cloud Slough. River miles 828 and 827 are shown in
circles. The dates of each dredge cut and associated deposition (spoil)
site are indicated (from Grimes 1980).

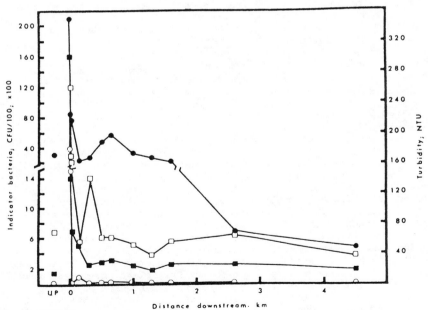

Figure 18.4
Mean turbidity (■), TC (●), FC (□), and FS (O) values in discrete water
samples, distributed according to distance downstream from the point of
discharge of dredged material (from Grimes 1980).

ranged from 6,600/100 ml to 640/100 ml, 10 m and 2.6 km downriver from the effluent pipe, respectively. Thus, while FC levels had returned to background values within a short distance downstream (2.6 km), there was, nevertheless, a significant increase in numbers of fecal indicator bacteria in the water column as a result of dredging. These bacteria had been lying "out of sight" in the sediment. There is no way of telling how many more fecal indicators escaped detection because they were in an injured or nonculturable state.

A project that is currently being investigated in our laboratory involves the effect of pharmaceutical wastes on the treatment of domestic wastewater and subsequent release of that mixed waste to the Atlantic Ocean (Grimes et al. 1984). The study site is the Barceloneta Regional Treatment Plant in Barceloneta, Puerto Rico. Domestic wastes account for 5.7% of the 3.5 MGD mean flow through the plant, with the remainder being comprised of pharmaceutical wastewater (43.5%), chemical industry wastewater (3%), and unidentified infiltration (25.8%). The effluent is highly contaminated with toxic chemicals, is not being chlorinated, and has an average fecal coliform content of greater than 10^6/100 ml. The effluent is released through a Y-shaped diffuser located in 25 m to 30 m of water, 800 m offshore of an area known locally as La Roca Beach. Beyond the 30 m contour, the ocean floor drops precipitously into the Puerto Rico Trench, reaching depths of more than 8000 m. Unfortunately, the sewage effluent does not appear to settle into these deeper waters. Instead, prevailing currents and heavy wave action tend to move the effluent westerly and toward the northern shore of Puerto Rico. FC counts at the outfall ranged from 480/100 ml at the bottom to 83/100 ml at the surface. Beach densities of FCs ranged from 110/100 ml to 430/100 ml. Bacterial pathogens have also been isolated from this study. Presumptive <u>Campylobacter</u> <u>jejuni</u> was isolated from sewage and from water samples, but the cultures inadvertently died before biochemical confirmations could be completed. <u>Salmonella</u> <u>enteritidis</u> serogroup C_2 was isolated from a sewage effluent sample, and <u>S. enteritidis</u> C_1 and C_2 were isolated from water samples near the outfall. Biochemically presumptive <u>Shigella</u> species were also isolated from water and sewage samples, but these cultures were not serologically confirmed.

In summary, the data that I have reported are all from aquatic environments, environments that have hydrologic features—that is, currents or wave action—that dilute and dissipate many of the entering pollutants, including bacteria. Wetlands, of course, do not have this advantage. Rather, they are lentic habitats that are only periodically flushed out by spring snowmelt and floods. Accordingly, introduction of improperly treated wastes could have severe deleterious impact on the health status of such wetlands. Both pathogenic bacteria and the nutrients to support those bacteria could accumulate, thereby creating a disease reservoir. There are numerous waterborne pathogens that I have not mentioned; some are infectious for humans, and some, as pointed out by Dr. Friend, are infectious for other animals (Friend, this conference). Perhaps most dangerous in this context are those that are infectious for both humans and animals. <u>Campylobacter</u> <u>jejuni</u>, for example, is emerging as one of the more common and cosmopolitan enteric pathogens of man; it has been shown to be carried in large numbers in the gut of migratory waterfowl. Therefore, while the problems I have addressed are, at this point, conceptual for wetlands, they are nevertheless of serious nature and must be dealt with in considering application of municipal wastes to wetlands.

REFERENCES

Grabow, W. O. K., I. G. Middendorff, and O. W. Prozesky, 1973, Survival in maturation ponds of coliform bacteria with transferable drug resistance, Water Res. 7:1589-1597.

Grabow, W. O. K., and O. W. Prozesky, 1973, Drug resistance of coliform bacteria in hospital and city sewage, Antimicrob. Agents Chemother. 3:175-180.

Grimes, D. J., 1980, Bacteriological water quality effects of hydraulically dredging contaminated upper Mississippi River bottom sediment, Appl. Environ. Microbiol. 39:782-789.

Grimes, D. J., F. L. Singleton, J. Stemmler, L. M. Palmer, P. Brayton, and R. R. Colwell, 1984, Microbiological effects of wastewater effluent discharge into coastal waters of Puerto Rico, Water Res. 18:(in press).

LaLiberte, P., and D. J. Grimes, 1982, Survival of Escherichia coli in lake bottom sediment, Appl. Environ. Microbiol. 43:623-628.

Mach, P. M., and D. J. Grimes, 1982, R plasmid transfer in a wastewater treatment plant, Appl. Environ. Microbiol. 44:1395-1403.

McFeters, G. A., G. K. Bissonnette, J. J. Jezeski, C. A. Thomson, and D. G. Stuart, 1974, Comparative survival of indicator bacteria and enteric pathogens in well water, Appl. Microbiol. 27:828-829.

Rosenberg, M. L., K. K. Hazlet, J. Schaefer, J. G. Wells, and R. C. Pruneda, 1976, Shigellosis from swimming, Am. Med. Assoc. J. 236:1849-1852.

Shuval, H. I., 1982, Health risks associated with water recycling, Water Sci. Technol. 14:6E1-6E10.

Xu, H.-S., N. Roberts, F. L. Singleton, R. W. Attwell, D. J. Grimes, and R. R. Colwell, 1982, Survival and viability of nonculturable Escherichia coli and Vibrio cholerae in the estuarine and marine environment, Microb. Ecol. 8:313-323.

DISCUSSION

Jackson: Do we have any idea about carrying capacities of wetlands for microorganisms? In other words, if you put in $x + y$ amount of organisms, do you end up only with x because of the ability of that environment to hold or maintain organisms?

Grimes: The answer to that question is yes and no. In terms of the natural habitat, no, I don't think we have any of this information. I am not aware of any in the literature. My yes answer involves some microcosm work that has been done and is now being done in our laboratory. It is at least a start in getting some answers to the question, What is the carrying capacity for some of these pathogens? Specifically, I am thinking about the work that we're doing with Vibrio cholerae in estuaries. We're trying to determine what is the carrying capacity of a given estuary for V. cholerae. How does it exist in the estuary? Is it truly infectious, or is it in a noninfectious state?

Odum: Gabriel Bitton in our lab did quite a bit on the cypress projects, both in the lab and in the field. I can't summarize his work very well, but I think his concept is that, for example, the viruses are being removed by some decomposition mechanism--I don't suppose it's known--in about two weeks. You do have a model for carrying capacity there, where you go at a certain rate. If they are held up for a certain period of time, then they are denatured in some way. The last two or three papers [presented at this workshop] are really not about wetlands. They are about the problems we have known for a long time: difficulty with water, particularly disturbed waters. There are lots of aquatic systems that are well established that process all of the bacterial-sized particles through animal guts, filter-feeders of one type or another, in the course of a couple of days, and the horror stories that you have been documenting come from places that are not well established. The question, I guess, is whether these wetlands that we have been dealing with are places where the stuff goes and does not get processed and put through stomachs--I mean, my inclination from our own studies is that the system, although the mechanisms are not all known, does process and denature these things pretty fast. So, is it right to take a sewage outfall somewhere in the ocean and say that's an example of what we should worry about in wetlands?

Grimes: I agree with you. I merely told my "horror stories" or "sea stories" so that this type of perturbation would not happen in wetlands. As Dr. Shiaris pointed out, we do not have any information, or at least any great amount of information, on wetlands. I would hate to see some of these kinds of things happening in wetlands. But I did not intend to make you think that this sort of thing has happened in wetlands or will happen.

Giblin: Is there any relationship between plasmids and virulence?

Grimes: We're looking at that right now. We are doing some membrane protein studies and some plasmid studies, and we are relating these to virulence, from toxigenicity to infectivity. We are, for example, trying to see what happens when these pathogens get into a natural environment and become "nonculturable," so to speak. Are they still infectious? Are they ready to die? Just exactly what is going on? We do not know yet, but we are looking at that.

Friend: I'd just like to elaborate on Dr. Odum's comment. The processing of pathogens in wetlands, I gathered from your comment, is possibly evolutionary in nature through the relationships that have developed. If that is so, then the comment I'd like to make is that the disease patterns that occur, and water problems today, are quite different from the disease patterns that occurred in waterfowl less than a hundred years ago. We did not have--at least, we have not had any previous experience with or documentation for--evidence of infectious disease in waterfowl being a problem. I am a believer in the theory that the evolution of migration of animals in part is due to this process in the environment. That is one of the problems we're getting into now. It relates to the degradation of the mountain environment and the lack of migration. I think we're undergoing a change in process here that is different than in past years.

19

Microbial Transformations of Detrital Carbon in Wetland Ecosystems: Effects of Environmental Stress

Robert E. Hodson, A. E. Maccubbin,
Ronald Benner, and Robert E. Murray

Much of the living plant biomass in wetland ecosystems is not eaten by grazing animals but rather enters the aquatic food webs as particulate organic detritus. A significant percentage of the detrital-derived carbon is assimilated into microbial biomass. In this way, the plant detritus, much of which is indigestible to animals in its original fibrous form, is converted to highly nutritive compounds that can be easily digested by detritivores (Teal 1962; Odum and de la Cruz 1967; Gallegher 1974). Thus, microbial degradation of detritus serves as an important link between primary and secondary production in wetlands (Odum and de la Cruz 1967).

In general, water-soluble, leachable compounds in the vascular plant detritus decompose rapidly, leaving a residue of fibrous material that is more resistant to degradation (Harrison and Mann 1975; Fallon and Pfaender 1976; Godshalk and Wetzel 1978). We have recently examined the refractory fraction of plant materials from a number of species of wetland plants and have found that it consists primarily of lignocellulose, a macromolecular complex consisting of polysaccharides (cellulose and hemicellulose) and lignin in intimate physical and probably covalent contact (Table 19.1). Lignocellulose accounts for between 50% and 90% of the plant dry weight, depending on species, plant part, age of the plant, and so forth (Hodson, Christian, and Maccubbin, Marine Biology, in press). In fact, most of the annual primary production in wetland ecosystems is in the form of lignocellulose. Hence, the factors that control the rates of lignocellulose-carbon mineralization, solubilization, conversion to microbial biomass, and ultimate assimilation into animal biomass also control the rates of overall secondary production in these systems. Any event, natural or anthropogenic, that alters the rate of transformation of lignocellulose in a wetland will eventually affect the abundance, diversity, and production at higher trophic levels as well. Figure 19.1 is a simplified model of the transformations of lignocellulosic detritus in wetland ecosystems. The model is derived from our results of studies in several wetlands as well as from analogous studies of terrestrial ecosystems. It can serve as a working hypothesis in the design of experiments. Using this hypothetical model, we are formulating methodology for assessing rates of transformation of both the labile and refractory components of

Table 19.1
Lignocellulose and Klason Lignin Contents in
Plant Detritus

Source	Extractive-Free Lignocellulose* (%)	Klason Lignin* (%)
Carex walteriana	85.4	9.9
Panicum hemitomon	87.7	14.8
Nymphaea odorata	54.8	7.2
Orantium aquaticum	54.9	4.1
Spartina alterniflora	75.6	15.1
Juncus roemarianus	75.3	18.7
Pinus elliottii	60.7	20.9

*Percent of total dry weight plant material on an
ash-free basis.

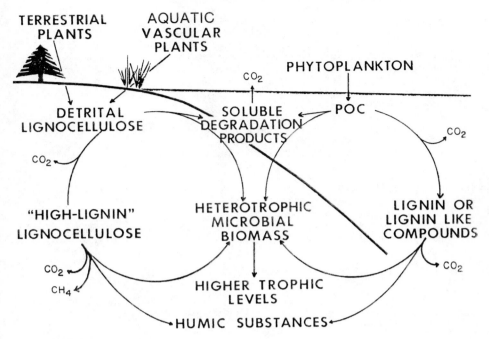

Figure 19.1
Model for the sources, transformations, and fates of carbon derived
from lignocellulosic detritus in aquatic environments.

wetland plant material. In this report, we describe the approach
used and present some of our results obtained from experiments
conducted in Georgia salt marshes and in a freshwater environment,
the Okefenokee Swamp. The approaches we have used in our basic
investigations of microbial processes in pristine wetland ecosystems
are directly applicable to similar studies of natural or artificial
wetlands used for wastewater treatment.

MATERIALS AND METHODS

MICROBIAL TURNOVER OF LABILE DISSOLVED
ORGANIC COMPOUNDS

Previous work has demonstrated that any of a variety of low-
molecular-weight, soluble organic compounds can be used to assess the
rates of microbial turnover of the easily degraded component of plant
material (e.g., see Hodson, Azam, and Lee 1977; Hodson et al. 1981).
Compounds such as simple sugars and amino acids are often found to
have similar rates of turnover in any given system under identical
environmental conditions. For our studies, we used D-glucose because
it is a common component of the dissolved organic carbon (DOC) of
natural waters. The basic procedure that was followed has been
described in Azam and Holm-Hansen (1973) and Murray and Hodson

(1984). Rates of turnover of dissolved D-glucose in water were determined by incubating triplicate 25 ml samples for 30 min at the in situ temperature in the presence of 1.3 nM ^{3}H-D-glucose (30.0 Ci mmol^{-1}). Incubations were terminated by the addition of formalin to a final concentration of 2%, and samples were filtered on 0.2 μm pore-size Nuclepore^R filters, washed twice with filter-sterilized water from the sampling site, and air dried. The assimilated radio-activity was combusted to ^{3}H$_2$O in an R. J. Harvey Biological Sample Oxidizer. The ^{3}H$_2$O was collected in 14 ml of Scintiverse^R (Fisher Scientific Co.) and assayed by liquid scintillation spectrometry. Samples of wetlands sediment were similarly assayed for rates of microbial turnover of D-glucose. Sediment samples were made into slurries using filtered water from the same habitat and incubated with labelled D-glucose. After incubation, samples were filtered and assayed as for water samples. As a correction for the respiration of ^{3}H-D-glucose to ^{3}H$_2$O during the incubation, 1 ml samples of the incubation liquid were removed, filtered, and the filtrates air dried. The volatile fraction of the radioactivity (^{3}H$_2$O) was taken as the respiration that had occurred during the incubation.

MICROBIAL TRANSFORMATIONS OF LIGNOCELLULOSIC DETRITUS

In order to assess the rates of mineralization, solubilization, and assimilation into microbial biomass of carbon derived from ligno-cellulosic detritus, specifically radiolabelled ^{14}C-(cellulose)-lignocellulose and ^{14}C-(lignin)-lignocellulose were prepared from a variety of wetland plants, including the salt-marsh grass, Spartina alterniflora, the salt-marsh rush, Juncus roemerianus, the freshwater sedge, Carex walteriana and the freshwater grass, Panicum hemitomon. In addition, ^{14}C-(lignin)-lignocellulose from slash pine, Pinus elliottii, was prepared as a representative lignin of terrestrial origin that could be deposited in aquatic environments from runoff. The procedures for preparation of radiolabelled lignocelluloses (Fig. 19.2) were modified from methodology described by Crawford and Crawford (1976) for labelling woody terrestrial plants. Plant cuttings for radiolabelling included young pine twigs and the entire aboveground portions of aquatic macrophytes (approximately 20 cm in height). Each cutting was placed in 1.0 ml of sterile, distilled water to which was added 5μ Ci of ^{14}C-(U)-L-phenylalanine or ^{14}C-(3-side chain)-cinnamic acid as lignin precursors or ^{14}C-(U)-D-glucose as cellulose precursor. Cuttings were incubated under constant illumination until 80% to 90% of the volume of liquid had been taken up. Additional liquid was added intermittently to maintain the volume at about 1.0 ml for 72 hours (Maccubbin and Hodson 1980; Benner, Maccubbin, and Hodson 1984a). After incubation, the plant material was dried to constant weight and extracted to remove unin-corporated label and other non-lignocellulose compounds (Maccubbin and Hodson 1980; Benner, Maccubbin, and Hodson, 1984b). The labelled extractive-free lignocellulose was characterized to ensure that the labelled moieties were, in fact, lignin and cellulose by strong acid extraction (Klason Lignin technique, as described in Browning 1967) and analysis of extracts by thin-layer and paper chromatography (Vomhof and Tucker 1965; Crawford and Crawford 1978). The resulting lignocellulose preparations are labelled in either the lignin or cellulose moiety and can be used to assess rates of

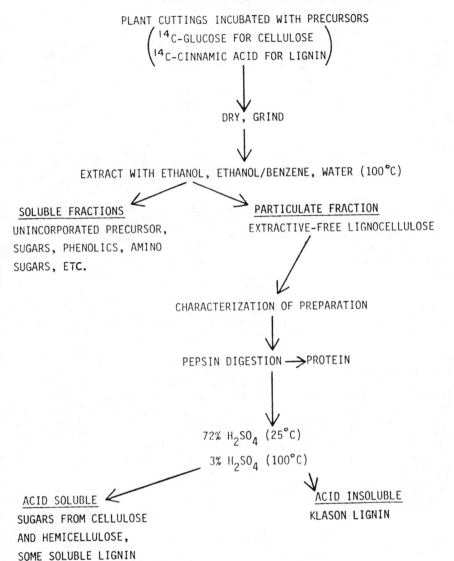

Figure 19.2
Procedures for preparing radiolabelled lignocellulose.

transformations of these macromolecules under natural wetlands conditions. The preparations are realistic representations of wetland detritus in that they contain lignocellulose in the natural matrix characteristic of the particular plant used. The labelled lignocelluloses were used in long-term incubations to assess rates of lignin and cellulose mineralization to CO_2 (and, hence, changes in the ratios of lignin and cellulose in the detritus with time), conversion to soluble metabolites, and incorporation into microbial biomass.

The labelled lignocelluloses were incubated for periods of up to one month with slurries of water and sediments from salt marshes or the Okefenokee Swamp. Samples (10 mg) of labelled material were added to 10 ml of sediment-water slurry in sterile 150 ml bottles sealed with butyl rubber stoppers fitted with gas ports. Samples were incubated in the dark with agitation and periodically flushed with CO_2-free, water-saturated air (for aerobic incubations) or oxygen-free nitrogen (for anaerobic incubations). Any $^{14}CO_2$ resulting from mineralization of the labelled lignocellulose was collected by passing the gas stream through CO_2-absorbing liquid scintillation counting medium (Maccubbin and Hodson 1980). For anaerobic incubations, the gas stream was subsequently passed through an R. J. Harvey Biological Sample Oxidizer especially modified to accept gases, trapping any evolved labelled methane after its combustion to $^{14}CO_2$ and dispensing it into a scintillation counting medium for radioassay (Benner, Maccubbin, and Hodson 1984b). After termination of the incubations, samples of the liquid were removed, filtered, and assayed for the presence of labelled bicarbonate and acid-stable, soluble, organic carbon derived from the lignocellulose. Rates of mineralization and solubilization were compared to those supported by formalin-fixed controls.

Assimilation of lignocellulose-derived carbon into microbial biomass was assessed after the incubations were terminated by extracting the contents of the incubation flasks with enzymes that selectively lyse bacteria and fungi without significantly solubilizing the lignocellulose preparations. After extraction, the sediments were centrifuged and the radioactivity in the supernatant determined by assaying a 1 ml subsample via liquid scintillation spectrometry.

MICROBIAL BIOMASS

Total microbial biomass in wetland waters and sediments can be estimated via extraction and determination of cellular adenosine triphosphate (ATP) using various modifications of the firefly luciferase assay developed by Holm-Hansen and Booth (1966). High concentrations of suspended sediment or humic substances in wetland waters complicate the recovery and determination of ATP and can necessitate the use of numerous correction factors to account for ATP losses at each step in the procedure (Hodson, Holm-Hansen, and Azam 1976, Hodson, Maccubbin, and Pomeroy 1981). We have found that with proper correction for losses, extraction of samples with bicarbonate buffer at 100°C (Bancroft, Paul, and Wiebe 1976) is an effective procedure for samples from wetlands (Murray and Hodson 1984).

Bacterial biomass is easily determined using direct examination with epifluorescence microscopy of acridine orange-stained samples collected on Nucleopore[R] filters (Hobbie, Daley, and Jasper 1977). Differential filtration of water samples on various size filters prior to staining the samples enables one to estimate the biomass of individual size classes of bacteria and, hence, to differentiate between biomass of free-living and attached bacteria (Hodson, Maccubin, and Pomeroy 1981). One possible shortcoming of the approach is the inability to differentiate between live and dead bacteria. However, it seems likely that dead microorganisms would have a relatively rapid turnover due to autolysis or grazing by microzooplankton. Bacterial biomass in sediments can be determined by similar procedures if sediment samples are first diluted.

SECONDARY PRODUCTION BY BACTERIA

No method for determining rates of microbial secondary production is as universally applicable as is the Steemann-Nielsen (1952) method for estimating primary production by phytoplankton. The source of carbon for autotrophs is CO_2, whereas the sources of carbon for microheterotrophs are numerous, including literally thousands of soluble organic compounds, and vary among species of microheterotrophs. However, recently, Fuhrman and Azam (1980) described a method that gives a reasonable estimate of rates of bacterial secondary production in aquatic systems. This method is based on the assimilation of ^{3}H-thymidine into cellular deoxyribonucleic acid (DNA). The authors have shown tight correlations between rates of incorporation of ^{3}H-thymidine into DNA and overall production of particulate organic matter by bacteria. This method should be highly useful in assessing temporal changes in production rates that might result from the effects of various anthropogenic perturbations to wetland ecosystems. We are presently using the method, with good results, to assess rates of bacterial production in waters of the Okefenokee Swamp (Murray and Hodson 1984). Basically, the method involves incubating the water sample with the labelled thymidine for one to several hours, filtering the sample onto Millipore[R] filters to collect the labelled bacteria, washing the filter with cold, dilute trichloroacetic acid to remove unincorporated label and thymidine taken up but not incorporated into DNA, and radioassaying the samples (filters) by liquid scintillation spectrometry. If the total amount of DNA in the sample can be estimated (e.g., from microscopic counts of bacterial density and conversion factors relating bacterial numbers with DNA concentrations), the rate of DNA production can be calculated. For relatively heterogeneous Okefenokee Swamp samples, we obtain very good results by combusting the filters and counting the incorporated radioactivity as ^{3}H$_2$O by liquid scintillation spectrometry. In this way, we avoid the problems associated with counting labelled samples in the presence of humic (colored) material (Murray and Hodson, 1984).

RESULTS AND CONCLUSIONS

The basic assumption made in our approach to the study of the impact of environmental stress on wetlands is that wetlands are detritus-based ecosystems in which much of the carbon and energy originating from primary production passes through "compartments"

consisting of detritus and microbial biomass before it is available
to the higher trophic level consumers in the food web. Thus, microbial transformations of plant detritus serve as the principal link
between primary and secondary production. The plant-derived carbon
first enters the aquatic environment as either dissolved organic
compounds or particulate, lignocellulosic detritus. Ultimately, the
microbial degradation of both these components results in either
incorporation of plant-derived carbon into microbial biomass or
respiration to CO_2. In the former case, the carbon and energy (in
the form of microbial biomass) are available to higher animals as
food, and the microbial processing can be considered to have served
as a link in the food web. In the latter case, the potential carbon
and energy are lost to the aquatic system and the microbial processing is, in effect, a terminal sink in the carbon flux through the
system (Pomeroy 1974). Any anthropogenic changes in either the rates
of these microbial processes or the relative percentages of carbon
assimilated into microbial biomass versus respired to CO_2 will affect
the flux of carbon and energy through the entire system.
    In selecting representative microbial processes in wetlands to
follow as indices of the overall natural rates of organic carbon
transformation, as well as the effects of perturbations on the rates,
one must keep in mind that the types and intensities of individual
reactions change over time. No single index will be sensitive under
all conditions. Results of several tests should be considered
together when possible. Some typical results of the analyses described above are given below as examples of the applicability of
these analyses to wetland ecosystem studies.
    Table 19.2 shows the results of a study that determined the
rates of D-glucose turnover in the water columns of several subhabitats in the Okefenokee Swamp ecosystem. Turnover times varied
between a few hours and approximately 300 hours. In contrast, Table
19.3 shows the turnover times for D-glucose in the sediments underlying the same habitats in the swamp. These values were on the order
of a few minutes, showing the overall much higher microbial processing rates in sediments than in water columns. This difference
was also seen when sediments and water columns were examined for
total microbial biomass (ATP) and bacterial biomass (acridine orange-
stained microscopic counts). Surface sediments had more biomass in
each case than water samples (data not shown). Interestingly, the
total microbial biomass in the waters and surface sediments of the
Okefenokee Swamp was found by us to be among the highest ever
reported (up to 6.6 μgl^{-1} in water and 28 μgg^{-1} dry wt in sediments).
In contrast, previous investigators had suggested that microbial
activity and biomass in the Okefenokee Swamp waters and surface
sediments were depressed and that this situation was an underlying
cause of the observed accumulation of organic material as peat.
    Bacterial secondary production in wetland waters and sediments
can be assessed by determining rates of incorporation of tritiated
thymidine into bacterial DNA. For instance, Table 19.4 presents
thymidine incorporation and bacterial production in Okefenokee Swamp
water. The rates of thymidine incorporation by bacteria in the
Okefenokee water column are comparable to similar measurements from
near-shore and eutrophic marine systems and are several orders of
magnitude greater than values from oligotrophic marine systems
(Fuhrman and Azam 1980; Fuhrman, Ammerman, and Azam 1980).

Table 19.2
Water-Column Glucose Turnover Time (hours) from Four Habitats in the Okefenokee Swamp

	September	December	February	March	April	May	June	July	August	Mean
Chesser Prairie	243.85 ±37.64	66.47 ±6.48	4.86 ±0.33	1.57 ±0.24	7.97 ±3.57	16.37 ±1.75	-	-	1.26 ±0.04	48.91 ±88.97
Chesser Shrub	39.77 ±1.15	701.25 ±251.73	37.97 ±4.02	15.42 ±0.93	8.95 ±0.73	-	-	-	-	160.67 ±302.50
Grand Cypress	-	509.85 ±31.23	105.87 ±6.92	8.87 ±0.58	-	-	-	-	-	41.53 ±55.72
Mizell Prairie	259.64 ±13.29	53.75 ±42.92	12.48 ±4.90	3.40 ±0.47	105.51 13.77	-	-	-	-	86.96 ±104.64

Note: Samples were collected during 1980-1981. Dashed line indicates no standing water on site.

Table 19.3
Surface-Sediment Glucose Turnover Time (minutes) from Four Habitats in the Okefenokee Swamp

	September	December	February	March	April	May	June	July	August	Mean
Chesser Prairie	2.46 ±0.36	4.62 ±0.78	5.46 ±0.72	12.54 ±6.00	26.04 ±7.32	8.76 ±2.94	2.04 ±0.48	3.36 ±0.42	5.76 ±0.84	7.89 ±7.56
Chesser Shrub	2.52 ±0.54	24.12 ±3.00	11.64 ±2.10	7.80 ±2.16	7.80 ±3.18	8.22 ±3.78	7.92 ±2.16	6.12 ±0.84	2.22 ±0.06	8.71 ±6.48
Grand Cypress	7.86 ±1.26	69.96 ±27.36	6.66 ±0.42	12.66 ±5.40	10.20 ±1.98	9.24 ±2.40	-	5.46 ±2.40	2.46 ±0.36	15.56 ±22.20
Mizell Prairie	2.88 ±0.24	4.68 ±1.20	5.58 ±0.60	22.74 ±2.34	24.30 ±2.34	8.10 ±2.52	4.80 ±0.66	5.10 ±0.12	2.34 ±0.42	8.95 ±8.43

Note: Samples were collected during 1980-1981. Dashed line indicates not sampled.

Table 19.4
Water-Column Thymidine Incorporation and Bacterial Production from
Four Habitats in the Okefenokee Swamp

	Thymidine Incorporation (moles l^{-1} d^{-1})	Bacterial Production (cells l^{-1} d^{-1})
Rookery	2.93×10^{-9} $\pm 0.41 \times 10^{-9}$	4.98×10^{9}
Rookery control	1.87×10^{-9} $\pm 0.23 \times 10^{-9}$	3.18×10^{9}
Little Cooter Prairie	1.01×10^{-9} $\pm 0.47 \times 10^{-9}$	1.72×10^{9}
Buzzards Roost Lake	0.76×10^{-9} $\pm 0.04 \times 10^{-9}$	1.29×10^{9}

Note: Each value represents the mean of nine replicates. Conversion
factor (moles of thymidine incorporated per cell produced)
from Fuhrman and Azam 1982.

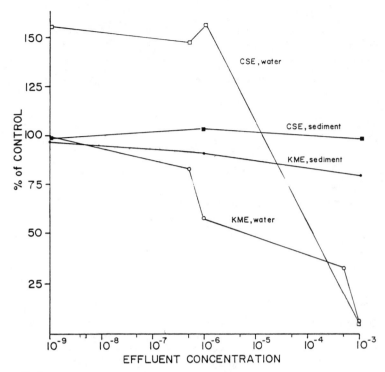

Figure 19.3
The effects of various concentrations (g ml^{-1}) of kraft
mill effluents (circles) and caustic soda effluents
(squares) on glucose uptake by microflora of estuarine
waters (open symbols) and sediments (solid symbols).

Rates of microbial activity, as estimated by our indices, are
highly sensitive to pollutional stress. For example, effluents from
pulp production, a major industry in Georgia and the Southeast, can
significantly inhibit the rates of glucose uptake by microbial populations from marsh waters and sediments. Rates of glucose uptake by
microbial populations preincubated 2 h in the presence of pulp mill
effluents (PME) were dependent on the type and concentration of
effluent (Fig. 19.3). Low concentrations (less than 1 ppm) of
caustic soda pulping effluent enhanced glucose uptake. At 1 part per
thousand (ppt), however, glucose uptake was inhibited by approximately 90% relative to controls. In contrast, all levels of kraft
mill effluent added inhibited glucose uptake, and inhibition
increased with effluent concentration. At 1 part per billion (ppb)
kraft mill effluent, glucose uptake was inhibited less than 10%,
while 1 ppt inhibited glucose uptake by 90% (Fig. 19.3). Sediment
samples were affected in a similar but less-marked manner. For
example, 1 ppt kraft mill effluent inhibited glucose uptake by only
20%, compared to 90% inhibition of water samples (Fig. 19.3).

Inhibitory effects of pulp mill effluents could not be accounted for solely by pH changes, although addition of effluents did increase sample pH slightly. Adjusting pH values of control samples with NaOH to match values of PME-treated samples had negligible effect on microbial activity as measured by glucose uptake.

Rates of microbial mineralization of the refractory (ligno-cellulosic) component of plant-derived particulate organic matter in wetland ecosystems are much slower than rates of mineralization of the soluble fraction of organic material. For example, Figure 19.4 shows the results of a representative experiment in which the mineralization of ^{14}C-(lignin)-lignocellulose and ^{14}C-(cellulose)-ligno-cellulose from the salt-marsh plants Spartina alterniflora and Juncus roemerianus and ^{14}C-(lignin)-lignocellulose from a representative tree of the coastal plain, Pinus elliottii, by salt-marsh sediment microflora was followed for nearly 600 h. Typically, the cellulose moieties were mineralized at rates 1.5 to 10 times those of the lignin moieties of lignocellulose from the same plant. The ratios of rates of cellulose mineralization to lignin mineralization for each lignocellulose decreased with incubation time. Thin-layer chromatography of the acid-soluble fraction of lignocellulose labelled with ^{14}C-D-glucose revealed that some of the label was associated with mannose and xylose, suggesting that hemicelluloses were labelled in addition to cellulose. Thus, the changing rates of cellulose degradation may reflect differential rates of mineralization of several polysaccharides present in the lignocellulosic detritus.

Rates of ^{14}C-(lignin)-lignocellulose mineralization were highest for S. alterniflora, intermediate for J. roemerianus, and lowest for P. elliottii (Fig. 19.4). The observed difference in the rates of mineralization of the lignin moieties of lignocelluloses from various plants is probably indicative of basic structural differences between lignins of grasses, rushes, and gymnosperms. Basically, lignins are three-dimensional polymers of phenylpropane units, derived from the oxidative polymerization of ρ-coumaryl, coniferyl, and sinapyl alcohols. The relative proportions of the individual alcohols and the types of bonds forming the polymers vary with plant species and can affect the relative resistance to microbial degradation of the resultant lignin.

Likewise, in degradation experiments using specifically radiolabelled lignocelluloses from Okefenokee Swamp plants, similar trends were observed (Fig. 19.5). The cellulose moiety from both Carex and Panicum was mineralized to CO_2 about 3.0 times faster than the lignin moiety (Fig. 19.5). After 500 h of incubation, similar amounts of ^{14}C-(cellulose)-lignin from Carex (2.5%) and Panicum (3.8%) had been mineralized to $^{14}CO_2$. However, the amount of $^{14}CO_2$ mineralized from ^{14}C-(lignin)-lignocellulose of both plants was only 1%. Rates of lignocellulose mineralization in Okefenokee Swamp sediment are generally lower than those in salt-marsh sediment at equal temperature. This difference could be due to the low pH of the swamp, the relatively low inorganic nutrient level in the swamp, or the more recalcitrant nature of the lignocelluloses from swamp plants. Recent experiments have indicated that pH is the major factor in the slower lignocellulose mineralization rates in the Okefenokee than in the salt marsh, and that polysaccharide mineralization is affected to a greater extent than lignin mineralization (R. Benner, M. Moran, and R. E. Hodson, in preparation).

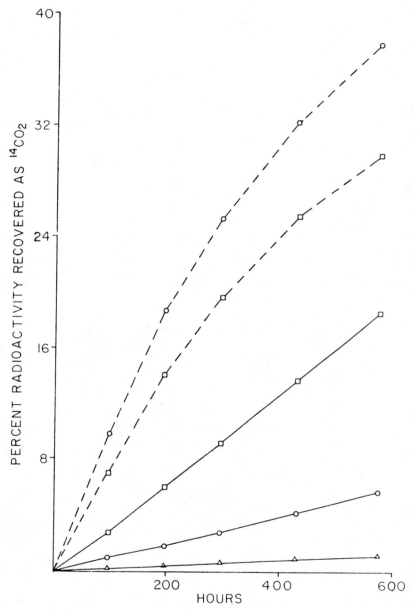

Figure 19.4
Microbial mineralization at 28°C of ^{14}C-(lignin)-lignocellulose (solid lines) and ^{14}C-(cellulose)-lignocellulose (dashed lines) from S. alterniflora (squares), J. roemerianus (circles), and P. elliottii (triangles).

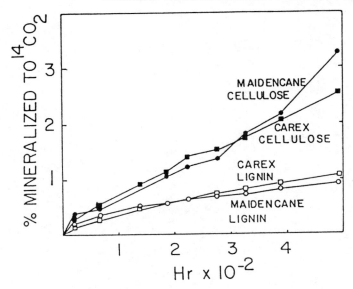

Figure 19.5
Mineralization at 20°C of specifically radiolabelled
^{14}C-(cellulose)-lignocellulose and ^{14}C-(lignin)-
lignocellulose from two Okefenokee Swamp plants.

Rates of mineralization of lignocellulosic detritus derived from
wetland plants are highly dependent on environmental conditions. For
example, we incubated ^{14}C-(lignin)-lignocellulose and ^{14}C-(cellulose)-lignocellulose from <u>Spartina alterniflora</u> with surface marsh
sediments over a range of temperatures typical of the marshes near
Sapelo Island, Georgia (Fig. 19.6). Sediments were collected in
March at an ambient temperature of 17°C. At this temperature, rates
of mineralization were six to eight times slower than those at 40°C
for both ^{14}C-(lignin)-lignocellulose and ^{14}C-(cellulose)-ligno-
cellulose. Rates of lignocellulose mineralization at 5°C were
negligible. Although data from a complete seasonal study are not yet
available, these results suggest that a high percentage of the annual
mineralization of lignocellulosic detritus occurs during the warmer
months. We are finding similar seasonal differences in lignocellulose mineralization rates in Okefenokee wetland environments as well.
    We have also compared the relative rates of lignocellulose
mineralization in wetlands under aerobic and anaerobic conditions.
Considering the reducing conditions in salt-marsh, Okefenokee, and
some other wetland sediments, it is likely that O_2 availability plays
an important role in lignocellulose degradation and transformation in
wetland ecosystems. In the absence of molecular oxygen, lignin
degradation has generally been believed not to occur (Zeikus 1980).
However, our data from this salt-marsh experiment (Table 19.5) and
similar ones with Okefenokee and mangrove swamp sediments (not shown)
all indicate that anaerobic sediment slurries can liberate $^{14}CO_2$ from
lignin-labelled as well as cellulose-labelled lignocellulose at low,

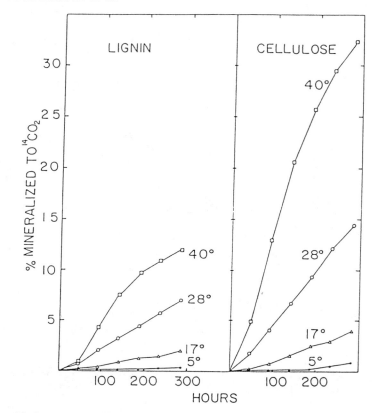

Figure 19.6
Microbial mineralization of radiolabelled
^{14}C-lignocellulose from <u>S. alterniflora</u> incubated with
marsh sediments at 5°C (solid circles), 17°C (triangles),
28°C (open circles), and 40°C (squares).

but significant rates. Periodically, incubations of labelled
<u>Spartina</u> lignocellulose with salt-marsh sediment were checked for
evolution of labelled methane and none was detected. Similar experiments in which specifically radiolabelled <u>Carex</u> lignocellulose
was incubated with anaerobic sediments from the Okefenokee Swamp indicated that anaerobic rates of lignocellulose degradation in this
freshwater wetland proceed at 20% to 25% of the aerobic rates.
Unlike the salt-marsh sediments, Okefenokee Swamp sediments produced
significant amounts of labelled methane from ^{14}C-lignocellulose. In
both the salt marsh (Table 19.5) and the Okefenokee (not shown),
purified cellulose was degraded much faster than was cellulose in the
natural lignocellulosic matrix, indicating that the lignin in lignocellulose limits the availability of the cellulosic moiety to microbial exoenzymes. Thus, contrary to the generally held belief that
lignin is not degraded anaerobically, results from our studies

Table 19.5
Comparison of the Aerobic and Anaerobic Rates of Cellulose and
Lignin Mineralization Incubated at 20°C for 480 Hours

^{14}C-Labelled Material	Aerobic Mineralization[*]	Anaerobic Mineralization[*]
Spartina		
^{14}C-lignin-LC	8.1	2.0
^{14}C-cellulose-LC	18.6	4.8
Pure-^{14}C-cellulose	51.6	8.5

[*]Percent radioactivity recovered as $^{14}CO_2$.

(Benner, Maccubbin and Hodson 1984b) indicate that both the lignin and
cellulose moieties of lignocellulose are significantly degraded in
the absence of molecular oxygen.

Specifically radiolabelled lignocellulose prepared from S.
alterniflora or other wetland plants can also be used to examine the
effects of pollutional stress on microbial mineralization of detrital
carbon in wetlands. For example, at KME concentrations of 500 ppm or
greater, the mineralization of both the lignin and cellulosic
moieties of labelled lignocellulosic detritus from Spartina was
inhibited by greater than 30% (Fig. 19.7). The extent of inhibition
gradually decreased with exposure time relative to controls. This
experiment suggested that lignin and cellulose mineralization were
directly inhibited by kraft mill effluents. However, it was also
possible that the effluent merely delayed colonization of detrital
particles, thus causing an apparent decrease in mineralization rate
initially. This possibility seems less likely in light of results of
an experiment (not shown) in which KME was added to ongoing incubations (48 h after addition of labelled detritus). Immediately after
addition of effluent, lignin and cellulose mineralization were
inhibited. The extent of inhibition relative to controls gradually
decreased over time. Thus the observed decrease with time in inhibition relative to controls is more likely due to microbial adaptation to, or detoxification of, the toxic components of KME.

To date our efforts at understanding microbial transformations
in wetland ecosystems have been directed toward measuring rates of
mineralization of soluble and particulate organic matter and incorporation of carbon from dissolved organic matter into microbial
biomass. In order to assess more completely the importance of the
various carbon fluxes postulated in our model (Fig. 19.1), we are now
expanding our measurements to include determinations of the rates and
efficiencies of incorporation of carbon derived from lignocellulosic
detritus into microbial and animal biomass. Preliminary data indicate that about 35% of lignocellulose-derived carbon is assimilated
into microbial biomass. Thus, although it is very slowly degraded,
lignocellulose, by virtue of its high proportion in plant detritus
and relatively efficient incorporation into microbial biomass, may be

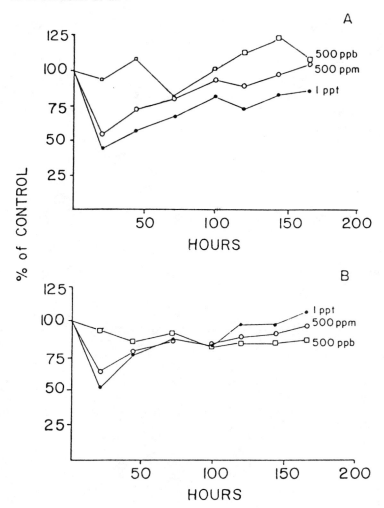

Figure 19.7
The effects of kraft mill effluents on microbial
mineralization of (A) ^{14}C-(lignin)-lignocellulose and
(B) ^{14}C-(cellulose)-lignocellulose. Kraft mill effluents
and labelled lignocellulose were both added to sediments
at time zero.

the single most significant carbon source to detritus-based wetland
ecosystems.

Management of natural or artificial wetlands for tertiary wastewater treatment involves either harvesting the macrophyte biomass or
promoting its degradation in situ by the natural microbial assemblages in the water and sediments. Harvesting and disposing of the
macrophyte biomass offsite can be expensive. Managing the wetland so
as to promote in situ degradation is more economical and has the
added advantage of potentially producing and harvesting commercially

marketable animals that feed directly or indirectly on the microbial biomass resulting from plant degradation. Thus any components in the wastewater that alter rates of microbial degradation of plant biomass could affect both the efficiency of water treatment and the animal secondary production in the wetland. Experiments such as those described herein can be used in both in vitro determinations of pollutant loading capacity and intermittent monitoring of microbial processes in situ.

ACKNOWLEDGMENTS

This is Contribution No. 477 of the University of Georgia Marine Institute, Sapelo Island, Georgia; Publication No. 37 of the Okefenokee Ecosystem Investigations. Funds for this research were provided by the National Science Foundation, Biological Oceanography Program Grant OCE-8117834 and Ecosystem Analysis Program Grants BSR-8114823 and BSR-8215587. Additional funding was provided by the U. S. Department of the Interior, Office of Water Resources and Technology and grant no. NA80AA-D-00091 from the U. S. Department of Commerce, Office of Sea Grant.

REFERENCES

Azam, F., and O. Holm-Hansen, 1973, Use of tritiated substrates in the study of heterotrophy in seawater, *Mar. Biol*. 23:191-196.
Bancroft, K., E. A. Paul, and W. J. Wiebe, 1976, The extraction and measurement of adenosine triphosphate from marine sediments, *Limnol. Oceanogr*. 21:473-479.
Benner, R., A. E. Maccubbin, and R. E. Hodson, 1984a, Preparation, characterization, and microbial degradation of specifically radio-labeled [^{14}C] lignocelluloses from marine and freshwater macrophytes, *Appl. Environ. Microbiol*. 47:381-389.
Benner, R., A. E. Maccubbin, and R. E. Hodson, 1984b, Anaerobic biodegradation of the lignin and polysaccharide components of ^{14}C-lignocelluloses and synthetic ^{14}C-lignin by sediment microflora, *Appl. Environ. Microbiol*. 47:998-1004.
Browning, B. L., 1967, *Methods of wood chemistry*, vol. 2, Interscience Publishers, New York, pp. 786-787.
Crawford, D. L., and R. L. Crawford, 1976, Microbial degradation of lignocellulose: The lignin component, *Appl. Environ. Microbiol*. 31:714-717.
Crawford, R. L., and D. L. Crawford, 1978, Radioisotopic methods for the study of lignin biodegradation, *Dev. Ind. Microbiol*. 19:35-49.
Fallon, R. D., and F. K. Pfaender, 1976, Carbon metabolism in model microbial systems from a temperate salt marsh, *Appl. Environ. Microbiol*. 31:959-968.
Fuhrman, J. A., and F. Azam, 1980, Bacterioplankton secondary production estimates for coastal waters of British Columbia, Antarctica, and California, *Appl. Environ. Microbiol*. 39:1085-1095.
Fuhrman, J. A., and F. Azam, 1982, Thymidine incorporation as a measure of heterotrophic bacterioplankton production in marine surface waters: Evaluation and field results, *Mar. Biol*. 66:109-120.

Fuhrman, J. A., J. W. Ammerman, and F. Azam, 1980, Bacterioplankton in the coastal euphotic zone: Distribution, activity and possible relationships with phytoplankton, Mar. Biol. 60:201-207.

Gallegher, J. L., 1974, Sampling macroorganic matter profiles in salt marsh plant root zones, Soil Sci. Soc. Proc. 33:154-155.

Godshalk, G. L., and R. G. Wetzel, 1978, Decomposition of aquatic angiosperms, III. Zostera marina L. and a conceptual model of decomposition, Aquat. Bot. 5:329-354.

Harrison, P. G., and K. H. Mann, 1975, Detritus formation from eel-grass (Zostera marina L.): The relative effects of fragmentation, leaching and decay, Limnol. Oceanogr. 20:924-934.

Hobbie, J. E., R. J. Daley, and S. Jasper, 1977, Use of Nuclepore filters for counting bacteria by fluorescence microscopy, Appl. Environ. Microbiol. 33:1225-1228.

Hodson, R. E., F. Azam, and R. F. Lee, 1977, Effects of four oils on marine bacterial population: Controlled ecosystem pollution experiment, Bull. Mar. Sci. 27:119-126.

Hodson, R. E., O. Holm-Hansen, and F. Azam, 1976, Improved methodology for ATP determination in marine environments, Mar. Biol. 34:143-149.

Hodson, R. E., A. E. Maccubbin, and L. R. Pomeroy, 1981, Dissolved adenosine triphosphate utilization by free-living and attached bacteria plankton, Mar. Biol. 64:43-51.

Hodson, R. E., F. Azam, A. F. Carlucci, J. A. Fuhrman, D. M. Karl, and O. Holm-Hansen, 1981, Microbial uptake of dissolved organic matter in McMurdo Sound, Antarctica, Mar. Biol. 61:89-94.

Holm-Hansen, O., and C. R. Booth, 1966, The measurement of adenosine triphosphate in the ocean and its ecological significance, Limnol. Oceanogr. 14:740-747.

Kirk, T. K., and H. Chang, 1971, Effects of microorganisms on lignin, Ann. Rev. Phytopathol. 9:185-210.

Maccubbin, A. E., and R. E. Hodson, 1980, Mineralization of detrital lignocelluloses by salt marsh sediment microflora, Appl. Environ. Microbiol. 40:735-740.

Murray, R. E., and R. E. Hodson, 1984, Microbial biomass and utilization of dissolved organic matter in the Okefenokee Swamp Ecosystem, Appl. Environ. Microbiol. 47:in press.

Odum, E. P., and A. A. de la Cruz, 1967, Particulate organic detritus in a Georgia salt marsh-estuarine ecosystem, in Estuaries, G. H. Lauff, ed., AAAS publ. no. 83, American Association for the Advancement of Science, Washington, D. C., pp. 383-388.

Pomeroy, L. R., 1974, The ocean's food web, a changing paradigm, Bioscience 24:499-504.

Steemann-Nielsen, E., 1952, The use of radioactive carbon (^{14}C) for measuring organic production in the sea, J. Cons., Cons. Int. Explor. Mer 18:117-140.

Teal, J. M., 1962, Energy flow in the salt marsh ecosystem of Georgia, Ecology 43:614-624.

Vomhof, D. W., and T. C. Tucker, 1965, The separation of simple sugars by cellulose thin-layer chromatography, J. Chromatogr. 17:300-306.

Zeikus, J. G., 1980, Fate of lignin and related aromatic substances in anaerobic environments, in Lignin biodegradation: Microbiology, chemistry, and applications, T. K. Kirk, T. Higuchi, and H. Chang, eds., CRC Press, West Palm Beach, Fla., pp. 101-109.

DISCUSSION

Brinson: Has any work been done with the potential phytoplankton source for lignin in terms of impact, end products, and so on?

Hodson: Several species have been examined by Martin Alexander and some of his co-workers several years ago. They found that certain phytoplankton species in freshwater have cell walls that are as refractory as lignocellulose and seem to be lignins; otherwise ligno-cellulose is not known to occur in nonvascular plants.

Ewel: You compared the terrestrial plant material with the vascular plant material from the wetlands. Do you have any feelings for the relative conservation of lignocellulose in the woody wetland material and the woody terrestrial material?

Hodson: The concentrations of lignocellulose are quite similar, it turns out, in the vascular plants, whether they are herbaceous or woody--slightly lower in herbaceous--but what we have is a lower lignin-to-cellulose ratio in the lignocellulose. So the ligno-cellulose is more labile in herbaceous plants. What we compared in Figure 19.4 was the primary input of detritus from the terrestrial system that in Georgia is slash pine, which is farmed along the coast, with the primary salt-marsh plants. In the salt marsh there are no woody plants, but in freshwater environments like the Okefen-okee, there are. We are only now looking at degradation of cypress and shrub lignin, so I cannot really answer directly.

Hemond: In a sphagnum-dominated wetland, a lot of the starting material is polyuronic acid. Is there any reason to think that there is a fundamental difference in degradation of polyuronic acids relative to the lignocellulose in the makeup of typical marshes and forested wetlands?

Hodson: In that the lignin content of sphagnum mosses is quite low, I might assume that degradation of this sort of polysaccaride-lignin complex would be more rapid. I would propose that it would not be difficult with the technology that has been developed over the last two or three years to prepare specifically labelled polyuronic acid material in its natural matrix from sphagnum mosses and examine its degradation kinetics. It has not been done to date.

Hemond: Is the presence of the dissolved organic acids in these low pH environments likely to have a major controlling influence?

Hodson: When we started about a year and a half ago to look at microbial processes in the Okefenokee, we started on the traditional wisdom that the low pH, peat-accumulating environment would have a low microbial biomass and low microbial activity relative to a normal pH system like the Georgia salt marsh. But we found that the micro-bial biomass and the activity, as measured by two or three indices, was just as high as in the pH 7 to pH 8 salt marsh. That result seems contradictory for a system that accumulates peat relatively rapidly. This shows how little we know. We do find that the low pH inhibits lignocellulose degradation in the Okefenokee, however.

SESSION VI

Long-term Effects

20

Some Long-term Consequences of Sewage Contamination in Salt Marsh Ecosystems

Ivan Valiela, John M. Teal, Charlotte Cogswell,
Jean Hartman, Sarah Allen, Richard Van Etten,
and Dale Goehringer

Salt marsh ecosystems are often located in coastal sites that
have become centers of human activity. Through direct emptying of
sewage wastes into salt marshes and by the more generalized contamination of tidal water with sewage effluents, salt marshes often
receive human wastewater. Most of these instances of contamination
are chronic, long-term events, but there are few studies available
that yield information on the consequences of such long-term
contamination.

We have been conducting an experimental study of long-term
consequences of sewage contamination since 1970 in Great Sippewissett
Salt Marsh on Cape Cod, Massachusetts. We have followed many of the
changes prompted by experimental additions of sewage-based fertilizer
on various aspects of this ecosystem. In this paper we will concentrate on the long-term changes that have become evident in the
vegetation of the experimental plots. These results demonstrate the
complex interactions and the degree of understanding of the natural
system that is necessary to predict the consequences of sewage
additions.

EXPERIMENTAL DESIGN AND METHODS

A commercially available fertilizer containing sewage sludge was
broadcast by hand to plots 10 m in radius. The fertilizer contained
10% nitrogen (N), 6% phosphorus expressed as P_2O_5, plus a wide
variety of other substances including heavy metals (see A. Giblin,
this volume). The fertilizer was added in three dosages (LF:
8 g m^{-2} wk^{-1}; HF: 25 g m^{-2} wk^{-1}; and XF: 75 g m^{-2} wk^{-1}), and
applications were carried out every two weeks from March to November.
Further details of the procedure are provided in Valiela, Teal, and
Sass (1975) and Valiela, Teal, and Persson (1976).

We carried out additional experiments to identify whether
nitrogen or phosphorus (P) were specific limiting elements. These
experiments consisted of separate additions of urea (an N source),
phosphate, or a combination of both urea and phosphate to salt-marsh
plots. The dosages of urea and phosphate were set to equal the N and
P provided by the HF dosage. In addition, there were untreated

control plots and each treatment was applied to two replicate plots.

The vegetation of the experimental plots consisted principally of the perennial grasses Spartina alterniflora, Spartina patens, and Distichlis spicata, and the annual glasswort Salicornia europaea. Estimates of biomass of these species were obtained either by harvest or by use of height measurements and height-weight regressions (Vince et al. 1981). Maps of the percent cover by the various species were also obtained for all the plots under each treatment.

LONG-TERM EXPERIMENTAL EUTROPHICATION AND SOME CONSEQUENCES

INCREASES IN BIOMASS AND PRODUCTIVITY

In any one growing season, the biomass of the principal species within fertilized plots increased relative to control plots (Valiela et al. 1982). For brevity, we consider here only S. alterniflora. Over the years, there were significant changes in the pattern of growth created by the various treatments. The peak biomass of control plots was remarkably constant over the 12-year period (Fig. 20.1, top). The lower two dosages (LF and HF) of fertilizer did not increase biomass during the first growing season but did so subsequently. The highest dosage (XF) prompted an increase in biomass even in the first year of treatment. Thus, the lag in response by the standing crop of grasses seems to depend on the dosage received, and the higher the dosage the shorter the lag.

The increase in biomass due to the fertilizer is associated with the added N, as demonstrated by the increased growth of grasses in plots to which N alone was added (Fig. 20.1, bottom). There was no response to phosphate addition (Valiela and Teal 1974). Similar results of enrichment experiments have been obtained by others elsewhere (Pigott 1969; Tyler 1967; Sullivan and Daiber 1974; Broome, Woodhouse, Jr., and Seneca 1975; Gallagher 1975; Patrick and Delaune 1976; Jefferies and Perkins 1977; Chalmers 1979). There is a secondary limitation due to P, once some additional N is provided (Figure 20.1, bottom).

The clear-cut nutrient limitation evidenced by the results of enrichment experiments is paradoxical in view of the high nutrient concentrations in interstitial water of salt-marsh sediments (Table 20.1) compared to seawater. One explanation for this situation is that nutrient uptake by plant roots is inhibited in anoxic sediments; the nutrient limitation is therefore mediated by anaerobic conditions. Anything that increases oxidation, such as plant activity or greater percolation rate (Howes et al. 1981; Mendelssohn, McKee, and Patrick 1981), will increase nutrient availability and therefore plant growth. These mechanisms prompt the tall stature of plants in creek banks, where percolation and plant biomass are greatest. Growth also increases if more nutrients are added to the interstitial water, as in our fertilized plots, since the greater concentrations of nutrients in the interstitial water may allow a greater uptake even if the redox potential is not changed.

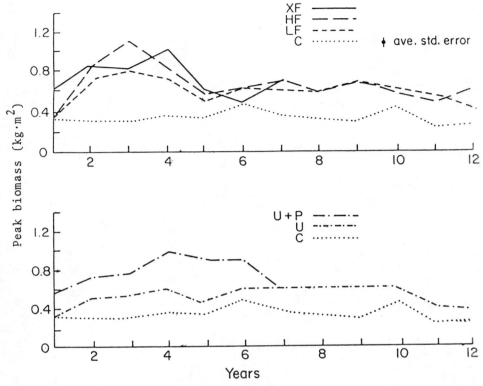

Figure 20.1
Long-term effects of various fertilization regimes on the annual
aboveground peak biomass of <u>Spartina alterniflora</u>. Standard errors
are omitted for clarity. The C, LF, and HF dosages were started in
1970; XF in 1974; and U + P in 1975. They are graphed according to
years from initiation for ease of comparison. Samples of biomass were
obtained by harvest; the values for peak biomass during each growing
season are shown.

CHANGES IN PLANT MORPHOLOGY AND SPACING

The enrichment treatments also changed the form of plant from
the typical short morphology to the tall form of <u>S. alterniflora</u>.
Short <u>S. alterniflora</u> consists of very dense, small-leaved, narrow-
stemmed plants. This condition is the most common in Great Sippewissett Marsh and is found in our control plots (Valiela, Teal, and
Deuser 1978). The effect of increasing doses of fertilizer is to
increase the size of plants, both leaves and stems. The stem diam-
eters, in fact, tend to converge on the sizes typical of the tall
form of <u>S. alterniflora</u>. In addition, the density of stems is
considerably lower, again converging on the lower values found in the
tall form (Valiela, Teal, and Deuser 1978).

One consequence of this change in plant form and spacing is that the enriched stands can be more easily penetrated by predatory fish, and that prey animals are more susceptible to these predators (Vince, Valiela, Backus, and Teal 1976).

INCREASED PERCENT NITROGEN

The fertilization treatments increased the N content of plant tissues. In S. alterniflora, there was a consistent increase in percent N in plants from fertilized plots compared to plants from control plots (Vince, Valiela, and Teal 1981). The increase in percent N was small, perhaps 1%. It turns out, however, that this small difference is very significant for the grazers that feed on S. alterniflora. A variety of herbivorous insects (Vince, Valiela, and Teal 1981), Canada geese (Buchsbaum, Valiela, and Teal 1982), and meadow voles (unpublished observations) fed on S. alterniflora, but usually their damage to stands has been minimal. In fertilized plots, however, grazing damage was considerably more important (Fig. 20.2) and may have consumed a substantial amount of the annual production (unpublished data). The plant biomass that was not consumed (typically, the bulk of annual production in salt marshes) became detritus. This detritus was also enriched in N, and the abundance of detritus-feeding invertebrates increased two- to fivefold in fertilized plots (unpublished data).

CHANGES IN SPECIES COMPOSITION

The experimental fertilization in all cases brought about a sequence of alterations in the taxonomic composition of the stands. Several subdominant species disappeared (Table 20.2); the number of species decreased from 11 in the controls or in initial conditions to about 4 in the highest dosage. There were also notable changes in the abundance of the species that remained in the plots. There are a number of possible mechanisms to account for these changes in species composition; one must be differential ability to take up nutrients and grow under differing rates of nutrient supply. S. alterniflora, for example, seems to reach an asymptote where yield is not increased beyond a certain rate of N supply (Fig. 20.3). D. spicata, on the other hand, continues to increase biomass at higher N inputs. The range of N inputs seen in Fig. 20.3 covers the span of inputs that enter wetlands, as reported by Kelly (this volume). Our treatments, therefore, covered the expected range of inputs and are appropriate comparisons to other instances of eutrophication. On the basis of these curves, one would expect that D. spicata would replace S. alterniflora and other species, and we, in fact, saw such replacement in the high marsh areas of the field plots. Other species have different responses to the increased input of nitrogen.

There are other mechanisms by which species replace one another. Notice in Figure 20.1 that after the initial increase in biomass during years 1 to 4, there is a decrease in standing crop. We are not sure what causes the drop but can suggest two possibilities. One alternative hypothesis is that the greater the amount of plant biomass, the greater the transpiration, and the greater the removal of freshwater from the sediment. In Great Sippewissett Marsh, the

Table 20.1
Rough Estimates of Modal and, in Parentheses, Range of
Concentrations of Nutrients in Seawater and in
Interstitial Water of Salt-Marsh Sediments

	Concentration (ugatom 1^{-1})		
	NO_3-N	NH_4-N	PO_3-P
Seawater	1(0-50)	1(0-50)	1
Interstitial water in salt-marsh sediments	5(0-50)	50(10-500)	10(5-20)

Table 20.2
Number of Species of Vascular Plants Found in the
Experimental Plots at Present

Treatment	Replicate Plot 1	Replicate Plot 2
XF	4	4
HF	4	7
LF	--	6
C	11	11
U	5	5

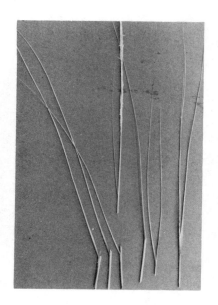

Figure 20.2
Top left: Patch of S. alterniflora that has been heavily grazed by
meadow voles. Although vole grazing occurs all over the marsh, damage
is not so marked as in fertilized plots, as shown here. Top right:
Remains of S. alterniflora shoots after feeding by voles. Bottom:
Typical evidence of grazing by Canada geese on S. alterniflora.

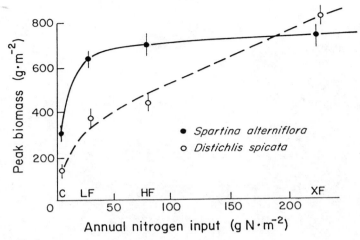

Figure 20.3
Peak aboveground biomass achieved by two salt–marsh grasses under increased nitrogen inputs. The nitrogen inputs provided by the fertilizer dosages are indicated along the horizontal axis.

salinity of intertidal water near the sediment surface rarely exceeds that of the flooding tidal water, about 32°/oo. This condition is in contrast to salt marshes in warmer climates, such as Georgia, where salinities may climb to 40°/oo to 60°/oo due to high rates of evapo-transpiration. Such higher salinities inhibit growth in laboratory cultures (Haines and Dunn 1976). In our experimental plots, the increased biomass could have transpired enough water to raise salin-ities of sediment above 40°/oo. The plants could thus inhibit their own growth, and perhaps this situation may lead to die-off in certain patches within plots fertilized at high dosages (Fig. 20.4). Perhaps because of local heterogeneities, patches rather than the entire sward are affected; slight differences in availability of nutrients from one specific patch of sediment to another may change the amount of biomass and therefore the loss of freshwater, for example, enough to push salinity over a critical threshold. This picture could be complicated by changes in sediment redox (Howes et al. 1981), that in turn would modify pH, perhaps through pyrite oxidation. We have no direct evidence as yet that the transpiration hypothesis is impor-tant, but we are in the process of studying the matter.

    A second hypothesis is that the enriched tissues attract herbivores, and the damage due to insects, voles, and other plant consumers may create or at least foster the appearance of bare patches (as can be seen in Fig. 20.2) or at least lower the density of S. alterniflora.

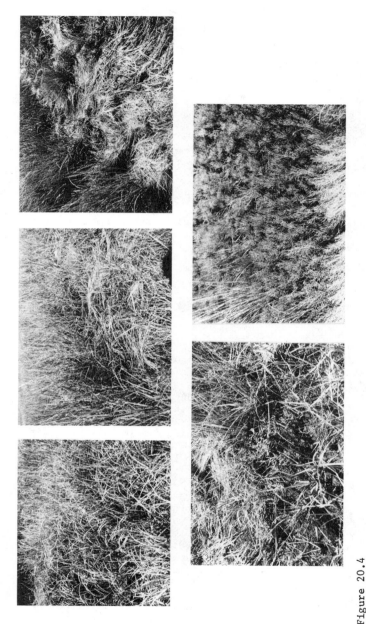

Figure 20.4
Composite of sequence of events in salt-marsh plots experimentally fertilized at a high (XF) dosage. The photos are not one sequence in the same site. Top left: Stand of S. alterniflora converted to the tall form during a year's growth. Middle top: Incipient formation of bare patch by yellowing and death of plants in a patch within a fertilized plot. Top right: Dead grasses at the end of the second or third growing season after fertilization. Bottom left: A few plants of the opportunist Salicornia europaea appearing in the bare areas during the following season. Bottom right: Fully developed patch of S. europaea about two years after the appearance of bare ground.

Figure 20.5
Density of seedlings of <u>Salicornia europaea</u> in early
spring in control plots (top) and fertilized (XF) plots
(below). The seedlings are 2 mm to 4 mm across the
longer dimension.

After a growing season, the opportunistic annual <u>Salicornia</u>
<u>europaea</u> occupied these newly bared patches. The expansion by <u>S.</u>
<u>europaea</u> may have occurred either through enhanced growth of the
small, scattered plants usually present or by the entrance of seed
from elsewhere. In the second or third year after the density of <u>S.</u>
<u>alterniflora</u> was lowered, the seed crop of <u>S. europaea</u> became very
large, and very dense growth of <u>S. europaea</u> occurred (Fig. 20.5).
This canopy may last one or two years.

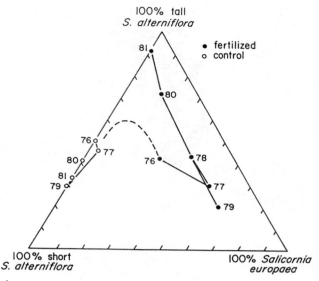

Figure 20.6
Distribution of percent cover between the three major plant types in
low marsh between 1976 and 1981 in a control and fertilized (XF) plot.
Dotted line shows presumed path based on other observations.

The changes in the composition of the plant community in low
marsh over the several years of our study can be shown using a
triangular graph where each corner represents the state where the
stand is 100% covered by each of short S. alterniflora, tall S.
alterniflora, or Salicornia europaea (Fig. 20.6). The variation in
control plots is caused by varying conditions (amount of light, rain,
wind, etc.) from one year to the next. That span of variation,
largely limited to small changes in the percent cover of short and
tall S. alterniflora, is the background variability in the absence of
new inputs of N. Eutrophied salt-marsh plots show a markedly different history. First, there was a trend toward a preponderance of the
tall form of S. alterniflora. Then, perhaps due to our postulated
increase of salinity with increased transpiration or to increased
grazing on the N-enriched plants, patches of S. alterniflora died and
bare patches appeared. Salicornia europaea then fairly quickly
invaded these vacated sites. There is little likelihood that S.
europaea actually outcompetes S. alterniflora since it does rather
poorly when in the company of almost any of the grasses. We believe
that this species is basically an opportunist that colonizes empty
space. Since it can tolerate the high salinity that may be present
in bare patches (if the transpiration hypothesis is correct), it may
be well suited to use the spaces vacated by S. alterniflora. There
are further changes in the species composition of these swards,
principally the disappearance of Salicornia europaea, that are
discussed in the next section. The evidence of this section suffices
to show, however, that eutrophication leads to significant changes in
species composition of vegetation.

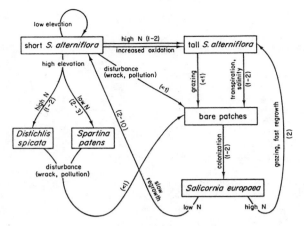

Figure 20.7
Processes that affect the structure of salt-marsh vegetation. The
mosaic of a salt marsh is made up of units shown as boxes. The arrows
show transformations from one to another kind of mosaic; the numbers
in parentheses show the years or growing seasons added to accomplish
the conversion.

RELATIONSHIPS THAT STRUCTURE SALT-MARSH PLANT COMMUNITIES

A New England salt marsh consists primarily of a mosaic of
patches of short and tall S. alterniflora, bare ground, S. europaea,
D. spicata, and S. patens (Fig. 20. 7, which should be referred to
for the remainder of this section). The results of the long-term
experiments provide clues as to how the balance among the various
components may be struck.
Much of the regularly flooded area of New England marshes is
covered by short S. alterniflora (Nixon 1982); this situation may
change if, as discussed above, either the N supply or oxidation of
sediments (as in creek banks) increases. If these changes occur, the
stand is converted to taller-growth S. alterniflora. If the taller
grasses do have more N available, their tissues will contain more N
and will therefore be more attractive to grazers. Grazing may lead
to the death of patches of plants that have been eaten. As also
mentioned earlier, perhaps the greater biomass--because of greater
transpiration--leads to higher interstitial salinity that may depress
growth or kill the grass. Thus, both these mechanisms may produce
bare patches where S. alterniflora used to be dominant.
Bare patches are soon colonized by S. europaea and, by the
second or third growing season after the appearance of bare ground,
this species dominates what used to be bare patches. This dominance
is transitory, especially if the high input of N continues. In such
a case, herbivores (mainly the chrysomelid beetle Erynephala maritima)
feed very intensively on the N-enriched S. europaea, and their
feeding leads to high mortality of the glasswort (C. Cogswell,
unpublished data). S. alterniflora by this time begins regrowth into

the patches now occupied by S. europaea and, in about two growing seasons, replaces the glasswort. We speculate that reinvasion of S. alterniflora may be facilitated by the fact that S. europaea does not transpire enough to maintain a high salinity in interstitial water. Thus, during the time that S. europaea occupies the space, salinity in the sediment resembles that of seawater. S. alterniflora that regrows in sites that continue to receive high N inputs becomes the taller version of this grass. In sites where nitrogen input is low, the whole process of replacement of S. europaea may be slower, and the resulting growth form of the grass is short S. alterniflora.

If all these events have taken place at fairly low elevations within the tidal range, the short form of S. alterniflora continues to exist. If, however, the events take place at somewhat higher elevation--that is, toward the top third of the tidal range--S. alterniflora is replaced by other grasses. If N inputs or availability are high, the species that dominates is D. spicata. If, on the other hand, the more common situation with lower N supply exists, the dominant species is S. patens.

Closure of this cycle of transformations is due to a process we have not yet mentioned. Throughout the salt marsh there is disturbance to stands of vegetation due to sea wrack. This material consists of plant litter that is floated and deposited by tides and may smother the vegetation underneath (Hartman, Caswell, and Valiela, in press). The stranding of wrack principally occurs high in the intertidal where, as depicted in Figure 20.7, stands of short S. alterniflora, S. patens, and Distichlis spicata may be thus turned into bare patches. We have unpublished evidence that the reinvasion of bare areas due to sea wrack disturbance may take 2 to 10 years (Cogswell and Hartman, unpublished data), depending on elevation, the length of time wrack is present, and nutrient input. This evidence is corroborated by studies of other types of disturbances in salt marshes (Beeftink 1979).

High marsh is much more subject to disturbances by wrack than is low marsh since the stranding of wrack tends to take place toward the high tide mark. High-marsh sites are considerably more diverse (at least 18 plant species in Great Sippewissett Marsh) than low marsh (about 8 species). The relative abundance of species individuals in high marsh is also much more evenly distributed than those of low marsh. As in other ecosystems--rainforests, coral reefs--some low level of disturbance may actually be associated with increased species diversity (Connell 1978).

The scheme of how a salt-marsh plant community may be structured (provided in Fig. 20.7) is preliminary and simplified, but it does provide some important notions in regard to long-term effects of sewage contamination:

1. Any eutrophication that occurs will affect the balance among the components of salt-marsh biota. Thus, to assess the consequences of wastewater contamination in any wetland, we need to know details of the interactions among the major species.
2. Many of the mechanisms involved in the transition from one kind of patch to another take time, from less than a year to perhaps several years. In the field, the changes that follow eutrophication may consist of a chain of such transitions such that there will necessarily be a continuing string of changes over considerably long periods of time. Assessment of the effects of eutrophication

therefore need to consider the possibility of such long-term changes.

3. Even at our highest dosage of sewage fertilizer, we have not found clearly detrimental effects to the plant species, even though the fertilizer used contains many elements and compounds that are potentially toxic. Perhaps salt-marsh plants, because of their adaptation to life in anaerobic substrates where, for example, metals are abundant, are tolerant of the presence of many kinds of toxic materials. Perhaps there are other reasons; the point is that the changes we see in Figure 20.7 were not brought about by any direct toxic effect but, rather, by alterations of common ecological processes: nutrient uptake, competition, grazing, and so forth.

4. Since our nutrient-enrichment experiments did not bring about the demise of organisms through toxic effects, it is very difficult to argue that this degree of eutrophication degrades a salt-marsh community. All the various combinations of components depicted in Figure 20.7 can be found in salt marshes in their natural state somewhere in New England and elsewhere. We can find places where any one of the components predominates or is very scarce. Our experimental eutrophication has merely shifted the balance among the components. Whether a particular parcel of marsh is degraded or not depends on anthropocentric criteria. For instance, if we value export of salt-marsh organic matter to coastal waters, we might feel that a marsh that has become dominated by Salicornia is a poor marsh since detritus from tissues of Salicornia is seldom exported. California marshes are typically dominated by glassworts and seldom export a significant amount of organic matter to coastal waters (Mauriello and Winfield 1978), in part because primary production is relatively low. Decay of Salicornia is rapid because the ligno-cellulose content of the plant biomass is low, so the pool of exportable detritus is low. Yet these Californian marshes have other important roles—for example, as stopovers along migratory lanes for shorebirds—and they are therefore no less valued by Californians.

There are many other important consequences of changes in vegetation type too numerous to deal with here. Suffice it to say that the rest of the food web—from grazers and predators to bacteria and fungi—are affected by the changes in vegetation documented here. Furthermore, the changes can be felt not only in the salt-marsh ecosystem itself, but also in adjoining coastal ecosystems since these exchanges of nutrients and particulate matter can be important (Valiela and Teal 1979; Valiela 1983).

Eutrophication can exert subtle but profound influence on the structure of communities, and many of the changes will take place over considerable periods of time, as shown above. Manipulative approaches, such as applied here, are the most effective way to gain insight into the mechanisms involved and furnish the evidence with which to predict at least some of the consequences of eutrophication.

ACKNOWLEDGMENTS

This research was carried out with support from National Science Foundation Grants GA-43009, OCE74-17859, DEB-7905127, DEB-8012437, NOAA Office of Sea Grant to the Woods Hole Oceanographic Institution (04-8-M01-149), the Victoria Foundation, and the Pew Memorial Foundation. This is Contribution No. 5202 from the Woods Hole Oceanographic Institution.

REFERENCES

Beeftink, W. G., 1979, The structure of salt marsh communities in relation to environmental disturbances, in Ecological processes in coastal environments, R. L. Jefferies and A. J. Davy, eds., Blackwell, Oxford, pp. 77-93.

Broome, S. W., W. W. Woodhouse, Jr., and E. D. Seneca, 1975, The relationship of mineral nutrients to growth of Spartina alterniflora in North Carolina, II: The effects of N, P, and Fe fertilizers, Soil Sci. Soc. Proc. 39:301-307.

Buchsbaum, R., I. Valiela, and J. M. Teal, 1982, Grazing by Canada geese and related aspects of the chemistry of salt marsh grass, Colonial Waterbirds 4:126-131.

Chalmers, S. G., 1979, The effects of fertilization on nitrogen distribution in a Spartina alterniflora salt marsh, Est. Coastal Mar. Sci. 8:327-337.

Connell, J. H., 1978, Diversity in tropical rainforests and coral reefs, Science 199:1302-1310.

Gallagher, J. L., 1975, Effect of an ammonium nitrate pulse on the growth and elemental composition of natural stands of Spartina alterniflora and Juncus roemerianus, Am. J. Bot. 62:644-648.

Haines, B. L., and E. L. Dunn, 1976, Growth and resource allocation responses of Spartina alterniflora Loisel. to three levels of NH_4-N, Fe, and NaCl in solution culture, Bot. Gaz. 137:224-230.

Hartman, J., H. Caswell, and I. Valiela, in press, Effects of wrack accumulation on salt marsh vegetation, Proceedings of 17th European marine biology symposium.

Howes, B. L., R. W. Howarth, J. M. Teal, and I. Valiela, 1981, Oxidation-reduction potential in a salt marsh: Spatial patterns and interactions with primary production, Limnol. Oceanogr. 26:350-360.

Jefferies, R. L., and N. Perkins, 1977, The effects on the vegetation of the additions of inorganic nutrients to salt marsh soils at Stiffkey, Norfolk, J. Ecol. 65:867-882. .

Mauriello, D., and T. Winfield, 1978, Nutrient exchange in the Tijuana estuary, Coastal Zone 78(3):2221-2238.

Mendelssohn, I. A., K. L. McKee, and W. H. Patrick, Jr., 1981, Oxygen deficiency in Spartina alterniflora roots: Metabolic adaptation to anoxia, Science 214:439-441.

Nixon, S. W., 1982, The ecology of New England high salt marshes: A community profile, FWS/OBS-81/55, Biological Sciences Program, U. S. Fish and Wildlife Service, Washington, D. C., 70p.

Patrick, W. H., Jr., and R. D. Delaune, 1976, Nitrogen and phosphorus utilization by Spartina alterniflora in a salt marsh in Barataria Bay, Louisiana, Estuar. Coastal Mar. Sci. 4:59-64.

Pigott, C. D., 1969, Influence of mineral nutrition on the zonation of flowering plants in coastal salt marshes, in Ecological aspects of mineral nutrition in plants, I. H. Rorison, ed., Blackwell, Oxford, pp. 25-35.

Sullivan, M. J., and F. C. Daiber, 1974, Response in production of cordgrass, Spartina alterniflora, to inorganic nitrogen and phosphorus fertilizer, Ches. Sci. 15:121-123.

Tyler, G., 1967, On the effect of phosphorus and nitrogen supplied to Baltic shore-meadow vegetation, Bot. Not. 120:433-447.

Valiela, I., 1983, Nitrogen in salt marsh ecosystems, in Nitrogen in the marine environment, E. J. Carpenter and D. G. Capone, eds., Nitrogen in the marine environment, Academic Press, New York, pp. 649-678.

Valiela, I., and J. M. Teal, 1979, The nitrogen budget of a salt marsh ecosystem, Nature 280:652-656.

Valiela, I., J. M. Teal, and W. G. Deuser, 1978, The nature of growth forms in the salt marsh grass Spartina alterniflora, Am. Nat. 112:461-470.

Valiela, I., J. M. Teal, and N. Y. Persson, 1976, Production and dynamics of experimentally enriched salt marsh vegetation: Belowground biomass, Limnol. Oceanogr. 21:245-252.

Valiela, I., J. M. Teal, and W. J. Sass, 1975, Production and dynamics of salt marsh vegetation and effect of sewage contamination, Biomass, production, and species composition, J. Appl. Ecol. 12:973-982.

Valiela, I., B. Howes, R. Howarth, A. Giblin, K. Foreman, J. M. Teal, and J. E. Hobbie, 1982, The regulation of primary production and decomposition in a salt marsh ecosystem, in Wetlands: Ecology and management, B. Gopal, R. E. Turner, R. G. Wetzel, and D. F. Whigham, eds., International Scientific Publications, Jaipur, India, pp. 151-168.

Vince, S. W., I. Valiela, and J. M. Teal, 1981, An experimental study of the structure of herbivorous insect communities in a salt marsh, Ecology 62:1662-1678.

Vince, S. W., I. Valiela, N. Backus, and J. M. Teal, 1976, Predation by the salt marsh killifish Fundulus heteroclitus (L.) in relation to prey size and habitat structure: Consequences for prey distribution and abundance, J. Exper. Mar. Biol. Ecol. 23:255-266.

DISCUSSION

Larson: I have a question about short and long forms of Spartina and the role of oxygen in making that determination. I am thinking of Mendelssohn's paper in Science recently [I. A. Mendelssohn, K. L. McKee, and W. H. Patrick, Jr., 1981, Science 214:439-441].

Valiela: Mendelssohn found very much what we reported in our article in Limnology and Oceanography--that is, that plants can oxidize sediments, maybe not by direct release of oxygen, but perhaps by the release of organic oxidants, perhaps glycolate. There is still some question as to what, in fact, the actual mechanism is. We have an increased oxidation in sediments where there are plant roots. And that, of course, changes the whole nature of the biogeochemistry of those sediments, as well as the plant growth form. At the same time, if you start off with an aerobic sediment, you get a much greater amount of growth.

Brinson: What do you suspect the effect of fertilization is on decomposition of peat accumulation?

Valiela: We have been studying the decomposition process in these plots for a number of years, and there's no single answer to that question. It's a very complicated set of processes. By and large, there's more production of biomass in our plots and that production

has to be decomposed. So, if you measure the active rate of decomposition—a tricky thing to do—you get a much faster decay rate in
fertilized plots. It doesn't necessarily mean that peat accumulation
increases because the peat accumulation depends primarily on the
phytochemical composition of the litter. There is some reduction in
the amount of lignin in fertilized grasses, but it would be hard to
see that difference after only two years worth of decay in a fertilized plot. It's a very slight amount; it's not a major thing.

Hemond: I want to point out that more vigorous plants may also
hydrologically increase the redox potential in sediments just through
creating negative partial pore pressures and developing the possibility of air entry. So, there may be an oxygenation of the sediments by that more or less passive mechanism associated with the more
vigorous plants.

21

Long-term Impacts of Agricultural Runoff in a Louisiana Swamp Forest

John W. Day, Jr. and G. Paul Kemp

This paper is a summary of a two-year research project on the dynamics of nutrient retention and release in a swamp receiving upland runoff (Kemp and Day 1981). The central objective was to estimate the capacity of this type of wetland for removing nutrients from upland runoff. We examined the role of redox in determining floodwater nutrient concentrations, both in the field and in laboratory microcosms, and tested the hypothesis that water quality deterioration in the region can be directly related to the cessation of overland water processing formerly performed by the swamp. An important aspect of the study was to determine the impacts of long-term loading by agricultural runoff.

STUDY SITE

The experimental site is a 320 ha section of swamp forest located in southern Louisiana (Fig. 21.1). Runoff enters the swamp system from the natural levees of the Mississippi River and Bayou Lafourche. Bald cypress (Taxodium distichium), water tupelo (Nyssa aquatica), and various bottomland hardwood species dominate the forest community. Swamp sediments are highly organic to a depth of 0.5 m to 1.0 m, where a laterally continuous alluvial clay lens is encountered that acts as a barrier to vertical groundwater movement. Locations for water sampling were chosen along a 2200 m transect extending from a field drainage ditch (Station 1) into a section of swamp ringed by levees to create two ponds that are managed for crayfish production. Water levels are artificially regulated in this two-pond system by pumps and weirs. From September to May, water is pumped into the 64 ha western pond from the Vacherie Canal (Station 2). It then flows through the 256 ha eastern pond before draining back into the Vacherie Canal. The ponds are drained during the summer.

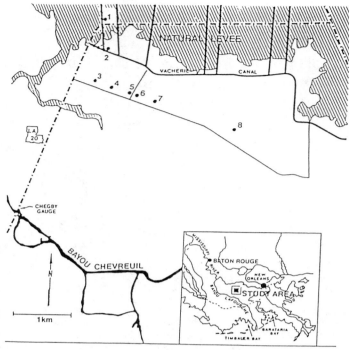

Figure 21.1
Study area-regional map showing relationship of crawfish farm to other
geomorphic features.

Table 21.1
Swamp Nutrient Input and Removal in g m^{-2} yr^{-1}

Nutrient	Pump	Precipitation	Total Input	Outflow	Out-In	Input Retained (-) or Exported (+) (%)
NH$_4$	0.46	0.16	0.62	0.25	-0.36	-58
NO$_3$	0.45	0.23	0.69	0.26	-0.43	-62
TIN	0.91	0.40	1.30	0.51	-0.79	-61
DON	3.16	0.24	3.40	4.13	+0.74	+22
PN	10.83	-	10.83	7.03	-3.80	-35
PO$_4$	0.71	0.02	0.73	0.83	+0.11	+15
TP	1.04	0.04	1.08	1.27	+0.20	+19
PP	3.12	-	3.12	1.24	-1.88	-60
ΣN	14.90	0.64	15.54	11.67	-3.87	-26
ΣP	4.16	0.04	4.20	2.51	-1.69	-41

METHODS

Water and nutrient budgets were calculated on the basis of measurements of water levels, pumping rates, precipitation, estimated evapotranspiration, and nutrient levels in the two ponds from September 1978 until August 1979. Water chemistry and litter decomposition were studied in microcosms under atmospheres of argon, air, and oxygen. Litter decomposition was also measured in the field. Details of methods are presented in Kemp and Day (1981).

RESULTS AND CONCLUSIONS

SWAMP NUTRIENT BUDGET

Approximately 10 metric tons of nitrogen (N) and 3 tons of phosphorus (P) were introduced into the 64 ha pond during the year studied. Of these totals, 70% of the N and 74% of the P were pumped from the canal in a particulate form (>0.45 μm) (Table 21.1). Atmospheric sources accounted for only 3% of the total N and 1% of the total P entering the pond; however, in terms of total inorganic N (TIN), the atmospheric contribution amounted to 30% (0.64 g N m^{-2}yr^{-1}) of the total input of these forms.

The swamp is a net nutrient sink, retaining 26% of the N and 41% of the P introduced annually. The settling of nutrient-rich particulate matter accounts for almost all of this effect. Working in an alluvial cypress swamp in southern Illinois, Mitsch and Ewel (1979) similarly found that more than 90% of the annual P input could be attributed to the deposition of P-rich sediment during periods of overbank flooding. In fact, although the swamp retains dissolved inorganic N (DIN) (0.79 g N m^{-2}yr^{-1}), it exports nearly the same amount of dissolved organic N (DON) (0.74 g N m^{-2}yr^{-1}), as well as total dissolved P (0.20 g P m^{-2}yr^{-1}).

The nutrient budget analysis indicates that the swamp removes dissolved inorganic N but adds dissolved PO_4 to waters passing over its surface. The atomic ratio of inorganic N and P, then, is a more sensitive index of swamp processing than concentration values alone. N:P ratios for swamp floodwaters ranged from 0.1:1 to 6.0:1 but clustered around a mean of 2:1 (SD = 1.8). Canal water, in comparison, ranged from 0.3 to 23.0:1 and showed considerable scatter around a mean of 6.0:1 (SD = 6.7). The swamp then acts to buffer the relative concentrations of inorganic N and P so that downstream systems receive water with a relatively stable inorganic nutrient composition.

MICROCOSM RESULTS

Litter decomposition was most rapid in the microcosms with an oxygen atmosphere. However, when nutrient levels are expressed on a per-gram basis, N and P increased while organic carbon (C) decreased (Fig. 21.2). The nutrient composition of the swamp sediment placed in the lab bottles did not change significantly over the 360 days of the experiment and can be considered the ultimate decompositional endpoint for the litter material. It is possible, then, to speculate

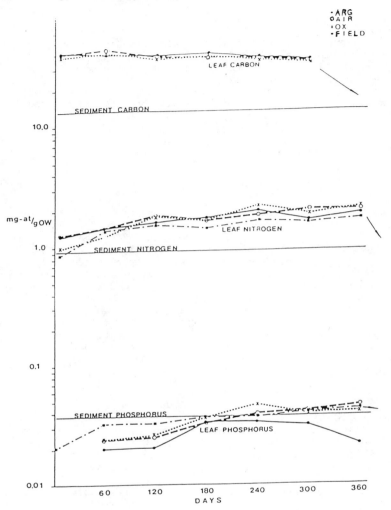

Figure 21.2
Carbon, nitrogen, and phosphorus concentrations of tupelo leaf litter
undergoing decomposition. Comparison between litter concentrations and
sediment values.

about long-term nutrient storage. Concentrations of N and C in the sediments are approximately half those of one-year-old litter material, whereas P concentrations in sediment and litter are roughly equivalent (Fig. 21.2). We might then expect, in a longer experiment, to see decreases in C and N concentrations in the litter material, but little change in P levels. It appears that in the swamp system, there is significant net microbial demand for N and C, but little for P.

EFFECTS OF DISSOLVED OXYGEN

There was a negative correlation between dissolved oxygen (DO) levels and P concentrations (Fig. 21.3). The exchange of PO_4 between the sediments and water column appears to be independent of any direct biological mediation. It is, however, indirectly influenced because biological activity controls sediment redox condition. Abiotic redox-dependent iron(II,III)-phosphate reactions have long been recognized as potential mechanisms for buffering PO_4 concentrations in natural waters and sediments (Mortimer 1941, 1942). Experiments with undisturbed sediment cores have quantitatively related exchange across the mud surface to redox condition, PO_4 concentration gradient, and sediment properties (Pomeroy, Smith, and Grant 1965; Stumm and Leckie 1971; Shukla et al. 1971; Li et al. 1972; Kamp-Nielsen 1974). Attention has focused recently on the reversible nature of PO_4 uptake and release by noncalcareous sediment suspensions (Patrick and Khalid 1974; Khalid, Patrick, Jr., and Delaune 1977; Holford and Patrick 1979). Our results, both in the field and in the laboratory microcosms, support this reversible mechanism. Swamp floodwater PO_4 concentrations respond rapidly and relatively predictably to redox-dependent changes in the capacity of the sediments to bind PO_4.

The swamp system exports excess PO_4. The lack of demand for PO_4 relative to NH_4 and NO_3 leads to the low N:P ratios of swamp-processed water. This situation explains why litter and sediment P levels are about the same. If the swamp forest is to be considered as a site for the processing of upland runoff, we need to understand, as best we can, (1) the nutrient pathways of the swamp under natural conditions, and (2) the long-term impacts of such a program.

THE SWAMP AS A BUFFER

The swamp acts as a buffer in time, concentration, and composition. Concentrations of DIN, particulate N, and particulate P were lower in swamp water than in input water, while levels of DON and dissolved P were somewhat higher in the swamp. The chemical composition of water overlying the swamp is different from that of the input water. The ratio of particulates to dissolved fractions for both N and P is much lower for the swamp water, and the composition of the particulate suspended matter in the swamp seems basically different from that in the input water. Particulates pumped into the swamp are inorganic silts and clays and phytoplankton cells. Particulates leaving the swamp appear to be made up primarily of organic materials generated within the swamp. Dissolved inorganic N:P ratios in swamp water are relatively constant at a value of about 2:1.

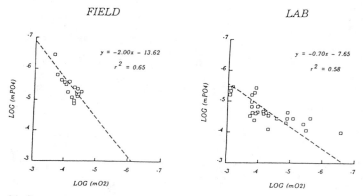

Figure 21.3
Comparison of phosphate/dissolved oxygen relationships developed from field and laboratory microcosm data.

Inorganic N:P ratios in channeled upland runoff that did not flow through the swamp averaged 6:1. Kemp and Day (1981) showed that this ratio is strongly influenced by rainfall and was as high as 22:1 following major runoff events.

It is apparent that a significant portion of N and P entering the swamp is retained. Nutrients exported to swamp bayous and lakes are released slowly over time.

LONG-TERM NUTRIENT PROCESSING

The results of this study are for one year. Can the rates of nutrient retention measured for this year be expected to hold for longer periods? Several workers have found that rates of retention and release may change or reach saturation after a short period of time. We believe that this swamp forest system, however, can continue to be a sink for N and P for long periods. The two major processes that act to enhance the role of the swamp as a sink are denitrification and sedimentation subsidence.

Recent research has shown that denitrification is an important avenue for "permanent" N loss in wetlands and shallow aquatic systems (Valiela and Teal 1979; Haines et al. 1977; Nixon 1979; Nixon et al. 1980; Boynton, Kemp, and Osborne 1980; Delaune and Patrick 1980). The closeness of the oxidized-reduced boundary to the sediment surface makes conditions for linkage between the nitrification and denitrification pathways favorable. The difference between N litter concentrations in one-year-old litter and sediments indicates that there is a further loss of N before the litter is incorporated into the sediments. Denitrification can explain this further loss.

The swamp forest, as part of the Mississippi River deltaic plain, is situated on a section of the Gulf Coast geosyncline that is undergoing rapid local and regional subsidence from the loading and compaction of alluvial sediments. Baumann (1980) measured a subsidence rate of 0.60 cm/yr at two sites in the swamp forest. Thus, sediments laid down in the swamp are rather rapidly incorporated into deep sediments and "permanently" lost in terms of current ecological dynamics.

CONCLUSIONS

1. Under overland flow conditions, the swamp can remove significant amounts of incoming nutrients: 21% of total N and 41% of total P were retained in the swamp.

2. Practically all of the removal takes place because of the settling of particulate N and P. Particulate nutrients entering the swamp are primarily associated with eroded upland soils and algal cells. Particulates leaving the swamp seem to be generated within the swamp.

3. For two reasons, it is not likely that the swamp will become saturated with N and P. First, the results indicate that denitrification is a significant pathway for the permanent loss of N. Second, the swamp is subsiding at a significant rate. This situation means that much of the suspended sediment that settles on the swamp floor will become buried. If these nutrients are not taken up by rooted vegetation, they will be permanently lost to the deep sediments.

4. In spite of nutrient retention in the swamp, significant amounts are still exported to swamp bayous and lakes. The swamp, however, acts as a buffer in time and composition, as well as in concentration. Nutrients are released slowly over a longer time, rather than in the erratic pulses associated with channeled runoff.

5. DO in the water column is the single most important factor determining sediment-water exchange of PO_4.

ACKNOWLEDGMENTS

The work on which this report is based was supported in part by funds provided by the U. S. Department of Interior, as authorized under the Water Research and Development Act of 1978, through the Louisiana Water Resources Research Institute, Grant No. A-043-2A.

REFERENCES

Baumann, R. H., 1980, Mechanisms of maintaining marsh elevation in a subsiding environment, M. S. thesis, Louisiana State University, Baton Rouge.

Boynton, W., W. Kemp, and C. Osborne, 1980, Nutrient fluxes across the sediment-water interface in the turbid zone of a coastal plain estuary, Estuarine perspectives, V. Kennedy, ed., Academic Press, New York, pp. 93-109.

Delaune, R., and W. Patrick, 1980, Nitrogen and phosphorus cycling in a Gulf Coast salt marsh, in Estuarine perspectives, V. Kennedy, ed., Academic Press, New York, pp. 143-151.

Haines, E. B., A. Chalmers, R. Hanson, and B. Sherr, 1977, Nitrogen pools and fluxes in a Georgia salt marsh, in Estuarine processes, vol. 2, M. Wiley, ed., Academic Press, New York, pp. 241-254.

Holford, I. C. R., and W. H. Patrick, Jr., 1979, Effects of reduction and pH changes on phosphate sorption and mobility in an acid soil, Soil Sci. Soc. Am. J. 43:292-297.

Kamp-Nielson, L., 1974, Mud-water exchange of phosphate and other ions in undisturbed sediment cores and factors affecting the exchange rates, Arch. Hydrobiol. 73:218-237.

Kemp, G. P., and J. W. Day, Jr., 1981, Floodwater nutrient processing in a Louisiana swamp forest receiving agricultural runoff, report A-043-LA, Louisiana Water Resources Research Institute, Louisiana State University, Baton Rouge, 60p.

Kemp, G. P., and J. W. Day, Jr., 1982, Nutrient dynamics in a Louisiana swamp receiving agricultural runoff, in Cypress swamps, K. E. Ewel and H. T. Odum, eds., University of Florida Press, Gainesville, in press.

Khalid, R. A., W. H. Patrick, Jr., and R. D. Delaune, 1977, Phosphorus sorption characteristics of flooded soils, Soil Sci. Soc. Am. J. 41:305-310.

Li, W. C., D. E. Armstrong, J. D. H. Williams, R. F. Harris, and J. K. Syers, 1972, Rate and extent of inorganic phosphate exchange in lake sediments, Soil Sci. Soc. Am. Proc. 36:279-285.

Mitsch, W. J., and K. C. Ewel, 1979, Comparative biomass and growth of cypress heads in north-central Florida, Am. Midl. Nat. 74:126-140.

Mortimer, C. H., 1941, The exchange of dissolved substances between mud and water in lakes: I and II, J. Ecol. 29:280-329.

Mortimer, C. H., 1942, The exchange of dissolved substances between mud and water in lakes: III and IV, J. Ecol. 30:147-201.

Nixon, S., 1979, Remineralization and nutrient cycling in coastal marine ecosystems, in Nutrient enrichment in estuaries, B. Neilson and L. Cronin, eds., Hamana Press, Clifton, N. J., pp. 111-138.

Nixon, S., J. Kelly, B. Furnas, and C. Oviatt, 1980, Phosphorus regeneration and the metabolism of coastal marine communities, in Marine benthic dynamics, K. Tenore and B. Coull, eds., J. South Carolina, Columbia, S. C.

Patrick, W. H., Jr., and R. A. Khalid, 1974, Phosphate release and sorption by soils and sediments, Science 186:53-55.

Pomeroy, L. R., E. E. Smith, and C. M. Grant, 1965, The exchange of phosphate between estuarine water and sediment, Limnol. Oceanogr. 10:167-172.

Shukla, S. S., J. K. Syers, J. D. H. Williams, D. E. Armstrong, and R. F. Harns, 1971, Sorption of inorganic phosphate by lake sediments, Soil Sci. Soc. Am. Proc. 35:244-249.

Stumm, W., and J. O. Leckie, 1971, Phosphate exchange with sediments: Its role in the productivity of surface waters, Proceedings of the fifth international water pollution conference, vol. 2, P. III-26/1-26/16, Pergamon Press, New York.

Valiela, I., and J. Teal, 1979, The nitrogen budget of a salt marsh ecosystem, Nature 280:652-656.

DISCUSSION

Brinson: What are the N:P ratios you have in the swamp water?

Day: 2 to 1.

Brinson: Is that total nitrogen to phosphorus?

Day: Inorganic. And they stay fairly close together, whereas they are highly variable in the agricultural runoff.

Brinson: Because the phosphorus is retained in the agricultural fields and the nitrogen is not?

Day: Yes.

Odum: Have you done anything with albedo? I noticed your aerial pictures. There was a pilot study done on vegetation at the Mississippi Test Center. Do you know if it was ever published or where it's available? I believe they characterized the signature for cypress versus other wetlands in Louisiana.

Day: I know who has the information, but it's never been published.

Odum: Do they discuss the possibility of determining the transpiration rate from the albedo?

Day: I think most of our wetlands fit at one end of the spectrum of what you find in Florida because of the climate. For example, we have more even rainfall and there are differences in soil types. We don't normally see continuous flooding.

Niering: I noticed that your slides had a different look in terms of the area where the water is going in and where it's going out. Have you checked the change in composition in this area?

Day: Yes, in fact, we have looked at that. Some of the difference in composition is due to the fact, in part, that the crayfish fishermen like to cut down some trees to try to encourage herbaceous growth, but I think they have stopped doing that in one end. Another thing is that there is a slight elevational difference across the study site and you can see the gradation from bottom-land hardwood species to cypress-tupelo. In general, the composition of that part of the forest looks very much like the control, because it's a mimic of a natural hydroperiod.

Niering: Do you see this as a compatible interrelationship between agriculture and the agriculture of wetlands?

Day: Yes. I think this could work for a long period of time. It's a close approximation of the type of hydroperiod that the swamps evolved with, and it is a reasonable way to treat this kind of nonpoint pollution that is practically impossible to treat in any other way.

Giblin: Do they use pesticides in the agricultural area?

Day: They use integrated pest management. The amount of pesticides put on the fields has been drastically reduced, and what is used disappears rather rapidly. We've done a couple of projects to look at that, and within three or four days of application, it is undetectable. A more important question is the herbicides. We haven't studied the long-term effects of those yet. The alternative to using this wetland is, of course, to just have upland runoff flow directly into water bodies. And I think that brings up a broader question that we may not have considered in our discussion. The alternative for this type of nonpoint pollution is to put it directly into receiving waters, as has been done in the past.

Niering: The potential long-term effects of herbicide use disturb me because findings from the Chesapeake Bay are pretty subtle and scary. I think that should really be looked at.

22

The Mississippi River Delta: A Natural Wastewater Treatment System

James G. Gosselink and Leila Gosselink

Wetlands have been shown to have considerable potential for wastewater treatment, but in only a few cases has it been possible to monitor the long-term effect of high nutrient-loading rates on wetland ecosystems (Kadlec and Kadlec 1979).

River deltas have evolved over millennia as the downstream receivers of runoff into rivers. Large rivers, especially, have high loading rates for sediments and sediment-related nutrients. The Mississippi River Delta is the largest natural wastewater treatment system in the United States; it receives runoff from more than one-half of the U. S. land area. Historically, as urban and industrial growth has occurred, the Mississippi River has become the conduit for elimination of wastes from a million households and thousands of industrial plants.

The present-day Mississippi River Delta region is about 10,000 years old. Its sediments contain a record of nutrient deposition that goes back to the end of the last glaciation. The hydrogen bomb fallout product Cesium-137 (^{137}Cs) has labelled the sediments and allows us to examine the last 30 years in detail. What can we learn from this record? Is the delta a useful analog for a municipal over-land flow system? If it is, what are the long-term characteristics of the delta, and what conditions make its activity effective? To answer these questions, we examined the nutrient concentrations and loading rates of river water flowing into the delta and the nutrient retention ability of several different kinds of delta environments.

CHARACTERISTICS OF INFLOWING MISSISSIPPI RIVER WATER

Figure 22.1 shows the seasonal concentrations of carbon (TOC), total nitrogen (N), total phosphorus (P) and suspended sediments (SS) in Mississippi River water close to the point at which it flows into the delta. One would expect the SS concentration to be closely related to water discharge, but the relationship appears to be casual. Perhaps this is so because 1980 was a low-water year without signif-icant flooding. Ordinarily, much of the suspended load is carried during the spring when currents associated with high water levels can carry large concentrations of sediment. The concentrations of

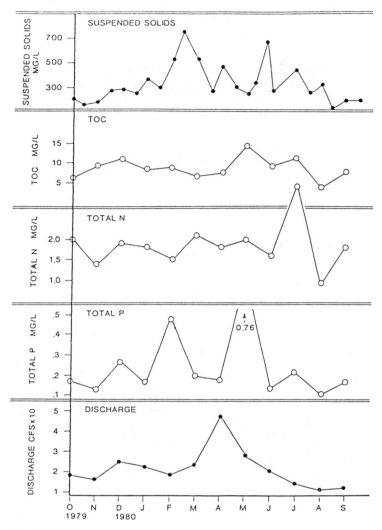

Figure 22.1
Seasonal water discharge into the Atchafalaya basin and
the concentrations of its suspended solids, organic
carbon, nitrogen and phosphorus.

nutrients are low compared with the inflow to selected municipal
overland flow systems (Table 22.1), and most of these nutrients are
adsorbed to suspended particles. Most treatment plants do not
process inorganic sediments so that the nutrients flowing into wetland treatment systems (after primary treatment to remove organic
solids) are primarily dissolved. The concentrations of nutrients in

Table 22.1
Concentrations of Chemicals in the Mississippi River at Melville,
Louisiana (Atchafalaya River), Compared to Selected Landflow
Treatment Systems

	Total N (mg/ℓ)	Total P (mg/ℓ)	TOC (mg/ℓ)	SS (mg/ℓ)
Atchafalaya River[a]	0.8-1.6	0.07-0.29	4-12	112-560
Domestic sewage treatment plants				
Michigan State Univ., Water Quality Management Program[b]	4.4-38	3.6-9.5	43-105	–
Tallahassee, Fla.[b]	25	10		–
Flushing Meadows, Ariz.[b]	30	10-15		–
Pennsylvania State Univ.[b]	2.6-30	0.25-4.75		–
Summary of 12, 2° treatment plants[c]	6.5-33.4	2.1-16.0		–
Rural storm-water runoff[c]				
Heavily fertilized, uncultivated	4.6-6.2	0.1-0.26		–
Unfertilized, uncultivated	0.05	0.17		–
Fertilized, plowed	84.6[4]	1.37[4]		–

[a] Water year 1980-81, USGS, 1981.
[b] USEPA 1976a.
[c] USEPA 1976b.
[d] 90%+ is associated with sediment.

Mississippi River water reflect the fact that most land runoff to the
river is from forested watersheds or uncultivated pastures.

In spite of the relatively low concentrations of nutrients, the
Mississippi River delivers large quantities of nutrients to the
delta, a reflection of the huge volume of water carried by the river.
For instance, about one-third of the combined flows of the Mississippi
and Red rivers runs through the Atchafalaya Basin, a 333,000-ha
bottomland swamp-aquatic system, before pouring into coastal Atchafalaya Bay. The loading rate to each hectare of this basin is shown
in Table 22.2 (assuming water flows across the whole basin), compared
with some more conventional, human-engineered systems. N and P loads
are considerably above those at the selected waste treatment systems.
Notice the enormous load of SS.

NUTRIENT RETENTION IN THE MISSISSIPPI RIVER DELTA

EVIDENCE FROM INPUT-OUTPUT BUDGETS: THE ATCHAFALAYA BASIN

Data for both water discharge and nutrient concentrations of
Mississippi River distributaries are inadequate for good input-output
budgets. The best data are for SS, as displayed in Figure 22.2 for
the Atchafalaya Basin. In the spring, if the annually retained
sediments were spread evenly over the basin, they would make a carpet
of about 3000 kg/ha, about 0.5-cm to 1-cm thick at a density commonly
found in coastal sediments. The basin has been steadily filling in
since the late 1800s, acting as a permanent sink. There is, of
course, considerable N and P associated with these deposited sediments, although input-output budgets indicate that less than 1% of
the nutrients are retained in the basin.

EVIDENCE FROM THE SOIL RECORD: ACCRETION STUDIES

Better evidence about nutrient retention in the delta comes from
sediment accretion studies. In recent years, a number of individuals
(Delaune, Patrick, Jr., and Buresh 1978; Baumann 1980; Redfield 1972;
Harrison and Bloom 1974; Muzyka 1976; Armentano and Woodwell 1975;
Richard 1978; Lord 1980) have studied the sedimentary record in
coastal wetlands. In many of these studies, accretion rates over the
last 30 years have been deduced from detection of the hydrogen bomb
fallout product ^{137}Cs in sediment cores. These data have been
supplemented by measurement of the rate of burial of marker horizons
over short periods of 1 to 5 years.

Salt Marshes

In the Mississippi River Delta, the most complete study of this
kind has come from the work of Delaune and his co-workers (Delaune,
Patrick, Jr., and Buresh 1978; Delaune, Reddy, and Patrick, Jr. 1981)
in the salt marsh. The measurements were made in an interdistrib-
utary basin (Barataria) that no longer receives the direct flow of
the Mississippi River. Nevertheless, the processes occurring here
seem characteristic for the Gulf Coast, although the rates may change
from location to location. Delaune, Patrick, Jr., and Buresh (1978),

Table 22.2
Loading Rates of Selected Chemicals to the Atchafalaya Basin Compared
with Selected Landflow Treatment Systems

	Total N (Kg/ha/yr)	Total P (Kg/ha/yr)	TOC (Kg/ha/yr)	SS (Kg/ha/yr)
Atchafalaya River Basin[a]	1060	150	5050	184,000
Michigan State Univ.[b]	325	113	–	–
Pennsylvania State Univ.[b]	198	44	–	–
Minnesota[c]	156	39	–	–
Houghton Lake, Mich.[d]	30	12	–	–

[a]Water year 1979-80. USGS 1981.

[b]USEPA 1976a.

[c]Farnham and Boelter 1976.

[d]Richardson et al. 1976.

331

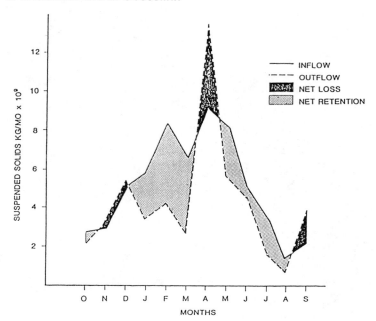

Figure 22.2
Seasonal influx and efflux of suspended sediments for the Atchafalaya
River Basin, 1980.

from ^{137}Cs studies, and Baumann (1980), from marker horizon studies,
found that accretion is occurring at the rate of about 0.75 cm/yr
(inland) to 1.35 cm/yr (streamside), and about 1.1 cm/yr in a nearby
shallow lake bottom. Deposited sediments change somewhat in nutrient
density as they are buried because of recycling by plants (Figure
22.3), and the dynamics of this recycling is element specific. In
about 30 years, nutrients deposited on the surface pass down through
and out of the root zone, into the permanent repository of the deep
marsh.
    Figure 22.4 summarizes N and P budgets for a streamside salt
marsh. Overflowing tidal water is the primary source of new
nutrients, although fixation of N from the atmosphere is also
significant. The surface flow of N and P through the system may be
much higher than indicated in Figure 22.4, but we have no way of
knowing this. Permanent retention of N and P amounts to about 210
kg/ha/yr and 15 kg/ha/yr, respectively.

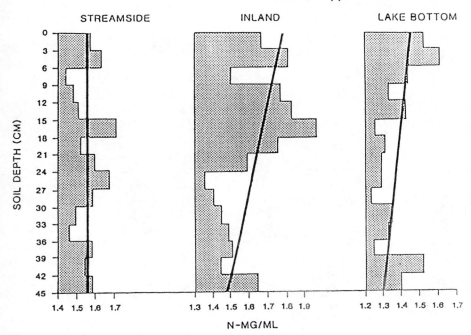

Figure 22.3
Depth profiles of nitrogen density in a Louisiana salt marsh.
(Data from Delaune et al. 1978.)

Other Wetlands

The salt marsh is not unique. Hatton (1981) measured accretion
rates in marshes along a salinity gradient. Table 22.3 summarizes
the nutrient accumulation rates he measured in salt, brackish,
intermediate, and fresh marshes. Accretion rates are fairly similar
in all these marshes. As salinity decreases, so does tidal activity.
Fresh marsh inundation is controlled more by local rainfall and winds
than by lunar tides. Flooding occurs less frequently in fresh
marshes than in salt marshes, but for longer durations (Gosselink,
Cordes, and Parsons 1979). Consequently, the flux of water across
the surface decreases with salinity, and this decrease is reflected
in much lower inorganic mineral transport and higher sediment organic
concentrations. While inorganic deposition decreases dramatically as
salinity decreases, N and P accumulation are much less affected.
Possibly, the longer contact time of waterborne nutrients with the
marsh surface in fresh and brackish marshes increases the efficiency
of their removal in comparison to salt marshes.

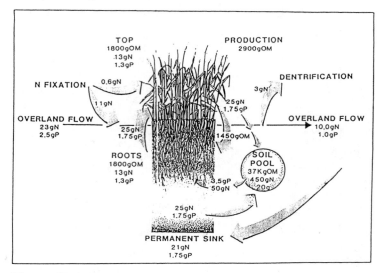

Figure 22.4
Nitrogen and phosphorus budgets of a Louisiana salt marsh.
(Data from Delaune et al. 1981.)

SUMMARY

The questions posed in the introduction to this article
concerned whether the Mississippi River Delta system is an analog for
municipal overland flow treatment systems and whether insights from
the river can be applied to human-engineered systems. The Mississippi River Delta does capture nutrients, and in significant quantities. The aptness of the analog, however, depends on how this
trapping occurs and, here, significant differences occur from most
conventional municipal overland flow treatment facilities.

1. The key to permanent, long-term retention of nutrients is
accretion. In the Mississippi River Delta, accumulation can occur
only because the coast is subsiding and the accretion rate is consequently high--somewhere around 1 cm/yr. This accretion cannot occur
indefinitely in the absence of subsidence because the wetland would
grow up out of the flood zone. Similarly, we surmise that in the
long run, a wetland wastewater treatment system must accrete to
continue to trap effectively any material other than C, N, and sulfur
(S).
2. In the Mississippi River Delta, the quantity of nutrient
retained is related both to the inorganic sediment input and to the
contact time of the flooding water with the marsh surface. The role
of inorganic sediments may be simply to act as a filter reservoir
that retains the nutrients. For a human-engineered system, then, the
question to be answered is how to use inorganic sediment efficiently

Table 22.3
Accretion Rates and Accumulation of Selected Chemicals
in Louisiana Marshes

Marsh Type	Mean Accretion Rate	Nutrient Accumulation (Kg/ha/yr)			
	(mm/yr)	TOC	N	P	Inorganic Minerals
Salt					
Streamside (S)	13.5	3,930	210	17	27,270[a]
Inland (I)	7.5	2,000	110	11	17,273
Brackish					
S	14.0	4,690	250	24	28,938
I	5.9	1,830	100	5	4,472
Intermediate					
S	13.5	4,150	280	15	16,416
I	6.4	1,540	110	4	2,464
Fresh					
S	10.6	2,500	160	10	7,134
I	6.5	1,450	90	5	4,290

Source: Data from Hatton 1981.
[a]Delaune et al. 1981.

in order to minimize the amount added each year and maximize its trapping effectiveness.

3. On a regional scale and long timeframe, the Mississippi River's strategy is to invade an area, trap nutrients and sediments as it grows, then abandon the site and start over again elsewhere. The old delta continues to function as it deteriorates, to support coastal geologic and biologic processes by a slow release of the sequestered sediments and nutrients. Eventually the same site is reinvaded by the river and the cycle starts over. At least two aspects of this strategy are interesting for human-engineered systems. First, would it be feasible or desirable to plan municipal overland flow systems in duplicate so that one field could be used to cleanse water while in the other, the water level is reduced to oxidize accumulated organics or farmed to use the trapped nutrients? Second, in natural deltaic systems, nutrient transformation is as important as permanent retention. In coastal areas, for instance, high fishery productivity depends on the transformation and release of nutrients to adjacent bays. Have we made the most of this possibility in engineered systems?

4. Finally, for Mississippi River Delta residents, there appear to be many opportunities for natural wetland wastewater treatment. Interior marshes in Louisiana need more sediments and nutrients since they are not presently maintaining themselves against the high rate of coastal subsidence (Gagliano, Meyer-Arendt, and Wicker 1981). The state has ambitious plans to nourish these marshes by diverting fresh river water into them. (Several potential sites are close to New Orleans.) Perhaps these diversions can be engineered with wastewater disposal for combined water treatment-marsh regeneration schemes.

REFERENCES

Armentano, T. V., and G. M. Woodwell, 1975, Sedimentation rates in a Long Island marsh determined by ^{210}Pb dating, Limnol. Oceanogr., 20:454-456.

Baumann, R. H., 1980, Mechanisms of maintaining marsh elevation in a subsiding environment, M. S. thesis, Louisiana State University, Baton Rouge.

Delaune, R. D., and W. H. Patrick, Jr., 1980, Nitrogen and phosphorus cycling in a Gulf Coast salt marsh, Proceedings 5th biennial international estuarine research conference, Jekyll Island, Ga., October 7-12, 1979, pp. 143-151.

Delaune, R. D., W. H. Patrick, Jr., and R. J. Buresh, 1978, Sedimentation rates determined by ^{137}Cs dating in a rapidly accreting salt marsh, Nature 275:532-533.

Delaune, R. D., C. N. Reddy, and W. H. Patrick, Jr., 1981, Accumulation of plant nutrients and heavy metals through sedimentation processes and accretion in a Louisiana salt marsh, Estuaries 4(4):328-334.

Farnham, R. S., and D. H. Boelter, 1976, Minnesota's peat resources: Their characteristics and use in sewage treatment, agriculture and energy, in Freshwater wetlands and sewage effluent disposal, proceedings of a national symposium, D. L. Tilton, R. H. Kadlec and C. J. Richardson, eds., May 10-11, 1976, University of Michigan, Wetlands Ecosystem Research Group, School of Natural Resources, Ann Arbor, Mich., pp. 241-255.

Gagliano, S. M., K. J. Meyer-Arendt, and K. M. Wicker, 1981, Land loss in the Mississippi River deltaic plain, Gulf Coast Assoc. of Geol. Soc. Trans. 39:295-300.

Gosselink, J. G., C. L. Cordes, and J. W. Parsons, 1979, An ecological characterization study of the Chenier Plain coastal ecosystem of Louisiana and Texas, 3 vols., FWS/OBS-78/9 through 78/11, U. S. Fish and Wildlife Service, Office of Biological Services, Washington, D. C.

Harrison, E. Z., and A. L. Bloom, 1974, The response of Connecticut salt marshes to the recent rise in sea level, Geol. Soc. Am. Abstr. with Programs 6:35-36.

Hatton, R. S., 1981, Aspects of marsh accretion and geochemistry: Barataria Basin, La., M. S. thesis, Louisiana State University, Baton Rouge.

Kadlec, R. H., and J. A. Kadlec, 1979, Wetlands and water quality, in Wetland functions and values: The state of our understanding, R. E. Greeson, J. R. Clark and J. E. Clark, eds., American Water Research Assoc., Minneapolis, Minn., pp. 436-456.

Lord, J. C., 1980, The chemistry and cycling of iron, manganese and sulfur in salt marsh sediments, Ph. D. dissertation, University of Delaware, Newark.

Muzyka, L. J., 1976, ^{210}Pb chronology in a core from the Flax Pond Marsh, Long Island, M. S. thesis, State University of New York (SUNY) at Stony Brook.

Redfield, A. C., 1972, Development of a New England salt marsh, Ecol. Monogr. 42:201-237.

Richard, G. A., 1978, Seasonal and environmental variations in sediment accretion in a Long Island salt marsh, Estuaries 1:29-35.

Richardson, C. J., W. A. Wentz, J. P. M. Channie, J. A. Kadlec, and D. L. Tilton, 1976, Plant growth, nutrient accumulation and decomposition in a central Michigan peatland used for effluent treatment, in Freshwater wetlands and sewage effluent disposal, proceedings of a national symposium, D. L. Tilton, R. H. Kadlec, and C. J. Richardson, eds., May 10-11, 1976, University of Michigan, Wetlands Ecosystem Research Group, School of Natural Resources, Ann Arbor, Mich., pp. 77-117.

U. S. Environmental Protection Agency, 1976a, Land treatment of municipal wastewater effluents--case histories, EPA tech. transfer seminar publication, Washington, D. C.

U. S. Environmental Protection Agency, 1976b, Land treatment of wastewater effluents--design factors--I, EPA tech. transfer seminar publication, Washington, D. C.

U. S. Geological Survey, 1981, Water resources data, Louisiana water year 1981, vol. 2, Southern Louisiana, U. S. Geological Survey water data report, LA-81-2, Baton Rouge.

23

Aging Phenomena in Wastewater Wetlands

Robert H. Kadlec

The treatment of wastewater by overland flow through a wetland has become a topic of scientific study during the last decade. Treatment systems are being established at natural wetlands and at wetlands specifically constructed for this purpose.

The prediction of the performance of a wetland treatment facility requires equations that describe both the response of the ecosystem to wastewater additions and the alteration of water quality. Since experience is somewhat limited, only the basic features of wetland processes are susceptible to meaningful analysis. These features include wetland hydrology and overland flow, removal rates for wastewater components, and the effects of nutrient additions on the continued ability of a wetland to treat wastewater.

Improved prediction techniques require consideration of individual phenomena and processes within the wetland. A larger body of reliable data is available on the function of wetland subsystems than on the performance of the wetland as a whole. Relatively simple models of significant ongoing processes make it possible to obtain further insight into the overall interactions between the wetland and applied wastewater. This procedure allows the synthesis of a conceptual model that, when represented in mathematical terms, can be used to evaluate aging phenomena.

To facilitate the use of a model over long periods of time (e.g., 20 to 50 years) a simple, specialized structure is desirable, as described in Figure 23.1. All transfers between the surface waters and the stationary ecosystem are taken as the annual net accumulation in each compartment. In this way, cycling of nutrients and other materials on a seasonal or even shorter-term basis does not complicate the model.

Removal of dissolved nutrients from surface waters is controlled by a two-step process: delivery and consumption. _Delivery_ is accomplished by convective mass transfer within surface waters, overland flow, or by downward flow due to water infiltration. _Consumption_ occurs principally at the surfaces of the soil, litter, plant stems, and algal mat. It consists collectively of a number of processes that initially are relatively fast, some of which slow considerably as wastewater treatment continues. Sorption will reach an equilibrium in the upper soil horizons, reducing the average areal uptake

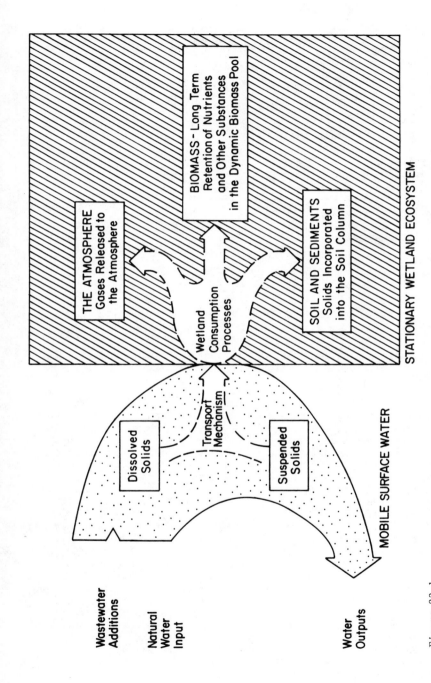

Figure 23.1
Simplified compartmental model for use in wetland treatment system design.

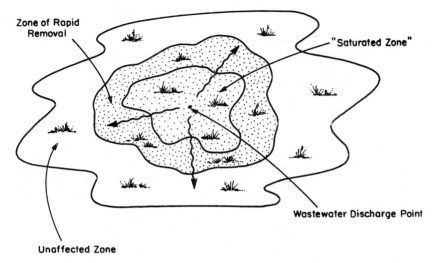

Figure 23.2
Schematic of the zone of affected soil and biomass.

rate. Biomass expansion, which offers a sink for nutrients, will
also reach a saturation condition where the release of nutrients due
to litter decay offsets any uptake in new growth. Woody biomass
production allows longer immobilization of nutrients and constitutes
a relatively permanent removal mechanism. Soil production also
represents a long-term removal process but is quite slow. While data
are extremely sparse, the same basic behavior can be anticipated for
heavy metals as well as nutrients.

Two regimes will exist in a wastewater wetland system for each
wastewater component considered, as shown in Figure 23.2. In the
vicinity of the wastewater discharge, a saturated region will exist.
Here, component removal rates will be quite slow, and a function of
the uptake rates, due to (1) sorption deep in the soil column, (2)
incorporation of material into new soil and woody plants, and (3)
microbial release of gases to the atmosphere. Outside this saturated
region, surface-water concentrations of wastewater components will
drop exponentially with distance. In this outer zone of rapid
removal, it is the transport of dissolved components through the
surface waters that limits the overall rate. The amount of wetland
area involved in this outer zone will be determined by mass transfer
considerations, and for stable discharge conditions (depth, velocity,
etc.), will not change. The inner zone in the saturated region will
continue removal at a rate that is slower but insensitive to modest
changes in water flow or depth. The expansion of the saturated
region, and consequently the surrounding unsaturated zone, will
continue until the total affected area is sufficient to allow all
incoming wastewater components to be removed by water infiltration,

incorporation into new soil and woody biomass, or release to the atmosphere. If the actual wetland area is less than that required for total retention of pollutants, breakthrough will occur. In this case, only a portion of the wastewater components fed to the wetland will be retained, and collection efficiency will drop sharply.

Harvesting plant biomass is a direct method of preventing saturation of the biomass compartment. Nitrogen, phosphorus and other wastewater components can be removed from the wetland system. Higher removal rates can be maintained indefinitely on a limited wetland area using this technique.

To employ this conceptual model in the evaluation of wetland system designs, it must be cast in mathematical terms.

THE MASS TRANSFER ZONE

When wastewater is caused to flow over the surface of a wetland, nutrients and other pollutants are removed, primarily by delivery to and consumption at solid surfaces. At the wetland surfaces, sorption and microbial processes may occur, as well as plant uptake. Algal and duckweed uptake may occur at the upper water surface. These processes require that each contaminant be transported to a bounding channel surface.

A typical relationship to describe this transport is:

$$\underline{N} = \underline{kA} \, (\underline{C}_w - \underline{C}_s)$$

(For a description of mathematical terms, see the Notation section following.) Such a rate expression must be coupled with mass balances for the contaminant and for the surface water to predict the distance or time for the removal of a dissolved substance. The contaminant mass balance is, for a linear-flow wetland:

$$\underline{v}\phi_s \underline{h} \, \frac{d \, C_w}{dx} = -\underline{k}\phi(\underline{C}_w - \underline{C}_s) - \underline{i} \, \underline{C}_s + \underline{p} \, \underline{C}_p - \underline{e} \, \underline{C}_e$$

At some upstream point, $\underline{x} = 0$, the concentration will be known ($\underline{C}_w = \underline{C}_f$). Presumably, this point is either the wastewater discharge point or, if the area around the discharge is saturated, it will be the outer edge of the slow removal zone. For example, the zone of rapid removal should begin near the discharge line for denitrification (NO_3^- removal) since saturation is unlikely. For an older facility, the rapid uptake zone for phosphorus would begin at a distance where saturation ends.

This model may be easily solved in a variety of special cases. It works well for those sites where transect data are available with which to validate it. The mass transfer parameter, k, and the depth-velocity relation are site specific.

THE LOADED ZONE

The addition of nutrients to a natural or newly constructed wetland will cause a zone of increased vegetative growth to appear. This zone will expand with time until either the permanent capacity of the zone equals the loading rate or the loaded zone reaches the boundaries of the wetland. The mathematical description of the advance of such a loaded zone for each substance of interest, such as phosphorus, consists of a mass balance on the zone and the rates at which the substance is taken up by the stationary ecosystem. These uptake rates fall in three general categories. The first is a permanent binding of the substance in question or a gaseous loss to the atmosphere, such as denitrification. In these cases, a component is permanently removed from surface waters. A second category consists of an increase in the sorbed quantity of the substance, which is the physical or chemical binding of the substance to the soil substrate within the wetland. Such processes are known to occur for phosphorus and ammonia, for example. The third category of nutrient consumption is storage in an expanding biomass compartment.

Other uptake/release processes also occur in the wetland ecosystem at rapid rates (compared to the annual averages considered here). An example is the uptake of phosphorus and nitrogen by algae. These algae grow during the summer months, die, and contribute a certain amount of algal litter to the sediment layers within the wetland. These algal sediments decompose and rerelease the nitrogen and phosphorus that was incorporated in the biomass. This process is fairly rapid in the summer months and, if one considers only year-to-year variations in area, too fast to be noticeable. There is little effect of such rapid cycling. Thus, in the model development that follows, only those processes that persist for a period greater than one growing season are considered as long-term consumers for nutrients. All quantities are expressed as rates, but these rates are averages over the period of one year or longer.

It is assumed that the loaded zone is interior to the wetland and the mass transfer zone. The zone of mass transfer limitation, described earlier, operates in addition to the loaded zone. A definite line of demarcation between the loaded and unloaded zones is presumed. This presumption is not entirely accurate since according to the principles of mass transfer, a zone must be present in which nutrient levels within surface water are decreasing.

In terms of the possible sinks, the mass balance equation for the loaded zone is, in words

addition rate = sorption rate + permanent removal rate

+ temporary binding rate

+ discharge rate

A variety of ways of expressing these terms exists; one set of choices leads to the following mass balance for the expanding loaded zone:

$$QC_i = KyC_\ell \frac{dA}{dt} + (r_s C_{sn} + r_A + r_w X_w + r_H X_H)A$$
$$+ X_L \int_0^t Fe^{-\alpha(t-\tau)} \frac{dA}{d\tau} d\tau$$

System parameters were estimated for operation of the Houghton Lake treatment site. Utilizing this mass balance equation, the

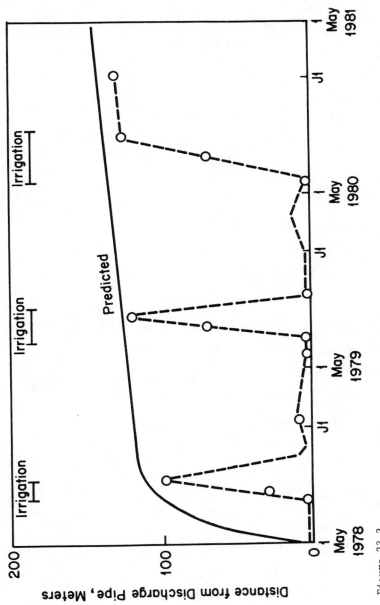

Figure 23.3
Movement of nitrogen (NH_4^+-N) concentration fronts in surface waters, $\underline{C}$ = 1 mg/1, Houghton Lake treatment site.

predicted nitrogen-front progression was calculated. The observed system behavior and the prediction for the nitrogen-front movement are shown in Figure 23.3.

The expansion of the saturated zones about the discharge point have been found to be much as predicted by the material balance, considering only the principal mechanisms. The aging of the Houghton Lake site can so far be described by this model.

ACKNOWLEDGMENT

This study was performed under Cooperative Agreement CR 807541-01-0, Robert S. Kerr Laboratory, U. S. EPA, Ada, Oklahoma.

NOTATION

$\underline{A}$ = area of surface, m^2

$\underline{C}_e$ = contaminant concentration in evapotranspired water, gm/m^3

$\underline{C}_f$ = contaminant concentration in wastewater, gm/m^3

$\underline{C}_i$ = mass average concentration of contaminant or nutrient in influent wastewater, gm/m^3

$\underline{C}_p$ = contaminant concentration in precipitation, gm/m^3

$\underline{C}_s$ = contaminant concentration in water at channel surface, gm/m^3

$\underline{C}_{sn}$ = contaminant concentration in new soil, gm/m^3

$\underline{C}_w$ = contaminant concentration in water, gm/m^3

$\underline{C}_\ell$ = average contaminant concentration in surface water, gm/m^3

$\underline{e}$ = evapotranspiration, m/day

$\underline{F}$ = average annual excess litter fall, $gm/m^2/yr$

$\underline{h}$ = depth, m

$\underline{i}$ = infiltration, m/day

$\underline{k}$ = mass transfer coefficient, m/day

$\underline{K}$ = sorption equilibrium constant, $(gm/m^3)_s$ / $(gm/m^3)_\ell$

$\underline{L}$ = amount of excess litter, mg/m^2

$\underline{N}$ = contaminant transport rate, gm/day

$\underline{p}$ = precipitation, m/day

$\underline{Q}$ = average annual wastewater addition rate, m^3/yr

$\underline{r}_A$ = excess average annual rate of loss to atmosphere, $gm/m^2/yr$

$\underline{r}_H$ = average annual harvest rate, $gm/m^2/yr$

$\underline{r}_s$ = excess average annual soil accretion rate, m/yr

$\underline{r}_w$ = excess average annual woody stem accumulation rate, $gm/m^2/yr$

$\underline{t}$ = time, years or days

$\underline{v}$ = actual velocity, m/day

$\underline{x}$ = distance, m

$\underline{X}_H$ = fraction contaminant in harvested biomass

$\underline{X}_L$ = fraction contaminant in remaining litter

$\underline{X}_W$ = excess average annual woody stem accumulation rate, $gm/m^2/yr$

$\underline{y}$ = average sorption depth, m

α = average annual specific litter decay rate, yr^{-1}

ϕ = flow porosity across the wetland surface

ϕ_s = storage porosity

τ = time, yr

REFERENCES

Association of Bay Area Governments, 1982, The use of wetlands for water pollution control, report to U. S. EPA.

Good, R. E., D. F. Whigham, and R. L. Simpson, eds., 1978, Freshwater wetlands, Academic Press, New York, 378p.

Greeson, P. E., J. R. Clark, and J. E. Clark, eds., 1978, Wetland functions and values: The state of our understanding, American Water Resources Association, Minneapolis, Minn., 674p.

Hammer, D. E., and R. H. Kadlec, 1982, Design principles for wetland treatment systems, PB83-188722, report to U. S. EPA, 244p.

Reed, S. C., and R. K. Bastian, eds., 1979, Aquaculture systems for wastewater treatment: Seminar proceedings and engineering assessment, 430/9-80-006, U. S. EPA, Office of Water Program Operations, Washington, D. C., 485p.

Richardson, B., ed., 1981, Selected proceedings of the Midwest conference on wetland values and management, Minnesota Water Planning Board, St. Paul, 660p.

WAPORA, Inc., in preparation, The effects of wastewater treatment facilities on wetlands in the Midwest, report to U. S. EPA, Region 5, Chicago, Ill.

Waste and Water International, in preparation, Emerging technology assessment: Wetlands wastewater treatment systems, Contract 68-03-3016.

DISCUSSION

Brinson: That original equation that you described, where you had three sinks: Which was the most important from the standpoint of permanent removal?

Kadlec: If you put in some numbers relevant to the Houghton Lake situation, what you would find is that in the first year, sorption is 20% or 30% of the total; expanding biomass is 70%. As time progresses, the value of the expanding biomass compartment is less and less. Sorption is less and less. Finally, you reach the state where the permanent removal mechanisms are the only thing remaining. Another point is that there's no guarantee in any of this that you have sufficient wetland area; and if you don't, then you get a break-through of a concentration front. I was pleased, in some ways, to note that at the Belaire site last year, that is precisely what happened. In other words, in previous years there was phosphorus removal at the level we had expected. But some of the characteristics of approaching that saturation condition from the interior measurements were observed and, lo and behold, no further removal, and the front broke out. That is also useful design information.

Hemond: In your first-order phosphorus removal model from the water column, you have a concentration in the water and a concentration associated with what I take to be a sediment surface. Is there a sensitivity that you observed or would expect; sensitivity to original sediment depth, not the increase resulting from expanded biomass department? Or, in your sites, does water not move into the sediments and carry things with it?

Kadlec: Another restriction of this talk that I didn't have time to announce is that we are dealing with overland flow. This could also be modified by infiltration, as is indicated. In those situations, we have found that only about the upper 5 cm of the soil column is affected--there is a litter layer and then a soil layer; out of 1 m to 2 m of peat, only about 5 cm penetration in five years is all we've gotten. That is why I made the simplification of a single depth absorption. Penetration is very slow by diffusion processes. So, in this system, downward flow is not important; I don't think downward diffusion is ever important in these tightly packed peat soils.

Hemond: Does this imply that there is a way of enhancing nutrient removal performance by redirection, by inducing infiltration, by creating a downward transport process so that you can utilize the rest of that peat?

Kadlec: Of course. Some of these ideas are explored in a paper that was done for EPA and that they will be issuing soon, but if you can get anything to go down, you are cutting your area requirements by whatever it is that you can move to lower soil horizons.

Bayley: How does this litter affect short-term sporadic events, let's say, a 1-in-20-year drought?

Kadlec: There are lots of things that this paper ignores; I think it safely ignores some things, but others you would have to calculate on the computer--for a 20-year drought, I think you'd have to do that. But, some other things--what about algal cycling? We know that algal cycling is probably just as important as the cycling through vascular plants; however, it also occurs within the growing season. When you're trying to depict things now and ten years from now, it's pretty safe to just compute annual averages. Yes, you will have picked up the annual cycle of things that decompose fast, but if they're here today and gone tomorrow, then the net effect on the biomass compartment is zero. But not the 20-year drought. So, when ignoring the seasonal variations in order to calculate long-term trends, it will be extremely difficult to try to get all the detail. In your 20-year drought, you'd have to program in the hydrology; you'd have to change velocities, and so forth, for that year.

Kaczynski: For your long-term steady state requirement for phosphorus removal, how did that compare to the area requirements for just water application? Water application is what we've had to use so far as our criteria in design, but here I see you can get at the area requirement for steady-state phosphorus removal.

Kadlec: The question in most years in this part of the world is not applicable because evaporation and transpiration just about balance precipitation and, consequently--if I understood you right--there is no dry-out area. The whole place is wet all the time and so all the water makes it through.

Kaczynski: Okay. But, the area where all the water makes it through--how did that area compare with the steady state area required?

Kadlec: At this particular site?

Kaczynski: For phosphorus removal over the long term, as an example.

Kadlec: As a matter of fact, the area is about two-thirds of the steady state area required, or some 800 or 900 acres for the Houghton Lake community if we understand the burial rate, and there's some question about that. By the way, for that burial rate, John Day mentioned 1 cm/yr. We've used the Cesium-137 technique, analyzed the Carbon-12 data for bog sediments and believe it's close to 1 mm/yr or 2 mm/yr, naturally. We do not know what effect, if any, we have had on that by fertilization. The odds are that it will go up a bit.

Ewel: These stochastic events, such as muskrats, fires, droughts, and so on, that we tend to regard as catastrophes when we look at a wetland for five years or so, are really cycles. It would be extremely interesting to put a stochastic element in a model like yours and see how much variability results, by incorporating two-year cycles or four-year cycles.

Kadlec: That's absolutely right. There are doctoral students preparing computer code to handle more complicated cases because it is not an easy computation and you do need to use the computer.

SESSION VII

Management

24

Ecological Analysis of Wastewater Management Criteria in Wetland Ecosystems

Curtis J. Richardson and Dale S. Nichols

Considerable interest has been focused on the use of wetlands to treat secondary municipal sewage effluent further (Valiela et al. 1975; Tilton, Kadlec, and Richardson 1976; Tilton and Kadlec 1979; Odum and Ewel 1980; Lyons and Benforado 1981). Reports on the effectiveness of wetlands in removing nitrogen (N) and phosphorus (P) from effluent have been encouraging (Richardson et al. 1976b; Tourbier and Pierson 1976; Kadlec 1979b; Kadlec and Tilton 1979; Whigham and Bayley 1979), but questions persist about sorption capacity, uptake rates (Richardson et al. 1978), and long-term capacities of wetland ecosystems to remove nutrients (Richardson 1981; Nichols 1983). In this paper, we present a series of ecological management criteria that should be addressed prior to the decision to use any wetland ecosystem for treatment of secondary municipal effluent. These criteria include the value of the effluent as a resource, the capacities and limitations of wetlands to accomplish wastewater treatment, wastewater management objectives, wastewater suitability for wetland discharge, and wetland values. Also presented are discussions of wetland hydrology, productivity, cycling of nutrients and heavy metals, and estimates of efficiencies of wastewater nutrient removal by wetlands and the wetland area needed for specific levels of nutrient removal.

EVALUATION CRITERIA

When considering municipal effluent application to a wetland, it is essential that the manager have a holistic view of the wastewater problem. We present some criteria that should be addressed before any decision is reached on a wastewater management system (see Table 24.1).

EFFLUENT VALUE

The first consideration is the value of the effluent itself. The use of wetlands to dispose of the wastewater or to remove nutrients (primarily N and P) from the effluent implies that these

Table 24.1
Management Criteria That Should Be Considered Prior to the Selection
and Utilization of a Wetland for Wastewater Discharge

1. Consider the value of the effluent as a resource.

2. Determine the wastewater management objectives of the treatment
   facility.

3. Determine the suitability of the wastewater for wetland
   discharge (consider presence of heavy metals, etc.).

4. Assess the capacity of the wetland to accomplish the desired
   treatment. In particular consider:

   Hydraulic loading,

   Nutrient loading, and

   Amount of wetland area required.

5. Compare wetland application to other treatment options
   (irrigation, overland flow, etc.) in terms of:

   degree of treatment accomplished,

   cost (not discussed in the paper, see Bastian
   and Reed 1979),

   energy requirements (see Bastian and Reed 1979), and

   legal implications (see Bastian and Reed 1979).

6. Determine all the values of the wetland system selected as a
   potential discharge site and the effects of disposal on
   these values.

7. Assess environmental impacts (ecosystem disruption, species
   losses, insect problems, disease vectors, odors, downstream
   impacts, etc.).

8. Select the best treatment system based on integrated management
   principles (ecological-economic-ethical) that maintain water
   quality, ecosystem value, sustained yield from ecosystem
   processes, option value to future generations and economic
   reality.

materials are detrimental rather than potentially valuable. It is well known that N fertilizer is costly to produce in terms of energy, P is a scarce resource, and water is in short supply in many areas of the country for agriculture and industry (Ehrlich, Ehrlich, and Holdren 1977). Hence, municipal effluent that is high in N and P content and low in heavy metals and other toxic materials may constitute a useful resource. Our contention is that the value of the wastewater first be considered in terms of fertilizer and irrigation potential on agricultural, forest, or marginal lands, following the recommendations of Sopper and Kardos (1973) and Sopper and Kerr (1979).

MANAGEMENT OBJECTIVES:
WHY PUT WASTEWATER IN A WETLAND?

The decision to discharge wastewater into a wetland should be preceded by an assessment of the treatment objectives and the capacity of the wetland to accomplish these objectives. If it is desired to reduce the effluent concentrations of biodegradable organic materials constituting a biochemical oxygen demand (BOD), suspended solids, nutrients, or some combination of these pollutants, it should be determined whether wetland size, hydrologic characteristics, vegetative community types, and adsorption capacities are amenable to the task. Overall effluent quality, nutrient and hydraulic loadings, and the presence of heavy metals, pesticides, and other toxic materials must also be considered. If the purpose of the wetland discharge is simply to put the effluent where it will do the least harm, then a quantitative evaluation should be made of the impacts of the discharge on the wetland compared to impacts on other potential discharge sites.

The following information on municipal effluent discharges gives some indication of the magnitude of current releases of wastewater into wetland areas in two regions of the United States. Discharges of municipal effluents into wetlands have recently been inventoried under the auspices of the U. S. Environmental Protection Agency (EPA) in Region 5 (Midwest) (U. S. EPA 1983) and Region 4 (Southeast) (Claude Terry and Associates, Inc. 1982). Some results of these inventories are given in Table 24.2. In the Midwest, 96 registered sites have discharged an average of 1.5 million litres per day (1/day) for an average of nearly 20 years, primarily in marsh and swamp wetlands averaging approximately 16 ha in size. The largest discharge site (284 ha) is the Houghton Lake research site in Michigan. The longest period of continuous discharge has been 60 years in Minnesota and Indiana. The quantity of discharge for most states has averaged nearly 1.5 million 1/day but the range has been from a few thousand litres per day at small package plants to over 3.8 million 1/day for larger municipal or industrial sites. Seventy-five percent of the 96 discharges to wetlands have been from municipal facilities, 17% from commercial, and 6% from a variety of specialized uses. In the Southeast, 224 registered sites were recorded in the eight states. Creek swamps were the most predominant discharge sites after secondary treatment. No data were given on the size of the wetlands used. The average number of years of discharge was 19, with a few sites having been in operation for nearly 100 years. The average quantity of discharge at 4.2 million 1/day has

Table 24.2
Survey of the Known Wetland Discharge Sites by Region and State

Region 5 (EPA)	Discharge Number	Wetland Type	Area of Wetland (ha) (maximum)	Mean Years of Discharge (longest)	Quantity of Discharge[a] (range)
Midwest states					
Illinois	6	Marsh, swamp, fen	11(32)	16(32)	1.1(0.002 – 4.2)
Indiana	9	Marsh, swamp, hardwood Forest	16(61)	21(60)	1.9(0.012 – 5.7)
Michigan	7	Marsh, swamp, fen	93(284)	7(10)	2.3(0.15 – 5.7)
Minnesota	35	Marsh, swamp, bog	19(81)	24(60)	0.8(0.01 – 4.5)
Ohio	3	Slough, marsh, bottomland	81(81)	33(42)	1.1(0.68 – 1.5)
Wisconsin	36	Marsh, swamp, bog	17(41)	15(50)	0.8(0.001 – 15.1)
Subtotal	96	Means	16	>20	1.5
Region 4 (EPA)					
Southeastern states					
Alabama	13	Creek swamps, marshes, cypress	—	18(40)	3.4(0.11 – 11.3)
Florida	54	Cypress domes, swamps, marshes	—	22(90)	6.0(0.002 – 18.9)
Georgia	10	Marsh, cypress, creek swamps	—	22(25)	4.2(2.5 – 6.0)
Kentucky	0	—	—	-	
Mississippi	40	Bayous, sloughs	—	18(33)	1.1(0.19 – 2.3)
North Carolina	61	River and creek swamps	—	14(26)	7.6(0.08 – 113)
South Carolina	34	Creek swamps	—	19(100)	3.8(0.06 – 18.9)
Tennessee	12	River and creek swamps	—	18(29)	2.6(0.11 – 10.2)
Subtotal	224	Means	—	19	4.2

Source: Data from U. S. EPA 1983; Claude Terry and Associates 1982.

[a]Million 1/day.

been more than double that found for the Midwest. A wide range of
discharge rates was also found for this region, with larger industrial discharges (>11.3 million l/day) being found in Alabama,
Florida, North Carolina, South Carolina, and Tennessee.

Wetlands have been and are being used for wastewater discharge,
sometimes by design with specific objectives in mind and sometimes
arbitrarily. A variety of reasons were cited (U. S. EPA 1983; Claude
Terry and Associates, Inc. 1982) for the use of these wetlands for
effluent discharge:

1. Discharge directly into a river not allowed by authorities;
2. No other ecosystem locally available;
3. Low-cost or free land, often owned by state or local
    government;
4. Wetland considered to be a worthless area;
5. Convenience; and
6. A belief that wetlands have high filtration and
    storage capacities for nutrients, heavy metals, and water.

Outside of the research sites in several states, few data existed on
the actual efficiency of these wetlands to further treat effluent.
Local information on the problems of wetland discharge were mostly
qualitative and included statements on:

1. Excessive odor,
2. Increased algal growth,
3. Decreased water quality of wetland outflow,
4. Heavy-metal accumulation problems,
5. Botulism in gamebirds, and
6. Nutrient discharge standards exceeded by nutrient
    overloading of wetlands.

In general, most local officials thought wetlands worked if low
loading rates and large enough wetland systems were available. Site-
specific problems were reported to each state. These problems might
have been reduced if a prior analysis had been undertaken following
suggestions given in Table 24.1.

SUITABILITY OF WASTEWATER FOR WETLAND DISCHARGE

Certain effluent characteristics may present problems of large
enough magnitude to preclude the use of a wetland ecosystem for
wastewater discharge. These include excessive volumes of water,
excessive nutrient loads, high pH, high concentrations of heavy
metals, and chlorination.

High volumes of wastewater with consequent rapid flow rates
through the wetland and short retention times lead to poor removal of
nutrients, BOD, SS, and the like, and poor effluent quality. Excessive hydraulic loadings can also raise the wetland water level significantly. Although wetland vegetation is adapted to wet conditions,
the requirements of certain species are often critical, and an
increase in the water level can damage or kill existing vegetation
and cause a shift to another, possibly less desirable or less
valuable, vegetative type. This sensitivity to water level can be
seen, for example, in conifer swamps (Thuja, Larix, and Picea

peatlands) in the northern Great Lake states, where hindered drainage due to topography (<1.5 m of slope/km), road or pipeline construction, beaver activity, and so forth, reduces or prevents timber growth (Boelter and Close 1974; Boelter and Verry 1977). Boelter (1972) has also reported that an increase in the water depth on northern peatlands causes a decrease in the abundance of woody species, such as Chamaedaphne calyculata (L.), and increases abundance of sedges, grasses, and sphagnum.

At high nutrient-loading rates, nutrient removal by wetlands is poor. In addition, excessive amounts of nutrients can adversely affect wetlands by causing changes in plant species composition (Richardson et al. 1976a; Wentz 1976; Goffeng, Saebo, and Haugen 1979; Guntenspergen and Stearns 1981), promoting nuisance growths of algae (Schwegler 1978), with resultant odor problems and oxygen depletion following die-off, and altering peat accumulation rates (Coulson and Butterfield 1978). High-pH wastewaters have the potential to dissolve organic wetland soils and leach them away. This phenomenon has been observed at some wetland wastewater discharge sites, such as near Kincheloe, Michigan, where significant amounts of the peat soil have been lost to dissolution (Kadlec 1981b). Heavy metals are toxic, are mobile to various degrees in wetland systems, and tend to accumulate in the food chain. Chlorine in wastewater can lead to production of chlorinated hydrocarbons, can reduce natural bacteria populations, and disrupt nutrient cycling.

Wetlands are extremely diverse in terms of hydrology, primary productivity, nutrient cycling, soil adsorption capacity, and the like. An understanding of the effects of these characteristics and processes on the capabilities of wetlands to remove wastewater nutrients and on the fate of heavy metals in wetland ecosystems is essential to determining wetlands potential for treatment and will be discussed in more detail in following sections. However, prior to this detailed level of analysis, it is important to compare wetlands application to other options (irrigation, overland flow, conventional treatment, etc.), determine the ecological values of the wetlands, and assess potential environmental impacts of the proposed discharge.

WETLAND WASTEWATER TREATMENT CAPABILITIES

Of the contaminants in municipal sewage (Table 24.3), BOD, SS, and pathogens are of the greatest concern, and their removal is the major objective in conventional sewage treatment. Conventional treatment systems are not usually designed to remove high levels of N, P, pesticides, refractory organics, or heavy metals (Tchobanoglous et al. 1979). Typical concentrations of various constituents of raw sewage and the amounts of these pollutants still present after secondary treatment are shown in Table 24.4. The typical removal of these remaining pollutants by wetland application and removal by other land application methods are compared in Table 24.5. It should be noted that wetland application typically does not remove greater percentages of SS, BOD, N, P, or heavy metals than are removed by other land treatment systems. Table 24.5 also shows the hydraulic loading rates that are compatible with satisfactory wastewater treatment in each system. For similar levels of treatment, permissible loading rates for wetland application systems are significantly less than for the other land application systems. Consequently, more land

Table 24.3
Contaminants of Concern in Wastewater Treatment

Suspended solids	Suspended solids can lead to the development of sludge deposits and anaerobic conditions in the receiving water.
Biodegradable organics	Composed principally of proteins, carbohydrates, and fats, biodegradable organics are measured most commonly in terms of (BOD) biochemical oxygen demand and (COD) chemical oxygen demand. If discharged to the environment, the biological stabilization of these organics can lead to the depletion of natural oxygen resources and to the development of septic conditions.
Pathogens	Bacteria and viruses capable of causing communicable disease can be transmitted by water routes.
Nutrients	The nutrients essential for growth include carbon, nitrogen, phosphorus, and trace elements. When discharged to the aquatic environment, these nutrients can lead to excessive growths of undesirable aquatic life.
Refractory organic compounds	These organic compounds tend to be toxic in relatively low concentrations. Some may also accumulate in the environment, biologically and on adsorptive surfaces, concurrent with the slow decay of these compounds. Typical refractory organics are surfactants, phenols, and agricultural pesticides.
Heavy metals	Heavy metals are often toxic in relatively low concentrations. These contaminants are elemental, (i.e., environmentally conservative). They tend to accumulate biologically and on adsorptive surfaces. Typical examples are mercury, lead, and cadmium.
Dissolved inorganic salts	Inorganic constituents such as calcium, sodium, boron, and sulfate may have to be removed if the wastewater is to be reused.

Source: From G. Tchobanglous, R. Stowell, R. Ludwig, J. Colt, and A. Knight, adapted from Metcalf and Eddy 1979.

Table 24.4
Composition of Municipal Sewage and Secondary Sewage Effluent

Constituent	Raw Sewage Range[a]	Raw Sewage Typical Value[a]	% Removal by Secondary Treatment	Secondary Effluent Range[a]	Secondary Effluent Typical Value[a]
Suspended solids	100–350[d]	220[d]	70–95[h]	13–62[e]	25[f]
Biological oxygen demand	110–400[d]	220[d]	80–95[h]	13–75[e]	25[f]
Chemical oxygen demand	250–1000[d]	500[d]	–	50–160[e]	70[f]
Nitrogen, total	20–85[d]	40[d]	45–70[k]	15–40[g]	20[f]
Phosphorus, total	4–15[d]	8[d]	40[k]	7–10[e]	10[f]
Coliform bacteria	10^5–10^{9d}	10^{7d}	90–98[h]	–	–
Refractory organics[b]	0.2–7.4[d]	1.4[d]	–	–	0.2[d]
Chlorides	30–100[d]	50[d]	–	40–100[e]	45[f]
Trace Metals	50%[c]	90%[c]			
Cadmium	–	0.02[j]	33[i]	<0.005–6.4[g]	<0.005[g]
Chromium	0.2[i]	3.6[i]	58[i]–67[j]	<0.05–6.8[g]	0.025[g]
Cobalt	–	0.05[j]		<0.05–0.05[g]	<0.05[g]
Copper	0.1[j]	0.4[j]	28[j]–50[i]	<0.02–5.9[g]	0.10[g]
Iron	0.9[j]	1.9[j]	47[j]	0.10–4.3[f]	
Lead	0.1[i]	0.2[j]	47[i]	<0.02–6.0[g]	0.05[g]
Manganese	0.14[j]	0.3[j]	13[j]	–	0.20[f]
Mercury	0.001[j]	0.0045[j]	26[j]–83[i]	<0.0001–0.125[g]	0.001[g]
Nickel	0.08[j]	0.2[j]	33[i]	< 0.02–5.4[g]	0.02[g]
Zinc	0.18[j]	1.0[i]	47[j]–50[i]	< 0.02–20[g]	0.15[g]

[a] all concentrations mg/l except coliform bacteria MPN/100 ml.
[b] primarily surfactants.
[c] 50th and 90th percentiles, higher than 50% and 90%, respectively, of samples taken.
[d] Metcalf and Eddy 1979.
[e] Buzzell 1972.
[f] Driver et al. 1972.
[g] Menzies and Chaney 1974.
[h] McGauhey 1968.
[i] Konrad and Kleinart 1974.
[j] Mytelka et al. 1973.
[k] McCaull 1974.

Table 24.5
Purification of Secondary Sewage Effluent by Wetland Application
and Various Land Treatment Methods

Constituent	Natural Wetlands	Land Treatment Irrigation[a]	Land Treatment Rapid Infiltration[b]	Land Treatment Overland Flow[c]
		% Removal		
Suspended solids	60–90[d]	>90[h,i]	>90[h]	40–90[i]
Biological oxygen demand	70–95[d]	>90[h,i]	>90[h,i]	40–90[i]
Nitrogen, total	10–90[e,f,g]	60–90[h,i]	30–80[h]	40–90[i]
Phosphorus, total	10–90[e,f,g]	>90[h,i]	50–80[h]	20–60[i]
Heavy metals	20–100[d,g]	>90[i]	50–90[i]	50–90[i]
		cm/yr		
Typical hydraulic loading of wastewater to achieve satisfactory treatment	40–150[j]	150–200[i]	2000–5000[i]	200–1000[i]

[a]1.5 m of soil.

[b]4.5 m of soil.

[c]40 m of slope.

[d]Tchobanoglous and Culp 1980.

[e]Kadlec and Tilton 1979.

[f]Nichols 1983.

[g]Removal efficiencies variable, depend upon loading rates, wetland hydrology, etc.

[h]Metcalf and Eddy 1979.

[i]U.S. E.P.A. 1981b.

[j]Estimated from Nichols (1983)(for 50% to 75% removal of N and P) and Kadlec and Tilton (1979).

is required to achieve the same effluent quality from wetland systems
than from other on-land systems.

WETLAND VALUES

It is important to determine all the values of a particular
wetland and the effects that disposal may have on these values prior
to the selection of that wetland for wastewater disposal. If a
wetland is found to be a refuge for a rare or endangered species, is
an important breeding ground for fish species, provides water retention and flood control for the area, contributes to the regional
groundwater supply, is heavily used for recreation and hunting,
and/or is a unique ecosystem (i.e., the only wetland of this type in
the area), then it should not be considered for disposal since
effluent additions usually have negative impacts on these values (see
Table 24.6).
The determination of the impacts of wastewater on wetlands will
require specific information on a number of ecological variables.
The key relationships and the required data sets as outlined in
Figure 24.1 demonstrate the complexity of selecting, utilizing, and
managing a natural wetland ecosystem for effluent treatment. Excellent reviews of required data collections, methods of analyses,
environmental impacts, and mitigation methods can be found in recent
EPA reports (Bastian and Reed 1979; U. S. EPA 1981, 1983; Claude
Terry and Associates, Inc. 1982). The next section details some of
the specific ecosystem processes that should be considered when
analyzing a wetland for nutrient-removal potential.

SOME WETLAND PROCESSES AND CHARACTERISTICS
AFFECTING WASTEWATER-TREATMENT CAPACITY

WETLAND HYDROLOGY

All natural wetland functions are a result of, or are closely
related to, wetland hydrology (Carter et al. 1979). The wastewater-
treatment capacity of wetlands is closely tied to various aspects of
wetland hydrology that must be taken into account when wastewater
discharge to a wetland is being considered. The removal of nutrients
and other pollutants from wastewater requires intimate contact
between the wastewater and the wetland soil, litter, and vegetation.
Two factors--water depth and retention time--greatly affect pollutant
removal. Water depths in excess of a few centimeters reduce this
contact and lead to poor removal. The rapid passage of wastewater
through a wetland reduces contact time and also results in poor
removal. Rapid flow rates can also flush suspendable matter through
a wetland to the detriment of effluent quality. Water may flow in a
broad, thin sheet across a wetland or be largely confined to a few
relatively narrow, deep channels. The degree of channelization has
an obvious effect on wastewater treatment by wetlands. Discharges of
wastewater to wetlands increase water depths, decrease retention
times, and can cause channelization. Consequently, wastewater
loading rates consistent with satisfactory treatment are low (see
Table 24.5). Kadlec and Tilton (1979) reviewed some aspects of

Table 24.6
Attributed Natural Wetland Values and the Effects of Secondary
Municipal Effluent Additions

Wetland Values	Increased or Decreased Value Due to Effluent Additions
Net primary production	+
Detritus production	+
Wildlife habitat	+-
Refuge (rare and endangered)	-
Breeding grounds (fisheries)	-
Water retention	-
Flood control	-
Groundwater recharge	?
Carbon sink	-
Filtration (nutrients, sediments and organics)	-
Human use (recreation, hunting, fishing, etc.)	-
Ecosystem value to region	-

Note: + = increase.
      - = decrease.
      +- = increase or decrease depending on particular species.

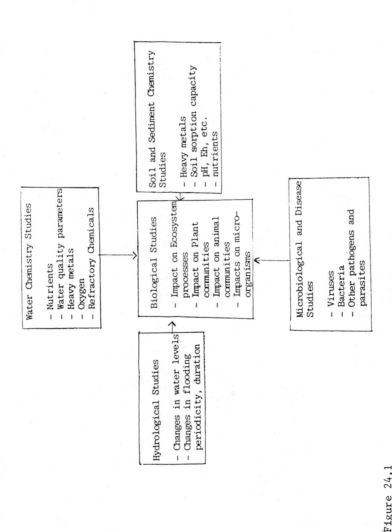

Figure 24.1
A brief overview of the key ecological variables that should be addressed in determining the effects of wastewater effluent on a wetland ecosystem. (After U. S. Environmental Protection Agency, 1983, The effects of wastewater treatment facilities on wetlands in the Midwest, tech. report, EPA-905/3-83-002, U. S. EPA, Center for Environmental Protection, Cincinnati, Ohio, p. 8-2.

wetland hydrology and wastewater treatment, and their information
demonstrates that the hydrologic characteristics of wetlands are
extremely diverse and greatly influence treatment capability. As an
example of this diversity, Verry and Boelter (1979) present a good
comparison between bogs (wetlands perched above the regional water
table and receiving water only from precipitation and runoff from
adjacent uplands) and fens (groundwater-fed wetlands). A large
portion of the annual runoff from perched bogs occurs in early spring
due to a combination of snowmelt, rain, high soil moisture, and low
evapotranspiration (Fig. 24.2). Although much of the annual precipitation occurs from late spring through early fall, bog runoff is low
during this period because precipitation is quickly lost by evapotranspiration. Streamflow from perched bogs is not well regulated;
flow duration curves are steep because the small perennial water
storage is depleted by streamflow and evapotranspiration. Streamflow
from fens is more uniform than from bogs (Fig. 24.2) because groundwater constantly flows into them from the surrounding landscape.
Flow duration curves from fens are almost flat, indicating a fairly
constant ratio of flow due to the relatively large amount of perennial storage in the groundwater system (Verry and Boelter 1979).

In wetlands not influenced by groundwater inputs, peak flows are
related to the depth to the water table (Verry and Boelter 1979)
(Fig. 24.3). Effluent discharges to wetlands should be minimized
when the wetland water table is high (Fig. 24.3) or during periods of
high rainfall and low evapotranspiration since runoff is highest
during these periods and effluent contact time with the wetland is
short (Fig. 24.4). The active growing season (late spring to early
fall) is optimum for wastewater discharge in terms of lower flows and
potential nutrient uptake by plants, soil, and microorganisms.
Effluent discharges to groundwater-fed wetlands should be reduced in
proportion to the larger natural hydraulic loading. Wetlands can
also feed into the groundwater system. Such wetlands should be
eliminated from consideration as wastewater discharge sites due to
potential groundwater contamination.

PRODUCTIVITY, ORGANIC MATTER, AND NUTRIENT ACCUMULATION

It has been suggested by a number of researchers (see Tilton et
al. 1976) that vascular plant nutrient uptake is an important factor
in removing nutrients from municipal effluent. Freshwater wetlands
are among the most productive ecosystems in the world (Richardson
1979) and therefore it might be assumed that highly productive wetlands like cattail marshes, reed marshes, or swamp forests (Table
24.7) are good prospects in terms of wastewater nutrient uptake and
storage. However, for emergent wetland vegetation, the soil rather
than the water is the major source of plant nutrients (Sculthorpe
1967). In addition, a high proportion of the seasonal nutrient
uptake is returned to the water compartment at the end of the growing
season (Spangler, Sloey, and Fetter, Jr. 1976). Klopatek (1975,
1978) calculated a nutrient budget for a stand of $\underline{Scirpus\ fluviatilis}$
in a Wisconsin marsh and found that about 17.5 g/m^2 of N and 3.8 g/m^2
of P per year were translocated from the wetland soil to the plant
shoots. At the end of the growing season, about 12% of this N and P
was transferred to the belowground portions of the plants and stored
over winter, 42% of the N and 58% of the P was leached into the

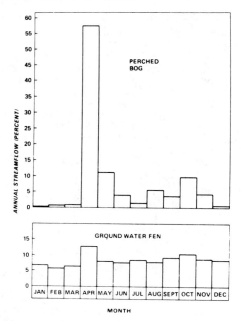

Figure 24.2
Monthly distribution of annual streamflow
from a perched bog and a groundwater fen in
1969. (After D.H. Boelter and E.S. Verry,
1977, Peatland and water, gen. tech. rep.
NC-31, USDA Forest Service, North Central
Forest Experiment Station, St. Paul, Minn.,
p. 14.

water, and the remainder was found in the dead plant material.
Prentki, Gustafson, and Adams (1978) reported a similar budget for P
in a Wisconsin cattail marsh. Even if a substantial part of the dead
plant tissue is incorporated into the soil, the net effect of rooted
emergent vegetation is to transfer nutrients from the soil to the
water. On the basis of the P loading in the sewage effluent from
Greenville, North Carolina, Brinson (this volume) reports that vegetation growth on 1 ha of highly productive swamp forest could only
remove the P from the effluent generated by 8 to 17 people. These
data indicate that under natural, unharvested conditions, the growth
and subsequent uptake of nutrients by wetland vegetation represents
only a minimal annual nutrient sink.
     Release from dead vegetation often results in a net export of
nutrients from wetlands at certain times of the year. The death of

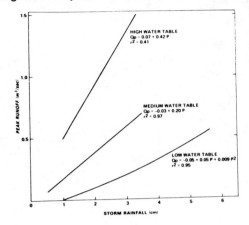

Figure 24.3
Relationship of peak flow (Qp) to storm rainfall (P) and
water table position in a perched bog watershed size is
9.7 ha. (After R.R. Bay, 1969, Runoff from small peatland
watersheds, J. Hydrol. 9: 100.

wetland vegetation is typically followed by the rapid release to the
water of 35% to 75% of the plant tissue P, and somewhat smaller but
still substantial amounts of N (Klopatek 1975, 1978; Boyd 1970; Davis
and van der Valk 1978). Lee, Bently, and Amundson (1975) concluded
that much of the P assimilated by two Wisconsin cattail marshes
during the growing season is flushed out during the fall and spring.
Cattail marshes adjacent to Lake Wingra, Wisconsin, retained 83% of
the P input from storm sewers during the summer, but only 1% in the
fall and 8% in the spring, for an annual retention of only 10%
(Loucks 1977). Verry and Timmons (1982) found N and P retention by a
Minnesota black spruce-sphagnum bog to be highest in the early
spring. In a Massachusetts marsh receiving secondary sewage effluent,
N and P were assimilated fastest early in the growing season and much
more slowly in late summer and fall; a net release of N and P
occurred in the winter (Yonika and Lowry 1979). In a study of small
artificial marshes to which sewage effluent was applied, almost all
of the P that accumulated in the marshes during the growing season
was lost during the fall (Spangler, Fetter, Jr., and Sloey 1977).
    However, the temporary, seasonal storage of wastewater N and P
by wetland vegetation may be beneficial to downstream water quality
since these nutrients are tied up during the growing season and

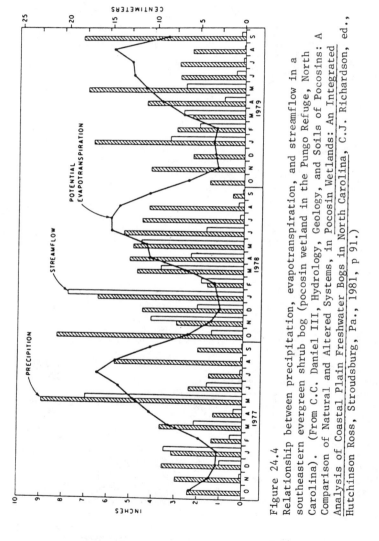

Figure 24.4

Relationship between precipitation, evapotranspiration, and streamflow in a southeastern evergreen shrub bog (pocosin wetland in the Pungo Refuge, North Carolina). (From C.C. Daniel III, Hydrology, Geology, and Soils of Pocosins: A Comparison of Natural and Altered Systems, in Pocosin Wetlands: An Integrated Analysis of Coastal Plain Freshwater Bogs in North Carolina, C.J. Richardson, ed., Hutchinson Ross, Stroudsburg, Pa., 1981, p 91.)

Table 24.7
A Comparison of Net Productivity Values from Wetland
and Terrestrial Ecosystems

Ecosystem Type	Net Productivity		Total of aboveground and belowground $gm^{-2} yr^{-1}$	$\underline{N}$[a]
	Aboveground	Belowground		
Cattail marsh	1270 ± 600	1430 ± 200	2740 ± 670	5
Reed-type marsh	–	–	2100 ± 580	5
Freshwater tidal	1290	460	1600 ± 200	2
Carex marsh	870 ± 90	180 ± 30	1040 ± 100	5
Bogs, fens, and muskeg	530 ± 230	400 ± 430	930 ± 460	9
Grasslands in U. S.	–	–	510 ± 150	9
Bog forest	480	190	670	1
Swamp forest	1050 ± 330	–	–	6
Boreal forest	900 ± 350	–	–	74
Pine forests	1040 ± 570	–	–	42
Temperate coniferous	1490 ± 900	–	–	67
Tropical rainforest	1650 to 2820	–	–	2

Source: Data from Richardson 1979.

[a] $\underline{N}$ = the number of ecosystem studies.

released during the nongrowing season, and because part of this N and P is converted from available to nonavailable forms. The eutrophication potential of wastewater may thus be lessened by its interaction with wetland vegetation. The only actual long-term nutrient sink associated with wetland plant growth is the process of organic soil development through the accumulation of partially decomposed vegetation.

Only a small portion of the total vegetative production is accumulated. Rates of peat accumulation in wetlands in Canada, Ireland, and Finland range from about 10 to 100 g dry matter/m^2/y (Reader and Stewart 1972; Pakarinen 1975; Tolonen 1979). Hemond (1980) estimates net peat accumulation in Thoreau's Bog in Massachusetts to be 180 g/m^2/y. Ranges of from 1.0% to 2.6% N and from 0.05% to 0.12% P in organic soils are typical (Stanek 1973; Richardson et al. 1978; Westman 1979). Thus the rates at which N and P are accumulated in the peat appear to range between 0.10 and 4.7 g N/m^2/y and between 0.005 and 0.22 g P/m^2/y in moderate to cold climates and possibly up to 10.0 g N/m^2/y and 0.50 g P/m^2/y in warm, highly productive areas. Schlesinger (1978) estimated annual accumulations of 3.8 g of N and 0.15 g of P/m^2 in Georgia's Okefenokee Swamp. These accumulation rates suggest a limited N and P removal capacity for wetlands if peat accumulation is the main mechanism being suggested for processing and storage.

MECHANISMS FOR REMOVAL OF WASTEWATER NUTRIENTS BY WETLAND SOILS

In a review, Nichols (1983) concluded that denitrification is the major mechanism for wetland removal of wastewater nitrogen. P is immobilized in soils by adsorption and precipitation reaction with aluminum (Al), iron (Fe), calcium (Ca), and clay minerals (Larsen, Warren, and Langston 1959; Fox and Kamprath 1971; Syers, Harris, and Armstrong 1973; Bloom 1981). The major differences that exist among wetland soils—for example, organic versus mineral substrates, aerobic versus anaerobic conditions, acid versus neutral pH's (Richardson et al. 1978)—can greatly affect N- and P-removal capacity.

Denitrification is an obvious mechanism for removing N from municipal effluent in wetlands since it occurs under anaerobic conditions. When oxygen is lacking, facultative anaerobic bacteria use NO_3^- in place of free O_2 as the terminal exogenous H acceptor in respiration. In this process NO_3^- is first converted to NO_2^-, then to gaseous N_2O and N_2. The rate of NO_3^- removal from surface waters is dependent on soil pH, redox potential, moisture status, and temperature (Avnimelech 1971; Van Cleemput, Patrick, and McIlhenny 1975). At pH of less than 6, the further reduction of N_2O to N_2 is strongly inhibited (Wijler and Delwiche 1954). Below pH 5, chemical rather than biochemical reactions can convert N to gaseous forms. At low pH, NO_2^- is unstable and will react with amino acids, ammonia, and urea to form N_2 gas. Soil organic matter, or some component of it, seems to increase NO_2^- instability. The disappearance of N from acid organic soils may be due as much to the chemical breakdown of NO_2^- as to microbial denitrification. Nitrate removal rates vary greatly, but laboratory studies at constant temperature (30°C) of intact flooded soil cores from a freshwater swamp and a salt marsh in

Louisiana showed NO_3^- removal rates of 2.50 and 7.64 mg N/l/d respectively (Engler and Patrick 1974).

Denitrification rates are related to the availability of organic matter that can furnish energy for the metabolic processes of denitrifying bacteria and serve as an H donor for the denitrification process (Broadbent and Clark 1965). In mineral soils, denitrification is usually limited by carbon (C) availability, and in laboratory studies, glucose or some other organic C source must often be added to the soil to maximize denitrification rates. In organic wetland soils, however, sufficient organic matter is available so that rapid rates of denitrification are typically obtained without adding a supplementary C source (Bartlett et al. 1979; Patrick and Reddy 1976; Reddy, Sacco, and Graetz 1980; Terry and Tate 1980).

Compared to the voluminous literature that exists on the fixation of P by mineral soils, little work has been done on P fixation by organic soils. However, the same adsorption and precipitation processes appear to be important. Organic soils low in Al, Fe, and Ca have very low P-fixing capacities (Larsen, Warren, and Langston 1959). The organic material itself appears to have almost no capacity to fix P (Richardson, unpublished data). In a recent study, Bloom (1981) related P fixation by organic soils to Al-organic matter complexes.

Adsorption and precipitation by soils are not necessarily permanent sinks for wastewater P; it is at least a partially reversible process. A soil functions to some extent as a "phosphate buffer" in regulating the concentration of phosphate in solution (Syers, Harris, and Armstrong 1973; Harter 1968; Williams et al. 1970). Soils that absorb P the least readily typically release P the most easily.

Phosphorus is fixed much more strongly by Fe in an oxidized state than in a reduced state. The oxidation-reduction potential of a wetland soil controls the oxidation state of Fe and thereby affects the ability of the soil to retain P (Patrick and Khalid 1974; Delaune, Reddy, and Patrick, Jr. 1981). The incorporation of oxygen (O_2) into the soil by diffusion or turbulent mixing and the formation of an oxidized surface layer in an otherwise anaerobic soil is of particular importance to P retention by wetland soils (Syers, Harris, and Armstrong 1973). The depletion of O_2 by the discharge into a wetland of high BOD wastewater or by the death and decay of wetland vegetation at the end of the growing season can release previously adsorbed P from the soil (Richardson et al. 1978).

Soil pH also affects P retention, and pH and redox effects interact. Bloom (1981) found less adsorption of P by Al-saturated peat at lower pH's, especially below 4.7. Less P was retained by estuarine sediments studied by Delaune, Reddy, and Patrick, Jr. (1981) as pH was reduced from 8.0 to 5.0. This effect was much more pronounced under anaerobic conditions (redox potential less than zero mV).

Organic soils are typically low in P and consequently have a high C:P ratio. Microbial immobilization has been suggested as a mechanism for wastewater P retention by peats with high C:P ratios (Brown and Farnham 1976). Microbial immobilization of P has been found to occur when crop residue containing less than 0.2% P is added to soil (Fuller 1956). Kaila (1956) believes that because much of the organic matter in peat is resistant to decomposition, immobilization of added P may not occur in peats with a P content of more than 0.1%. The P content of peats is commonly less than 0.1%, and 0.05% is not unusual. Consequently, microbial immobilization of P is likely to occur in response to the initial applications of wastewater to peats. However, because the C:P ratio of wastewater is low compared to that of peat, continued applications would soon satisfy microbial P requirements. Therefore, immobilization by micro-organisms is not likely to play a significant role in the long-term fixation of P by peats, but may be important initially if levels of addition are not excessive.

WASTEWATER NUTRIENT REMOVAL EFFICIENCY OF WETLANDS AND LOADING CAPACITY ESTIMATES

The removal efficiencies of wastewater N and P by wetlands vary over a wide range and depend on a number of factors, including wetland type, season of application, nutrient loading rates, hydro-logic conditions, and the number of years of applications. Few extensive studies exist from which to analyze efficiencies or compare wetlands to terrestrial applications (Kelly, this volume). A list of ten of the most complete research sites, which includes data on size of the wetland, years of application, hydraulic loading, N and P loading and nutrient removal, is given in Table 24.8. For those wetlands that have received wastewater for many years, the loading rates and N- and P-removal efficiencies given are recent measurements rather than long-term averages. The nutrient loadings include contributions from runoff and precipitation, as well as from wastewater. Contributions from groundwater or N fixation were not measured in any of these studies. To obtain a general estimate of the effects of loading rates and years of application on P removal, a hand-fit curve (dashed line) was drawn to the data in Table 24.8. At low loading rates, wetlands have the capacity to remove a high percentage of the P applied, but this capacity can decrease substantially in a few years, depending on the wetland type (Fig. 24.5). As the loading rate is increased, the efficiency of P removal also declines rapidly. At a loading rate of 1.5 g P/m^2/y, about 68% or 1.0 g/m^2/y would be removed from the wastewater (Fig. 24.5). A 10-fold increase in P loading to 15 g/m^2/y increases P removal only 4.5 times to 4.5 g/m^2/y, or about 30%. Application above 20 g/m^2/y reduces efficiency to below 30% (Fig. 24.5).

Caution must be used when making generalizations about wetland nutrient-removal efficiencies from such a diverse and sparse data set. This is especially true of this data set that includes such a variety of wetland types and a wide range of years of application of wastewater (1 to 69 years). However, trends from the most complete studies in terms of extensive year-by-year analysis (site 2 and site 3, where loading and removal efficiency is shown for 5 and 3 years, respectively) show the general pattern of decreased nutrient-removal efficiency with time and with higher loading rates. Initially the Michigan swamp (site 2 in Fig. 24.5), a peat soil system high in Ca (23,000 ± 13,100 µg/g dry weight of peat) and Fe content (8600 ± 9200 µg/g dry weight of peat) removed 91% of the P loadings (Fig. 24.5). By the fifth year, the efficiency level had been reduced to 65%, even though loading rates of less than 10 g/m^2/y were applied (Fig. 24.5).

The Irish Bog site (site 3 in Fig. 24.5) received fertilizer additions and initially retained 96% of the P; however, by year 3,

Table 24.8
Removal of N and P from Wastewater (Secondary Effluent) and Fertilizer Applied to Natural Wetlands

Types of Wetland	Location	Size (ha)	Years Nutrients Were Applied	Hydraulic Loading (cm/yr)		Nutrient Loading (g/m²/y)		Nutrient Removal (%)		Reference
				Waste-water	Other	Total P	Total N	Total P	Total N	
Shrub-sedge fen	Michigan	1[a]	1[b]	70[b]	-	1.7[b]	1.9[b,c]	95[b]	96[b,c]	Tilton and Kadlec (1979)
Forest-shrub fen	Michigan	18.2	1[d]	36.8[d]	-	0.9[d]	1.5[c,d]	91[d]	75[c,d]	Kadlec and Tilton (1977) (1978)
			2[e]	74.1[e]	205[e]	2.6[e]	6.5[c,e]	88[e]	80[c,e]	Kadlec (1979a)
			3[f]	65.2[f]	183[f]	1.8[f]	9.3[e,f]	72[f]	80[e,f]	Kadlec (1980)
			4[g]	55.7[g]	116[g]	1.8[g]	6.2[e,g]	64[g]	77[e,g]	Kadlec (1981a)
			5[g]	57.3[g]	97[g]	1.7[g]	9.3[e,g]	65[g]	75[e,g]	
Blanket bog	Ireland	-	1	-[h]	-	5.0	7.4[e]	96	82[e]	Burke (1975)
			2	-[h]	-	13.1	15.4[e]	72	87[e]	Burke (1975)
			3	-[h]	-	8.1	10.3[e]	43	68[e]	Burke (1975)
Hardwood swamp	Florida	204	20	10.2	83	0.9	-	87	-	Boyt et al. (1977)
Cattail marsh	Wisconsin	156	55	23.4	558	15.2	-	32	-	Spangler et al. (1977)
Cattail marsh	Massachusetts	19.4	69	684	159	7.1	53.6	47	31	Yonika and Lowry (1979)
Cattail	Massachusetts	2.4	69	5526	-	63.6	428	20	1	Yonika and Lowry (1979)
Deepwater marsh	Ontario	162	55	231	5569	11.6	78.6	58[i]	41[i]	Semkin et al. (1976)
Glyceria marsh	Ontario	20	55	1870	-	77	404	24[i]	38[i]	Semkin et al. (1976)
Cypress dome	Florida	1.0	4	130	127	17.2	-	41[j]	-	Ewel and Odum (1979)

[a] Area affected by study; entire wetland is 710 ha.
[b] May-September.
[c] Inorganic N only, organic N not measured.
[d] August-October.
[e] March-November.
[f] April-November.
[g] June-November.
[h] Chemical fertilizers, not wastewater, applied.
[i] Wastewater applied year-round, but percent removal measured during the growing season only. Percent removal would likely have been much less if calculated on a year-round basis.
[j] Infiltration accounts for 50 % (8.6 g/m²) of output, while runoff accounts for 9.3 % of output.

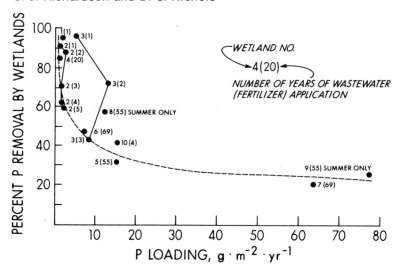

Figure 24.5
Phosphorus-removal efficiency by wetland types as affected by loading
rate and years of application. (After D.S. Nichols, 1983, Capacity of
natural wetlands to remove nutrients from wastewater, Water Pollut.
Control Fed. J. 55:499.)

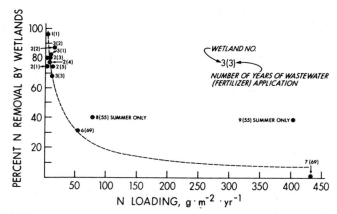

Figure 24.6
Nitrogen-removal efficiency by wetlands types as affected by loading
rate and years of application. (After D.S. Nichols, 1983, Capacity of
natural wetlands to remove nutrients from wastewater, Water Pollut.
Control Fed. J. 55:501.)

this wetland retained only 43% of the P additions (Table 24.8).
Excessive P losses have also been noted from fertilized bogs in North
Carolina (Richardson 1981) and mucklands in New York (Duxbury and
Peverly 1978). These findings suggest a low capacity for organic
soils (especially bog systems) to retain P.

It is not known how long a wetland can continue to remove P from
wastewater. It is known that if sufficient P is added, the soil can
be P saturated (Fox and Kamprath 1971). The capacity of a particular
wetland to remove P from wastewater should depend to a large degree
on those properties that determine the P adsorption capacity of the
soil--that is, Al, Fe, and Ca content and the relative amounts of
organic and inorganic materials. Current research indicates that Al
and Fe content, especially in peat substrates, may be the most important variables in determining P-holding capacity in undrained wetlands (Richardson, in preparation).

Points 8 and 9 (Fig. 24.5), representing wetland areas that have
received wastewater for 55 years, are located well above the curve
and seem to indicate that these wetlands have unusually high P-
removal efficiencies relative to their loading rates. However,
wastewater was applied to these two wetlands throughout the year, but
P-removal efficiency was only measured during the growing season when
the vegetation was actively taking up nutrients. The year-round P-
removal efficiencies for these areas would probably be much lower.

The N-removal pattern of wetlands (Fig. 24.6) is similar to that
for P with high removal efficiency--70%--at low loading rates--10 g
$N/m^2/y$--and rapidly declining efficiency as loading rates increase.
The loading rates and removal efficiencies for wetlands 1, 2, and 3
were calculated on the basis of inorganic N forms only; organic N was
not reported. Verry (1979) found that 85% of the N leaving a black
spruce-sphagnum wetland was in organic forms. Similar results for
another wetland were reported by Crisp (1966). Thus, the N-removal
efficiencies of wetlands 1, 2, and 3 would probably be lower if
calculated on the basis of total N. As with P, sites 8 and 9 (Fig.
24.5) show higher N-removal efficiency during the growing season.

Even though the natural rate of N accumulation in peat is about
20 times that of P, peat formation does not seem to be a significant
wastewater N sink except at low loading rates (Nichols 1983). The
major mechanism for removing N from wastewater applied to wetlands
appears to be denitrification. Unlike the case with P removal, no
reduction in N-removal efficiency occurs at a given loading rate with
continued application of wastewater (see data for wetland site 2 and
3, Fig. 24.6 and Table 24.8). Results from wetlands receiving applications of wastewater for many years and wetlands to which wastewater
has been applied for only a year or two also seem to plot along the
same hand-fit curve (Fig. 24.6), indicating little difference in
removal efficiency. As long as the supply of NO_3^- is maintained,
denitrification should continue at the same rate, unless the vast
supply of organic C available in a typical wetland soil becomes
exhausted.

The rapidly declining N-removal efficiency with increasing
loading rates seen in Figure 24.6 may be due to limits on the rate of
denitrification, nitrification, O_2 availability, or NH_4^+ or NO_3^-
diffusion. Some of this decrease may also be due to N fixation
induced by wastewater P. One of the wetlands (no. 6, Table 24.8)
achieved only a 31% N reduction, averaged over the whole year, at a
loading rate of 54 $g/m^2/y$, and another wetland (no. 7, Table 24.8)

removed only a small percentage when loaded at the very high rate of 428 g/m^2/y. Yet, in the laboratory under constant anaerobic conditions, and with continuous stirring so that diffusion was not a factor, soils from these same wetlands denitrified more than 90% of added NO$_3$-N at N concentrations as high as typically found in sewage effluent (Bartlett et al. 1979).

Hydrologic conditions in a wetland can also affect the removal of wastewater N and P. Higher N and P loading rates are generally accompanied by higher hydraulic loadings (Table 24.8), so that retention times in the wetland are reduced and less time is allowed for N- and P-removal reactions to occur. At very high loading rates, nutrient removal may be limited primarily to the sedimentation of particulate forms. For the wetlands in Table 24.8, sufficient data are not available to separate the effects of hydraulic loading rates and nutrient loading rates. Wetland morphology is also important. As the depth of water in a wetland increases, the chance for reactions between wastewater nutrients and the wetland soil decreases. On the other hand, a deep-water wetland will have a longer retention time than a shallow-water wetland, given the same hydraulic loadings.

In spite of the shortcomings of the previous data set, some patterns and trends are evident, and some preliminary estimates can be made of the capacity of wetlands to remove N and P from wastewater. Vollenweider (1968) estimated the average per-capita loading of N and P from sewage and sewage effluents to be about 2.2 g P/d and 10.8 g N/d. These values were used to convert P and N loading rates (Table 24.8) to numbers of people in order to estimate the P and N removal that might be expected if the sewage produced by various numbers of people was applied to 1 ha of wetland (Nichols 1983). According to these estimates, about 1 ha of wetland is needed for every 60 people for 50% N and P removal and approximately 1 ha is needed for every 20 people for 75% removal (Fig. 24.7). Of course, the relationship between nutrient removal and loading rate will differ from one site to another according to local conditions and wetland type. In general, however, it appears that wetland application can be an effective method of removing N and P from secondarily treated wastewater if large wetland areas are available, populations and loading rates are low, nongrowing-season applications are avoided, and long-term high removal efficiencies are not required. To maximize removal efficiency, it may also be necessary to consider two wetland sites and provide a "resting period" by alternating disposal between them. This management technique has proved itself in land application at a number of sites (Sopper and Kardos 1974; Sopper and Kerr 1979).

HEAVY-METAL CYCLING IN NATURAL WETLANDS

It is well known that municipal sewage effluents often contain elevated concentrations of toxic heavy metals, primarily from industrial sources (see Table 24.4). The accumulation of toxic metals in the food chain is a major problem in wetland disposal of wastewater (and in other on-land disposal systems as well). It is not feasible in this paper to do an in-depth review of the effects and interactions of heavy metals in wetlands, but a brief overview of some aspects of heavy-metal cycling, along with case examples, should provide insight into the ecological issues involved with discharge of

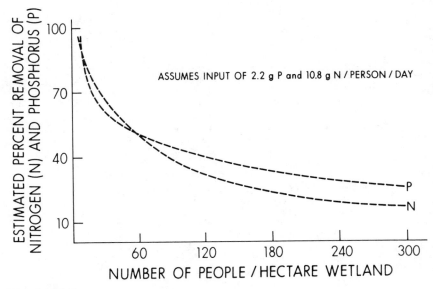

Figure 24.7
Estimated removal efficiency for nitrogen and phosphorus by 1 ha of
wetland receiving the average discharge of 2.2 g P and 10.8 g
N/person/day. (After D.S. Nichols, 1983, Capacity of natural wetlands
to remove nutrients from wastewater, Water Pollut. Control Fed. J.
55:502.)

metal-laden effluent into wetlands.
    Wetland soils generally serve as an initial sink for heavy
metals such as cadmium (Cd), chromium (Cr), copper (Cu), lead (Pb),
mercury (Hg), nickel (Ni), and zinc (Zn) (Bunzl et al. 1976; Giblin
et al. 1980; Gallagher and Kibby 1980; Hazen and Kneip 1980; Giblin,
this volume). Metals dissolved in the soil solution or weakly
adsorbed to soil particles, however, are readily available to the
biota. Many studies have shown that heavy metals are taken up by
freshwater and marine wetland plants through their roots and translocated to other parts of the plants (Murdoch and Capobianco 1979;
Giblin et al. 1980; Gallagher and Kibby 1980; Welsh and Denny 1980;
Cushing and Thomas 1980; Ragsdale and Thorhaug 1980; Drifmeyer et al.
1980; Hazen and Kneip 1980; Folsom and Lee 1981; Lee, Sturgis, and
Landin 1981). Various amounts of these metals are then passed on to
animals feeding on live vegetation, litter, and detritus. Heavy
metals differ in their mobility in wetland environments. Pb is
largely retained in the soil and finds its way into the food chain to
only a small degree. Cd and Cr are highly mobile; Cu and Zn somewhat
less mobile and are quickly passed on to consumer organisms.
    The mobility of heavy metals varies with different conditions of
water and soil chemistry. For example, increasing the salinity of
overlying water increases the amount of dissolved Hg from contaminated sediments (Feick, Horne, and Yeaple 1972). Long and Angino
(1977) report that highly soluble chloride complexes of Zn and Cd

will form in saline waters. Strong reducing conditions in sediments may promote the retention of Cd, Cr, Cu, Pb, Fe, and Zn as insoluble sulfides (Giblin et al. 1980). Folsom and Lee (1981) found that added Zn and Cd were taken up by marsh plants from the soil in greater amounts under oxidized than under reduced conditions. On the other hand, Giblin et al. (1980) believe that reduced manganese (Mn) may be released from the soil and taken up by plants.

An analysis of the cycle of Hg, a well-recognized toxic element (Wood 1974) that has been discharged in substantial amounts from chlor-alkali plants into rivers, estuaries, and wetlands (Windom et al. 1976), gives some insight into potential problems with heavy-metals disposal in wetlands. Hg has three oxidation states: the native metal Hg^o, the mercurous form Hg^+, and the mercuric form Hg^{++} (Wood 1974). Hg may be adsorbed to sediments in a variety of compounds including HgS, $HgSO_4$, HgCl, $HgCl_2$, HgO, and $Hg(OH)_2$. Conversion of Hg to methylmercury, CH_3Hg^+ (or MeHg), a highly toxic form, and subsequent bioaccumulation is a major concern in ecosystems (Wood 1974). Once the free Hg ion is formed, it is quickly methyl-ated and taken up by biota, with bacteria playing a role in this process (Windom et al. 1976). Under anaerobic conditions frequently encountered in wetlands, mercuric Hg as $HgCl_2$ and $Hg(OH)_2$ is already in the proper valence state for formation of the free ion, so methyl-ation is facilitated. While rarely detectable in water and present as only a small fraction of sediment Hg (<0.01%), methylmercury is preferentially soluble in lipids. Thus, it accumulates through the food chain and is the main form found in fish (Gardner et al. 1978).

Gardner et al. (1978) and Windom et al. (1976) measured total and methylmercury concentrations in various compartments of a salt-marsh ecosystem near Brunswick, Georgia. A chlor-alkali plant had discharged 1 kg/y of inorganic Hg into the estuary for 6 years ending in 1972. Table 24.9 shows data obtained for total and methylmercury in sediments, Spartina (salt-marsh grass) root and stem, soft tissue of Littorina irrorata (periwinkle snail), Uca spp. (fiddler crabs), and Modiolus demissus (mussel). The periwinkle snail feeds on benthic algae and epiphytes on salt-marsh grasses. Fiddler crabs are detritus feeders and burrow in the sediments; mussels are filter feeders. Together they constitute over 90% of the production at the primary consumer level, the first level at which methylmercury is consistently detectable (Windom et al. 1976).

The highest sediment concentrations occurred at the chlor-alkali plant station 4 and at a point upstream, possibly moved there by upriver tidal flow (Table 24.9). As distance from the plant increased, sediment Hg concentrations generally decreased, but not in a linear fashion. Water circulation, sediment deposition patterns, and periodic flooding and drying patterns that would change redox conditions probably are important factors in determining sediment Hg levels (Windom et al. 1976). Contamination of sediments up to 12 km distant is evident (Table 24.9), but decreasing levels show that the marsh has either filtered or diluted a substantial amount of Hg at that distance. Hg in the methyl form in the soil was undetectable by the methods used (<0.001 µg/g).

Total Hg in Spartina roots was found to be correlated with surface sediment concentrations. But the root appears to bar trans-port to the rest of the plant, as stems and leaves did not show any increased levels (except at the highly contaminated station 4). Methylmercury was rarely detectable in plants and soils, but traces

Table 24.9
Concentrations of Total Hg and Methylmercury in Various Components of the Contaminated Salt-Marsh
Ecosystem Studied by Gardner et al. (1978) and Windom et al. (1976)

Total Hg (μg/g) dry weight

Station Number	Distance from Plant[b] (km)	Sediment[a] 0–5 cm	Sediment[a] 0–10cm	Spartina Root	Spartina Stem	Littorina irrorata Total Hg	Littorina irrorata MeHg	Uca sp. Total Hg	Uca sp. MeHg	Modiolus demissus Total Hg	Modiolus demissus MeH
1	4 (upstream)	1.34	0.15	1.47[d]	0.06[d]	6.1	0.61	0.5	0.36	–	–
2	2 (upstream)	0.34	0.08	0.07	0.06	6.4	0.51	–	–	–	–
3	1 (downstream)	–	0.06	0.14	0.03	2.5	0.25	0.5	0.22	–	–
4	0	1.70	0.48	1.17	0.12	9.4	0.56	0.9	0.40	2.5	0.2
5	4	0.48	0.33	0.74	0.07	1.6	0.13	0.4	0.18	–	–
6	4	0.27	0.08	0.40	0.06[d]	1.8	0.13	–	–	–	–
7	8	0.55	0.82	0.23	0.03	1.5	0.10	–	–	1.0	0.0
8	12	0.41	0.30	0.21	0.03	0.6	–	–	–	–	–
9	8–12[c]	0.28	0.25	0.09	0.03	2.5	0.12	–	–	–	–
10	6	0.27	0.23	0.11	0.03[d]	2.6	0.08	0.3	0.20	–	–

[a]Average (Hg) in salt-marsh sediments in the southeastern United States is 0.07 μg/g (Windom et al. 1976).
[b]Distances estimated from a map in Gardner et al. (1978).
[c]Distance is 8 km if measured across land (mainly marsh), 12 km if measured along the river course.
[d]Traces of MeHg (0.001 to 0.002 μg/g) detected.

were found in seeds at four out of five stations where they were sampled. The conversion of Hg to organic forms in the roots would also make it available for movement through the food web by detritus feeders. The primary consumers were found to accumulate large concentrations (0.08 µg/g dry weight to 0.61 µg/g dry weight in snails and 0.18 µg/g dry weight to 0.36 µg/g dry weight in fiddler crabs) of methylmercury, the most toxic chemical form (Table 24.9). Total Hg was also higher in the snails. Uptake by these species was used to estimate annual methylmercury production in the marsh to be 50 µg/g total Hg in the upper surface sediments (Windom et al. 1976). Concentration of total Hg and methylmercury in these species was correlated to sediment concentrations.

Bioaccumulation was shown at the primary consumer level by the large percentage of total Hg present as MeHg (3% to 10% in snails and 43% to 72% in fiddler crab tissues) as opposed to the small percentage (<0.01% as MeHg) in sediments and plants. Fish collected in the river showed even higher concentrations of total Hg and methylmercury that concentrated in the liver and, to a slightly lesser extent, in muscle tissue. Fish at high trophic levels generally accumulated more Hg than those lower in the food web. Annelids, mollusks, and echinoderms found lower on the food chain had fairly low concentrations of MeHg, possibly due to close contact with sediments that contain very little MeHg as a source of Hg (Windom et al. 1976). Birds found in the area routinely contained an order of magnitude more methylmercury than primary consumers, with top carnivores, such as heron and egrets, showing the highest accumulations. Carnivorous mammals held similar concentrations.

This case study of a salt-marsh system contaminated with Hg shows the importance of sediments as both a sink and a source of Hg and the seriousness of methylmercury transport into food webs. In light of the continuing contamination of the food web from sedimental Hg, the discharge of that metal in natural marsh ecosystems should not be permitted. Similar problems of uptake of Cd by plants, vertebrates, and invertebrates, especially crabs, was noted as a potential human health problem in a New York marsh by Hazen and Kneip (1980). Giblin et al. (1980) found elevated levels of Cd, Cr, and Cu in mussels, Cd, Cu, and Mn in crabs, and Cd and Cr in fish in a saltwater marsh experimentally treated with sewage sludge. The uptake of significant amounts of Pb, Zn, Cr, Cd, and so forth, by plants such as Lemna minor (duckweed), Myriophyllum spp. (water milfoil) and Glyceria grandis (reed-meadow grass) in a freshwater marsh (Ontario, Canada) affected by discharge from an inefficient wastewater treatment plant also suggests food web contamination problems (Mudroch and Capobianco 1979).

From the previous studies, it is apparent that wetlands can sorb many different heavy metals, but they also recycle, release, and often move some toxic metals up the food chain and into downstream ecosystems. This fact suggests the possibility of a number of ecological and human health problems from these and similar toxic metals released into wetlands. The movement of these contaminants also indicates that natural wetlands are not a final sink for heavy metals and thus should not be used for treatment of effluent laden with these toxic materials.

However, one possible way to take advantage of the high initial adsorptive capacity of organic soils for heavy metals is the use of artificial peat systems for tertiary treatment. This alternative was tested by Coupal and Lalancette (1976), who found that wastewater containing heavy metals such as Hg, Cd, Cu, Fe, Ni, Cr^{6+}, Cr^{3+}, Ag, Pb, and Sb, or cyanide, phosphates, and organic materials such as oil, detergents, and dyes, can be treated efficiently after filtration through peat moss.

In the treatment process, precipitation with sulfide or hydroxide and crude settling are followed by passage of the supernatant over a mat of peat at a pH of 6 to 8. The peat can absorb up to 4% of its dry weight in metals. After two passes through the mat, even highly contaminated waters have been purified to meet EPA standards. Disposal of the contaminated peat in an environmentally sound way is then a problem, but alternatives are available (Coupal and Lalancette 1976). Sulfuric acid treatment is reported to increase the capacity of peat to adsorb additional cations (Smith et al. 1977). The successful removal of acid dye from effluent by peat has also been reported by Poots, McKay, and Healy (1976).

CONCLUSIONS AND MANAGEMENT SUGGESTIONS

From the preceding discussion of values, ecosystem criteria, removal efficiencies for N and P, and heavy metals in wetlands, it is suggested that consideration of any wetland ecosystem as a treatment system first entail a complete analysis of:

1. Value of wastewater as a resource;
2. All wetland values;
3. Suitability of wastewater for wetland discharge;
4. Wastewater treatment objectives;
5. Capacity of wetland to accomplish desired treatment on the basis of hydraulic loading, nutrient loading, and wetland area needed;
6. Comparison with other treatment options (irrigation, overland flow, and so forth) by degree of treatment accomplished, cost, and energy requirements;
7. Environmental impacts (insect problems, disease vectors, odors, species loss, community change, etc.); and
8. Legal aspects of wetland utilization.

The key to utilizing any wetland ecosystem for N, P, BOD, and organic removal is low loading rates coupled with sufficient wetland area. A minimum of 1 ha/60 people is suggested for a 50% removal efficiency of N and P. Management of natural wetlands for treatment is extremely difficult, but the following management suggestions (taken in part from Sloey [1978] and Tchobanoglous and Culp [1980]) may improve treatment performance.

Pretreatment: Suppose a wetland system being analyzed for effluent disposal has a low P-removal efficiency but will remove other wastewater pollutants. It may become an acceptable system by removing most of the P in the influent by chemical precipitation prior to wetland discharge (e.g., alum treatment). Ozonation may be the preferable disinfection method for wastewater effluent that is to be disposed of in wetlands because of chlorinated hydrocarbon problems and wetland bacterial dieback, both of which may be associated with the use of chlorine.

Seasonal application: Removal of nutrients has been shown to be highest during the growing season. This finding suggests that storage of wastewater during the winter months may be required if maximum uptake is required.

Flushing: Periodic flushing of wetlands might be used to control the buildup and/or release of specific materials, providing downstream systems can handle the constituents without eutrophication problems. For example, P could be flushed from the system during high flow (often winter and spring periods) when effects on dormant downstream aquatic systems are likely to be minimal. Silt-laden water could also be passed through the wetland system to increase or restore the adsorption characteristics of the wetland for P or other constituent removal.

Outflow regulation: The hydraulic detention time could be controlled by regulating the depth of water in the wetland. This technique would be especially important for the control of N (nitri-fication-denitrification).

Changing application locations and methods: It may be advan-tageous to vary the locations (surface versus submerged water appli-cation) and method of disposal (spray versus irrigation pipe versus trickle system, and so on) to achieve higher removal of wastewater constituents. It may also be necessary to develop two wetland areas and alternate discharges between sites to obtain a "recharge effect"--that is, restoration of removal efficiency due to chemical changes under aerobic versus anaerobic soil conditions.

Harvesting of vegetation: The harvesting of biomass may allow for greater uptake of nutrients depending on the plant species and constituents of concern. Plant uptake by itself does not constitute the major nutrient sink in most wetland systems but is important in elemental allocation to other system components (e.g., heavy metals, N, and so on). Plant removal may be useful in increasing uptake and preventing metal movement, but treatment costs will rise significantly.

Finally, it appears that management of wetlands for effluent treatment has potential in only some specific wetland types under low loading conditions of N and P, but the methodology has not reached the stage of development for routine utilization by municipalities. The diversity of wetland types and the wide ranges of reported efficiencies dictate that considerable site-specific research is needed before any community considers an investment in a natural wetland system. Also, the value of water for irrigation and of N and P as fertilizers on agricultural, forestry, or marginal land may prove to be the best utilization of wastewater for many areas of the United States.

ACKNOWLEDGMENTS

The following people provided invaluable assistance in the preparation of this paper. Jay Benforado (U. S. Fish and Wildlife Service) supplied key references and unpublished data, Gregory Bourne (Claude Terry and Associates, Inc.) provided data from their EPA report on wetland disposal in the southeastern United States. Catherine Garra sent information from the EPA report on wastewater treatment in the Midwest. Ruth Cupery aided in the review of heavy metals in wetlands. Sam Matthews aided greatly in the review of the manuscript and Cheryl Spann graciously typed the various drafts.

REFERENCES

Avnimelech, Y., 1971, Nitrate transformation in peat, Soil Sci. 111:113-118.
Bartlett, M. S., L. C. Brown, N. B. Hanes, and N. H. Nickerson, 1979, Denitrification in freshwater wetland soil, J. Environ. Qual. 8:460-464.
Bastian, R. K., and S. C. Reed, 1979, Aquaculture systems for wastewater treatment, U. S. EPA 430/90-80-006, Environmental Protection Agency, Washington, D. C., 485p.
Bay, R. R., 1969, Runoff from small peatland watersheds, J. Hydrol. 9:90-102.
Bloom, P. R., 1981, Phosphorus adsorption by an aluminum-peat complex, Soil Sci. Soc. Proc. 45:267-272.
Boelter, D. H., 1972, Preliminary results of water level control on small plots on a peat bog, Proceedings 4th international peat congress, I-IV, Helsinki, Finland, International Peat Society, pp. 347-384.
Boelter, D. H., and G. E. Close, 1974, Pipelines in forested wetlands: Cross drainage needed to prevent timber damage, J. For. 72:561-563.
Boelter, D. H., and E. S. Verry, 1977, Peatland and water, gen. tech. rep. NC-31, USDA Forest Service, North Central Forest Experiment Station, St. Paul, Minn., 22p.
Boyd, C. E., 1970, Losses of nutrients during decomposition of Typha latifolia, Arch. Hydrobiol. 66:511-517.
Boyt, F. L., S. E. Bayley, and J. Zoltek, Jr., 1977, Removal of nutrients from treated municipal wastewater by wetland vegetation, J. Water Pollut. Contr. Fed. 49:789-799.
Broadbent, F. E. and F. E. Clark, 1965, Denitrification, in Soil nitrogen, Agronomy 10, W. V. Bartholomew and F. E. Clark, eds., American Society of Agronomy, Madison, Wis., pp. 344-359.
Brown, J. L., and R. S. Farnham, 1976, Use of peat for wastewater filtration, principles and methods, vol. I. Proceedings 5th international peat congress, Poznan, Poland, International Peat Society, pp. 349-357.
Bunzl, K., W. Schmidt, and B. Sansoni, 1976, Kinetics of ion exchange in soil organic matter, IV. Adsorption and desorption of Pb, Cu, Cd, Zn, and Ca by peat, J. of Soil Sci. 27:32-41.
Burke, W., 1975, Fertilizer and other chemical losses in drainage water from a blanket bog, Irish J. Agric. Res. 14:163-178.
Buzzell, T., 1972, Secondary treatment processes, Wastewater management by dispersal on the land, special rep. no. 171, U. S. Army Corps of Engineers, Cold Regions Research and Engineering Laboratory, Hanover, N. H., pp. 35-47.
Carter, V., M. S. Bedinger, R. P. Novitzki, and W. O. Wilen, 1979, Water resources and wetlands, in Wetland functions and values: The state of our understanding, P. E. Greeson, J. R. Clark, and J. E. Clark, eds., tech. pub. TPS 79-2, American Water Research

Association, Minneapolis, Minn., pp. 344-376.

Claude Terry and Associates, Inc., 1982, Wetlands disposal of treated wastewater: Environmental impact statement phase I report to U. S. EPA region 4, Claude Terry & Assoc., Inc., and Gannett, Fleming, Corddry, & Carpenter, Inc., Atlanta, Ga., 397p.

Conner, W. H., J. G. Gosselink, and R. T. Parrondo, 1981, Comparison of the vegetation of three Louisiana swamp sites with different flooding regimes, Am. J. Bot. 68:320-331.

Coulson, J. C. and J. Butterfield, 1978, An investigation of the biotic factors determining the rates of plant decomposition on blanket bogs, J. Ecol. 66:631-650.

Coupal, B., and J. M. Lalancette, 1976, The treatment of wastewaters with peat moss, Water Res. 10:1071-1076.

Crisp, D. T., 1966, Input and output of minerals for an area of Pennine moorland: The importance of precipitation, drainage, peat erosion, and animals, J. Appl. Ecol. 3:327-348.

Cushing, C. E., Jr., and J. M. Thomas, 1980, Cu and Zn kinetics in Myriophyllum heterophyllum Michx. and Potamogeton richardsonii (Ar. Benn.) Rydb., Ecology 61:1321-1326.

Daniel, C., III, 1981, Hydrology, geology, and soils of pocosins: A comparison of natural and altered systems, in Pocosin wetlands, C. J. Richardson, ed., Hutchinson Ross, Stroudsburg, Pa., pp. 69-108.

Davis, C. B., and A. G. van der Valk, 1978, Litter decomposition in prairie glacial marshes, in Freshwater wetlands: Ecological processes and management potential, R. R. Good, D. F. Whigham, and R. L. Simpson, eds., Academic Press, New York, pp. 99-113.

Delaune, R. D., C. N. Reddy and W. H. Patrick, Jr., 1981, Effect of pH and redox potential on concentrations of dissolved nutrients in an estuarine sediment, J. Environ. Qual. 10:276-279.

Drifmeyer, J. E., G. W. Thayer, F. A. Cross, and J. C. Zieman, 1980, Cycling of Mn, Fe, Cu, and Zn by eelgrass, Zostera marina L., Am. J. Bot. 67:1089-1096.

Driver, C. H., B. F. Hrutfiord, D. E. Spyridakis, E. B. Welch, and D. D. Wooldridge, 1972, Assessment of the effectiveness and effects of land disposal methodologies of wastewater management, wastewater management rep. 72-1, U. S. Army Corps of Engineers, Cold Regions Research Laboratory, Hanover, N. H., 147p.

Duxbury, J. M., and J. H. Peverly, 1978, Nitrogen and phosphorus losses from organic soils, J. Environ. Qual. 7:566-570.

Ehrlich, P. R., A. H. Ehrlich, and J. P. Holdren, 1977, Ecoscience: Population, resources, environment, W. H. Freeman, San Francisco, Calif., 1051p.

Engler, R. M., and W. H. Patrick, Jr., 1974, Nitrate removal from floodwater overlying flooded soils and sediments, J. Environ. Qual. 3:409-413.

Ewel, K. C., and H. T. Odum, 1979, Cypress domes: Nature's tertiary treatment filter, in Utilization of municipal sewage effluent and sludge on forest and disturbed land, W. E. Sopper and S. N. Kerr, eds., Pennsylvania State University Press, University Park, pp. 103-114.

Feick, G., R. A. Horne, and D. Yeaple, 1972, Release of mercury from contaminated freshwater sediments by runoff of road deicing salt, Science 175:1142-1143.

Folsom, B. L., Jr., and C. R. Lee, 1981, Zinc and cadmium uptake by the freshwater marsh plant Cyperus esculentus grown in contaminated sediments under reduced (flooded) and oxidized (upland) disposal conditions, J. Plant Nutrition 3:233-244.

Fox, R. L., and E. J. Kamprath, 1971, Adsorption and leaching of P in acid organic soils and high organic matter sand, Soil Sci. Soc. Am. Proc. 35:154-156.

Fuller, W. H., D. R. Nielsen, and R. W. Miller, 1956, Some factors influencing the utilization of phosphorus from crop residues, Soil Sci. Soc. Proc. 20:218-224.

Gallagher, J. L., and H. V. Kibby, 1980, Marsh plants as vectors in trace metal transport in Oregon tidal marshes, Am. J. Bot. 67:1069-1074.

Gardner, W. S., D. R. Kendall, R. R. Odom, H. L. Windom, and J. A. Stephens, 1978, The distribution of methylmercury in a contaminated salt marsh ecosystem, Envir. Pollut. 5:242-251.

Giblin, A. E., A. C. M. Bourg, I. Valiela, and J. M. Teal, 1980, Uptake and losses of heavy metals in sewage sludge by a New England salt marsh, Am. J. Bot. 67:1059-1068.

Goffeng, G., S. Saebo, and L. E. Haugen, 1979, Action of effluents from a tourist centre on a peatland ecosystem in Norway, in Classification of peat and peatlands, E. Kivinen, L. Heikurainen, and P. Pakarinen, eds., International Peat Society, Helsinki, Finland; pp. 332-340.

Guntenspergen, G., and F. Stearns, 1981, Ecological limitations on wetland use for wastewater treatment, Selected proceedings of the midwest conference on wetland values and management, June 17-19, St. Paul, Minn., B. Richardson, ed., Freshwater Society, Navarre, Minn., pp. 273-284.

Harter, R. D., 1968, Adsorption of phosphorus by lake sediments, Soil Sci. Soc. Am. Proc. 32:514-519.

Hazen, R. E., and T. J. Kneip, 1980, Biogeochemical cycling of cadmium in a marsh ecosystem, in Cadmium in the environment, J. O. Nriagu, ed., Wiley, New York, pp. 399-424.

Hemond, H. F., 1980, Biogeochemistry of Thoreau's Bog, Concord, Massachusetts, Ecol. Monogr. 50:507-526.

Kadlec, R. H., 1979a, Monitoring report on the Bellaire wastewater treatment facility, 1978, utilization report no. 3, University of Michigan Wetlands Ecosystem Research Group, College of Engineering, University of Michigan, Ann Arbor.

Kadlec, R. H., 1979b, Wetlands for tertiary treatment, in Wetland functions and values: The state of our understanding, P. E. Greeson, J. R. Clark, and J. E. Clark, eds., tech. pub. TPS 79-2, American Water Resources Assoc., Minneapolis, Minn., pp. 490-504.

Kadlec, R. H., 1980, Monitoring report on the Bellaire wastewater treatment facility, 1979, utilization rep. no. 4, University of Michigan Wetlands Ecosystem Research Group, College of Engineering, University of Michigan, Ann Arbor.

Kadlec, R. H., 1981a, Monitoring report on the Bellaire wastewater treatment facility, 1980, utilization rep. no. 5, University of Michigan Wetlands Ecosystem Research Group, College of Engineering, University of Michigan, Ann Arbor.

Kadlec, R. H., 1981b, How natural wetlands treat wastewater, in Selected proceedings of the midwest conference on wetland values and management, June 17-19, 1981, St. Paul, Minn., B. Richardson, ed., Freshwater Society, Navarre, Minn., pp. 241-254.

Kadlec, R. H., and D. L. Tilton, 1977, Monitoring report on the
    Bellaire wastewater treatment facility, 1976-77, utilization rep.
    no. 1, University of Michigan Wetlands Ecosystem Research Group,
    College of Engineering, University of Michigan, Ann Arbor.
Kadlec, R. H., and D. L. Tilton, 1978, Monitoring report on the
    Bellaire wastewater treatment facility, 1977, utilization rep.
    no. 2, University of Michigan Wetlands Ecosystem Research Group,
    College of Engineering, University of Michigan, Ann Arbor.
Kadlec, R. H., and D. L. Tilton, 1979, The use of freshwater
    wetlands as a tertiary wastewater treatment alternative,
    CRC Crit. Rev. Environ. Control 9(2):185-212.
Kaila, A., 1956, Phosphorus in virgin peat soils, Sci. Agr. Soc.
    Finland J. 28:142-167.
Klopatek, J. M., 1975, The role of emergent macrophytes in mineral
    cycling in a freshwater marsh, in Mineral cycling in southeastern
    ecosystems, F. G. Howell, J. B. Gentry, and M. H. Smith, eds.,
    ERDA Symposium Series (CONF-740513), U. S. Energy Research and
    Development Administration, Aiken S. C., pp. 367-393.
Klopatek, J. M., 1978, Nutrient dynamics of freshwater riverine
    marshes and the role of emergent macrophytes, in Freshwater
    wetlands: Ecological processes and management potential, R. E.
    Good, D. F. Whigham and R. L. Simpson, eds., Academic Press, New
    York, pp. 195-216.
Konrad, J. G., and S. J. Kleinert, 1974, Surveys of toxic metals in
    Wisconsin, tech. bull. no. 74, Wisconsin Dept. Natural Resources,
    Madison, pp. 2-7.
Larsen, J. E., G. F. Warren, and R. Langston, 1959, Effect of iron,
    aluminum, and humic acid on phosphorus fixation by organic soils,
    Soil Sci. Soc. Proc. 23:438-440.
Lee, C. R., T. C. Sturgis, and M. C. Landin, 1981, Heavy metal
    uptake by marsh plants in hydroponic solution cultures, J. Plant
    Nutrition 3:139-151.
Lee, G. F., E. Bently, and R. Amundson, 1975, Effects of marshes on
    water quality, in Coupling of land and water systems, A. D.
    Hasler, ed., Springer-Verlag, New York, pp. 105-127.
Long, D. R., and E. E. Angino, 1977, Chemical speciation of Cd, Cu,
    Pb, and Zn in mixed freshwater, seawater, and brine solutions,
    Geochim. Cosmochim. Acta 41:1183-1191.
Loucks, O., R. Prentki, U. Watson, B. Reynolds, P. Weiler, S. Bartell,
    and A. B. D'Allessio, 1977, Studies of the Lake Wingra watershed:
    An interim report, report 78, Center for Biotic Systems, Institute
    for Environmental Studies, University of Wisconsin, Madison.
Lyons, L. A., and J. Benforado, 1981, Wetlands for wastewater treat-
    ment: An annotated bibliography on ecological impacts, U. S.
    Dept. of the Interior, Fish and Wildlife Service, Kearneysville,
    W. Va., 203p.
McCaull, J., and J. Crossland, 1974, Water pollution, Harcourt Brace
    Jovanovich, New York, 206p.
McGauhey, P. H., 1968, Engineering management of water quality,
    McGraw-Hill, New York, 295p.
Menzies, J. D., and R. L. Chaney, 1974, Waste characteristics in
    factors involved in land application of agricultural and
    municipal wastes, USDA-ARS, National Program Staff, Soil, Water
    and Air Sciences, Beltsville, Md., 200p.
Metcalf and Eddy, Inc., 1979, Wastewater engineering:
    Treatment/disposal/reuse, McGraw-Hill, New York, 920p.

Mudroch, A., and J. A. Capobianco, 1979, Effects of treated effluent on a natural marsh, Water Pollut. Control Fed. J. 51:2243-2256.

Mytelka, A. J., J. S. Czachor, W. B. Guggino, and H. Golub, 1973, Heavy metals in wastewater and treatment plant effluents, Water Pollut. Control Fed. J. 45:1859-1864.

Nichols, D. S., 1983, Capacity of natural wetlands to remove nutrients from wastewater, Water Pollut. Control Fed. J. 55:495-505.

Odum, H. T., and K. C. Ewel, eds., 1980, Cypress wetlands for water management, recycling, and conservation, fifth and final report, Center for Wetlands, University of Florida, Gainesville, 291p.

Pakarinen, P., 1975, Bogs as peat-producing ecosystems, Internat. Peat Soc. Bull. 7:51-54.

Patrick, W. H., Jr., and R. A. Khalid, 1974, Phosphate release and sorption by soils and sediments: Effects of aerobic and anaerobic conditions, Science 186:53-55.

Patrick, W. H., Jr., and K. R. Reddy, 1976, Nitrification-denitrification reactions in flooded soils and water bottoms: Dependence on oxygen supply and ammonium diffusion, J. Environ. Qual. 5:469-472.

Poots, V. J. P., G. McKay, and J. J. Healy, 1976, The removal of acid dye from effluent using natural adsorbents, I: Peat, Water Res. 10:1061-1066.

Prentki, R. T., T. D. Gustafson, and M. S. Adams, 1978, Nutrient movements in lakeshore marshes, in Freshwater wetlands: Ecological processes and management potential, R. E. Good, D. F. Whigham, and R. L. Simpson, eds., Academic Press, pp. 169-194.

Ragsdale, H. L., and A. Thorhaug, 1980, Trace metal cycling in the U. S. coastal zone: A synthesis, Am. J. Bot. 67:1102-1112.

Reader, R. J., and J. M. Stewart, 1972, The relationship between net primary production and accumulation for a peatland in southeastern Manitoba, Ecology 53:1024-1037.

Reddy, K. R., P. D. Sacco, and D. A. Graetz, 1980, Nitrate reduction in an organic soil-water system, J. Environ. Qual. 9:283-288.

Richardson, C. J., 1979, Primary productivity values in freshwater wetlands, in Wetland functions and values: The state of our understanding, P. E. Greeson, J. R. Clark, and J. E. Clark, eds., tech. pub. TPS 79-2, American Water Resources Assoc., Minneapolis, Minn., pp. 131-145.

Richardson, C. J., 1981, Pocosins: Ecosystem processes and the influence of man, in Pocosin wetlands, C. J. Richardson, ed., Hutchinson Ross, Stroudsburg, Pa., pp. 135-154.

Richardson, C. J., J. A. Kadlec, W. A. Wentz, J. P. M. Chamie, and R. H. Kadlec, 1976a, Background ecology and the effects of nutrient additions on a central Michigan wetland, in Proceedings of the Third Wetlands Conference, M. W. LeFor, W. C. Kennard, and T. B. Helfsolt, editors. Institute of Water Resources, University of Connecticut, Storrs, pp. 34-72.

Richardson, C. J., W. A. Wentz, J. P. M. Chamie, J. A. Kadlec, and D. L. Tilton, 1976b, Plant growth, nutrient accumulation and decomposition in a central Michigan peatland used for effluent treatment, in Freshwater wetlands and sewage effluent disposal, proceedings of a national symposium, D. L. Tilton, R. H. Kadlec, and C. J. Richardson, eds., May 10-11, 1976, Wetlands Ecosystem Research Group, University of Michigan, Ann Arbor, pp. 77-118.

Richardson, C. J., D. L. Tilton, J. A. Kadlec, J. P. M. Chamie, and
W. A. Wentz, 1978, Nutrient dynamics of northern wetland ecosystems, in Freshwater wetlands: Ecological processes and management potential, R. E. Good, D. F. Whigham, and R. L. Simpson,
eds., Academic Press, New York, pp. 217-241.

Schlesinger, W. H., 1978, Community structure, dynamics, and nutrient
cycling in the Okefenokee cypress swamp forest, Ecol. Monogr.
48:43-65.

Schwegler, B. R., 1978, Effects of sewage effluent on algae dynamics
of a northern Mighigan wetland, M. S. thesis, University of
Michigan, Ann Arbor, 47p.

Sculthorpe, C. D., 1967, The biology of aquatic vascular plants,
St. Martin's Press, New York, 610p.

Semkin, R. G., A. W. McLarty, and D. Craig, 1976, A water quality
study of Cootes Paradise, Ontario Ministry of the Environment,
West-Central Region, Toronto, Canada.

Sloey, W. E., F. L. Spangler, and C. W. Felter, Jr., 1978, Management
of freshwater wetlands for nutrient assimilation, in Freshwater
wetlands: Ecological processes and management potential, R. E.
Good, D. F. Whigham, and R. L. Simpson, eds., Academic Press, New
York, pp. 321-340.

Smith, E. F., P. MacCarthy, T. C. Yu, and H. B. Mark, Jr., 1977,
Sulfuric acid treatment of peat for cation exchange, Water Pollut.
Control Fed. J. 49:633-638.

Sopper, W. E., and L. T. Kardos, eds., 1973, Recycling treated
municipal wastewater and sludge through forest and cropland,
Pennsylvania State University Press, University Park, 479p.

Sopper, W. E., and S. N. Kerr, eds., 1979, Utilization of municipal
sewage effluent and sludge on forest and disturbed land, Pennsylvania State University Press, 537p.

Spangler, F. L., C. W. Fetter, Jr., and W. E. Sloey, 1977, Phosphorus
accumulation discharge cycles in marshes, Water Resour. Bull.
13:1191-1201.

Spangler, F. L., W. E. Sloey, and C. W. Fetter, Jr., 1976, Experimental use of emergent vegetation for the biological treatment of
municipal wastewater in Wisconsin, in Biological control of water
pollution, J. Tourbier and R. W. Pierson, Jr., eds., University of
Pennsylvania Press, pp. 161-171.

Stanek, W., 1973, Classification of muskeg, in Muskeg and the
northern environment in Canada, N. W. Radforth and C. O. Brawner,
eds., University of Toronto Press, Toronto, Canada, pp. 31-62.

Syers, J. K., R. F. Harris, and D. E. Armstrong, 1973, Phosphate
chemistry in lake sediments, J. Environ. Qual. 2:1-14.

Tchobanoglous, G., and G. L. Culp, 1980, Wetland systems for wastewater treatment: An engineering assessment, in Aquaculture
systems for wastewater treatment: An engineering assessment,
S. C. Reed and R. K. Bastian, eds., 430/9-80-007, U. S. EPA,
Office of Water Program Operations, Washington, D. C., pp. 35-55.

Tchobanoglous, G., R. Stowell, R. Ludwig, J. Colt, and A. Knight,
1979, The use of aquatic plants and animals for the treatment of
wastewater: An overview, in Aquaculture systems for wastewater
treatment: Seminar proceedings and engineering assessment, R. K.
Bastian and S. C. Reed, eds., 430/9-80-006, U. S. EPA, Office of
Water Program Operations, pp. 35-55.

Terry, R. E., and R. L. Tate, III, 1980, Denitrification as a pathway
for nitrate removal from organic soil, Soil Sci. 129:162-167.

Tilton, D. L., and R. H. Kadlec, 1979, The utilization of a fresh-water wetland for nutrient removal from secondarily treated wastewater effluent, J. Environ. Qual. 8:328-334.

Tilton, D. L., R. H. Kadlec, and C. J. Richardson, eds., 1976, Freshwater wetlands and sewage effluent disposal, Proceedings of a national symposium, May 10-11, 1976, Wetlands Ecosystem Research Group, University of Michigan, Ann Arbor, 343p.

Tolonen, K., 1979, Peat as a renewable resource: Long-term accumulation rates in north-European mires, in Classification of peat and peatland, E. Kivenen, L. Heikurainen, and P. Pakarinen, eds., International Peat Society, Helsinki, Finland, pp. 282-296.

Tourbier, J., and R. W. Pierson, eds., 1976, Biological control of water pollution, University of Pennsylvania Press, 340p.

U. S. Environmental Protection Agency, 1981, Process design manual: Land treatment of municipal wastewater, U. S. EPA, Center for Environmental Protection; Rep. 625/1- 81-013; 458 pp.

U. S. Environmental Protection Agency, 1983, The effects of waste-water treatment facilities on wetlands in the Midwest, tech. report, EPA-905/3-83-002, U. S. EPA, Center for Environmental Protection, Cincinnati, Ohio.

Valiela, I., J. M. Teal, and W. J. Sass, 1975, Production and dynamics of salt marsh vegetation and the effects of experimental treatment with sewage sludge, J. Applied Ecology 12:973-982.

Van Cleemput, O., W. H. Patrick, and R. C. McIlhenny, 1975, Formation of chemical and biological denitrification products in flooded soil at controlled pH and redox potential, Soil Biol. Biochem. 7:329-332.

Verry, E. S., and D. H. Boelter, 1979, Peatland hydrology, in Wetland functions and values: The state of our understanding, P. E. Greeson, J. R. Clark, and J. E. Clark, eds., tech. pub. TPS 70-2, American Water Resources Assoc., Minneapolis, Minn., pp. 389-402.

Verry, E. S., and D. R. Timmons, 1982, Waterborne nutrient flow through an upland-peatland watershed in Minnesota, Ecology 63:1456-1467.

Vollenweider, R. A., 1968, Scientific fundamentals of the eutrophication of lakes and flowing waters, with particular reference to nitrogen and phosphorus as factors in eutrophication, Organization for Economic Cooperation and Development, Paris, France, 255p.

Welch, R. P. H., and P. Denny, 1980, The uptake of lead and copper by submerged aquatic macrophytes in two English lakes, J. Ecol. 68:443-455.

Wentz, A. W., 1976, The effects of simulated sewage effluents on the growth and productivity of peatland plants, Ph. D. dissertation, University of Michigan, Ann Arbor.

Westman, C. J., 1979, Climate-dependent variation in the nutrient content of the surface peat layer from sedge pine swamps, in Classification of peat and peatlands, E. Kivinen, L. Heikurainen, and P. Pakarinen, eds., International Peat Society, Helsinki, Finlad, pp. 160-170.

Whigham, D. F., and S. K. Bayley, 1979, Nutrient dynamics in fresh-water wetlands, Wetland functions and values: The state of our understanding, P. E. Greeson, J. R. Clark, and J. E. Clark, eds., tech. pub. TPS 79-2, American Water Resources Assoc., Minneapolis, Minn., pp. 468-478.

Wijler, J. and C. C. Delwiche, 1954, Investigations on the denitrifying process in soil, Plant Soil 5:155-169.

Williams, J. D. H., J. K. Syers, R. F. Harris, and D. E. Armstrong, 1970, Adsorption and desorption of inorganic phosphorus in lake sediments in a 0.1 M NaCl system, Environ. Sci. Technol. 4:517-519.

Windom, H., W. Gardner, J. Stephens, and F. Taylor, 1976, The role of methylmercury production in the transfer of mercury in a salt marsh ecosystem, Estuar. and Mar. Sci. 4:579-583.

Wood, J. M., 1974, Biological cycles for toxic elements in the environment, Science 183:1049-1052.

Yonika, D., and D. Lowry, 1979, Feasibility study of wetland disposal of wastewater treatment plant effluent, res. proj. 78-04, Final report to the Massachusetts Water Resources Commission, Division of Water Pollution Control, Westborough, Mass.

DISCUSSION

Ewel: I'm curious about the peat accumulation rate in warmer climates. I would suggest that productivity rates may be a bit higher, but decomposition rates are, too. This is supported by the observation that most of what we would call peat-based wetlands, certainly in the South and probably in the North, are steady-state systems rather than accumulating systems. I was surprised to see your accumulation rate so much higher than the loading rate.

Richardson: If you look at net decomposition rates, this indicates systems that are accumulating peat, but there are, clearly, a number of systems that are not accumulating. Pocosins may currently be an example of an accumulating system.

Ewel: Well, it's arguable that the pocosin areas are accumulating systems.

R. Kadlec: I would like to comment on some of your graphs. Another data point is now available for Figures 24.5 and 24.6; it is for Belair, Michigan. It falls below the horizontal axis because we had a breakthrough with that system. In that case, instead of following the design curve that you suggest, we should probably be far more conservative. On the other side of the spectrum, the Houghton Lake data points (of which there are now five) are registering 99+ for N and P removal, if you go far enough down-gradient, because there's plenty of area available. One must be careful about drawing one curve for N or P and inferring design principles from them.

Richardson: That's an excellent point. We discussed the drawing of those curves. Alternatively, we could look at clusters, although that is difficult. However, from the management point of view, people have been making faulty estimates and we need to quantify the various boundaries as best we can. I think you'll agree that with an areal requirement of 1 ha for 60 people, these will necessarily be modest systems with large wetland areas. As a result, when these wetlands have to be bought or permission obtained for use, few will get permits in the first place.

Berger: I have a question on the word "management." In your inventory of wetlands and in your contacts with municipal officials, what

would you say about their attitude with regard to the use of wetlands, and what would you say about the attitude of the consulting
engineers on whom they depend for advice?

Richardson: Municipal operators are comfortable with a traditional
system, although a number of municipalities are looking at cheaper
alternatives. In our part of the country, some consulting firms will
say that wetlands provide an alternative. As a result, there are a
few systems around. It's a regional issue. On the Southeastern
Coastal Plain, for example, communities may have three choices:
(1) some type of land system, assuming there is adequate space; (2) a
standard municipal system; or (3) wetlands treatment. Given the cost
of the first two and the fact that the towns may be near some
thousand acres of wetlands, officials are looking at the possibility
of using some type of wetland system. It's hard to give you one
answer, but I think the municipal wastewater people like the traditional system because they can manage it and, consequently, they feel
more comfortable with it. Some engineering firms are trying to sell
wetlands treatment, but only on a regional and a site-by-site basis.
Many consultants are <u>not</u> trying to promote wetlands treatment.

Bastian: In your list of states and numbers of projects, you didn't
distinguish between those systems designed to treat wastes versus
systems serving simply as convenient disposal sites for effluent. I
think you'll find that most are in the latter category.

Richardson: Seventy percent in the North are municipal treatment
systems.

Bastian: But for municipal, industrial, and agricultural applications, how many are actually trying to manage the system versus
discharging into a wetland and forgetting about it? Are they really
trying to use wetlands as part of a treatment system?

Richardson: Good point. I think that should be evaluated because it
is likely that the use of wetlands is mostly by default rather than
by planned design of treatment.

Bastian: Possibly 4 or 5 at most out of all 14 states listed have
managed wetland systems for treatment.

Richardson: You mean using the total wetland for treatment?

Bastian: No. As a treatment mechanism in any way, shape, or form.
Does the state regulatory agency recognize this as part of the treatment system or merely as a convenient place to dispose of the waste?
I think that what you really have is an inventory of discharges, most
of which are not monitored and have never been monitored. Whether or
not the wetland does in fact treat the waste is incidental to the
purpose of most managers.

Richardson: Well, Greg Bourne might be able to answer that question,
since he has visited a number of sites. Greg, how many of these
systems are monitored?

Bourne: To tell you the truth, most of them are as Bob Bastian says. When you ask the operators how large the wetlands are, the response is that they don't really know the boundary. If you ask if they monitor, they say they don't really know where to monitor. To specifically answer your question, I would say that the only ones really monitored in detail are the experimental ones, in Florida and in the Southeast, which the University of Florida has studied at various points.

Richardson: I know of two places in the Midwest where people use wetlands for treatment and do standard monitoring. But many wetlands are used just for discharge of effluent.

Garra: Of those 93 sites in our inventory, 3, I think, were designed as treatment systems. The rest "just happened." Another really interesting aspect is the process of locating records for wetlands that receive waste. Most states, except for perhaps one or two, don't keep records on this sort of thing. My guess would be that in Michigan, for example, there are more sites than are listed; the number seems very low considering the number of wetlands we have in Michigan. In our experience, it's hard to gather this information.

Richardson: But the important thing is that these wetlands are being used, maybe not for a scientific reason, but for reasons of cost and lack of other suitable sites for discharge. The states may not even know about many of these discharges.

Bastian: I'll bet only a handful of the sites have a permit that clearly specifies that the discharge location is a wetland.

Richardson: Oh, I'm sure of that. However, some states are getting a little better. Nevertheless, Kentucky officials, for example, say they have no discharges into wetlands in the entire state. Zero!

Bourne: I think most of the wetlands that we've seen in EPA Region 4 actually do have a permit that at least says they're going to a "creek swamp" or a "river swamp." However, these are nebulous terms. Of the many sites that we've examined, we saw maybe four or five that were stressed, or at least were visibly stressed; for example, the trees were either damaged or destroyed. These were either bottomland hardwoods or cypress strands. They are particularly interesting sites from which we can learn much. This may naturally lead us to the topic of the next session where we might discuss this concept of stress. Have we done enough research on system stress to gain information to aid in proper design? The fact of the matter is, as we pointed out, that small communities are going to use wetlands, especially if they are in the coastal plain; they hardly have a choice because the only nearby body of water (or quasi-water body) is a wetland. So, I think we need to develop some tentative design criteria, even though we still have questions about the appropriateness of wetland use for some areas.

Valiela: In regard to the question on peat accumulation, decomposition obviously varies with temperature, but what determines the amount of remaining peat would be the amount of lignin. That amount varies geographically with plant species composition.

Richardson: Yes. And there is variation among sites. Accumulation is related also to water level, temperature, and nutrients. The figures presented give you some rough idea of the phosphorus removal rate by organic accumulation. It is clearly not 100 g/m^2-yr or 200 g/m^2-yr, but 0.2 g/m^2-yr to 0.4 g/m^2-yr. That's not very much phosphorus. I don't care if you move the figures up or down a little; it's still very low for P retention per year in organic matter.

25

An Ecological Evaluation Procedure for Determining Wetland Suitability for Wastewater Treatment and Discharges

Donald M. Reed and Timothy J. Kubiak

The following observation was made by Sloey, Spangler, and Fetter, Jr. (1977) at the Symposium on Freshwater Marshes: Present Status, Future Needs, held at Rutgers University in New Brunswick, New Jersey:

> In the past, we caused the deterioration of the quality of our surface waters by using them to treat our wastes. When the practice was initiated, we marvelled at the remarkable ability of water to "self-purify." We based our decisions on short-term observations and immediate economics. Years later, the results of long-term overloading became evident. Lest we make the same mistake in handling our valuable and diminishing wetlands, it is mandatory that we carry out long-term, carefully monitored experiments at a severely limited number of sites. It is also important that those conducting the experiments document changes very carefully in the natural system that could signal future problems.

The questions raised in this statement are enormously complex and extremely important if we are to proceed with discharging waste-water into wetlands. Probably the most important question to be addressed by this workshop is: Have we, in the five years since 1977, institutionalized a formal wetland/wastewater evaluation process that provides adequate guidance and criteria for the consultants and various units and agencies of government to allow for wastewater to be discharged into natural wetlands without unacceptable environmental effects and economic costs to society? The answer seems to be a qualified no. Certainly, we have accumulated a relatively large information base on the topic, but, except for some general principles, it is disjunct and not easily applied in a generic sense (Tilton, Kadlec, and Richardson, eds. 1976; DeWitt and Soloway 1978; Good et al. 1978; Bastian and Reed 1979; Greeson, Clark, and Clark, eds. 1979; Sutherland and Kadlec 1979; Council on Environmental Quality and U. S. Fish and Wildlife Service 1980; Chan et al. 1981; Richardson 1981). The dynamic nature of wetlands

(Weller 1981), the variability of wetland types nationwide, and the ecoregion in which wetlands occur (Cowardin et al. 1979) all influence the function, nature, and societal importance we place on them (Horwitz 1978). These facts have frustrated the attempts of resource managers to perform meaningful ecological evaluations on a site-specific basis.

This situation does not mean that a comprehensive wastewater/wetland evaluation and decision-making process cannot be institutionalized. Indeed, the institutional framework already exists for wastewater issues within the Clean Water Act. This framework includes the Section 201 Facility Planning process, Section 303 Water Quality Standards process, and Section 402 National Pollution Discharge Elimination System permit process. Wastewater discharge evaluations involving this framework have been institutionalized within the last ten years. These evaluations have shown the direct integration of Section 201, 303, and 402 requirements. For the most part, the evaluations of and criteria for wastewater discharges have been directed primarily at stream discharges because of historic practice and technical understanding. However, the lack of integration in practice of the aforementioned sections of the Clean Water Act will continue to be a deterrent to the establishment of a technically sound and socially acceptable process for the discharge of wastewater into our nation's wetlands. In this regard, it is the purpose of this paper to discuss the need for, and considerations to be addressed in, a formal ecological evaluation process directed at determining wetland suitability to accept and treat wastewater effluent.

GENERAL EVALUATION PROCESS FOR WETLAND USE

Although a detailed discussion of the entire wastewater facility planning process is beyond the scope of this paper, we have attempted to provide, in Figures 25.1 and 25.2, a schematic presentation of the process that emphasizes wetlands as receiving waters. Figure 25.1 is a general flow diagram of the overall process. It emphasizes only treatment alternatives and alternative receiving waters. Figure 25.2 provides additional information concerning the environmental assessment of wastewater/wetland discharge alternatives utilizing the water-quality standards process and effluent limitations of the Section 402 permit process.

Any evaluation process (Fig. 25.1) for natural or artificial wetland use should be instituted at the onset of the facility planning process. Consideration of appropriate screening criteria for effluent quality and quantity and cost-effectiveness should be made. If these initial considerations suggest that a natural wetland would not be feasible, other surface or groundwater discharges would need to be identified and assessed as receiving waters. If an artificial wetland could be created and properly managed as a treatment system, an evaluation of the impact of the wetland on the surrounding environment should be made. Effluent, if any, discharged from the wetland treatment system would be assessed to assure compatibility with receiving-water requirements similar to conventional systems (artificial wetlands created for wastewater treatment should not be considered as receiving waters, but should have sufficient design requirements to avoid adverse impacts to the surrounding environment).

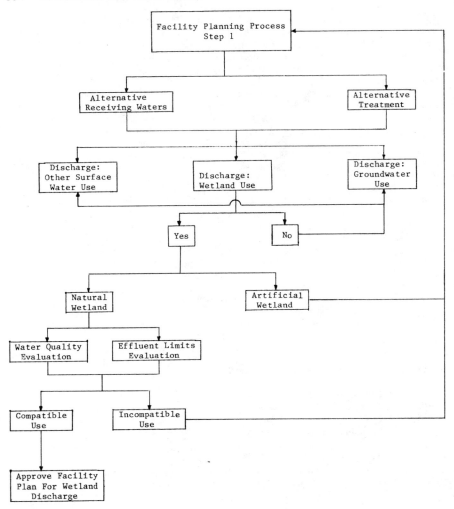

Figure 25.1
Generalized schematic of the facility planning process.

If receiving-water requirements preclude the wetland system effluent from being discharged, other treatment alternatives would have to be considered.

Use of a natural wetland as a receiving water requires an ecological evaluation of the subject wetland and its relationship to the surrounding landscape. As illustrated in Figure 25.2, the ecological evaluation of the wetland should precede decisions relative to approving the potential use of the wetland and should be used to determine compatibility with water quality standards (or other requirements) and effluent limitations to meet those standards. Where high quality values of the wetland could be adversely affected or the risks of proceeding are unknown but deemed unacceptable, use of the wetland could be precluded by appropriate nondegradation criteria.

Those wetland-use proposals that satisfactorily pass the test for nondegradation requirements would be candidates for regulated (controlled degradation or enhancement) wastewater discharge using selected criteria in the standards and appropriate effluent limitations. If the discharge can meet these requirements, effluent discharge would be approved and monitoring undertaken. In this manner, the predictive capability of the preapproval analysis can be validated and aid in the feedback of information into the water-quality standards process for further refinement. This enables the standards to remain flexible and revisable over time to allow for wetland use where determined acceptable and to preclude unwise use when monitoring or other new information shows unacceptable effects that cannot be eliminated. This whole illustration, of course, does not operate in a vacuum. Original research, as well as social, economic, and political considerations, influences the entire process and must also be factored into decisions.

DISCUSSION OF THE ECOLOGICAL EVALUATION PROCESS

The ecological evaluation of any wetland area being considered for the receipt of wastewater discharges should include consideration of: (1) the existing wetland functions and uses and their importance to the watershed as a whole; (2) the plant and animal communities present and a determination of their quality; (3) existing management activities occurring within the area; and (4) the existing regulatory functions that apply to the management of the area. This ecological evaluation should be completed in narrative form rather than being based on a checkoff list. Such a narrative would enable the more subtle functions and uses to be more fully considered in the determination of a wetland's suitability for the receipt of wastewater discharges.

Determination of existing wetland functions and uses and their significance within a watershed would enable a full characterization of the wetland and an evaluation of wetland effects to be made on both surface and groundwater quality; water quantity, particularly as it relates to storm-water runoff management; flood control and water storage and yield during low-flow events; groundwater recharge; shoreline erosion protection; and overall ecological health and diversity. Included within the ecological health and diversity evaluation should be a review of the existing and potential use of the subject wetland for recreation, research, and education, support

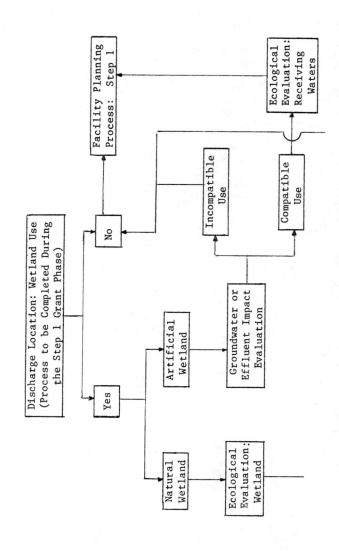

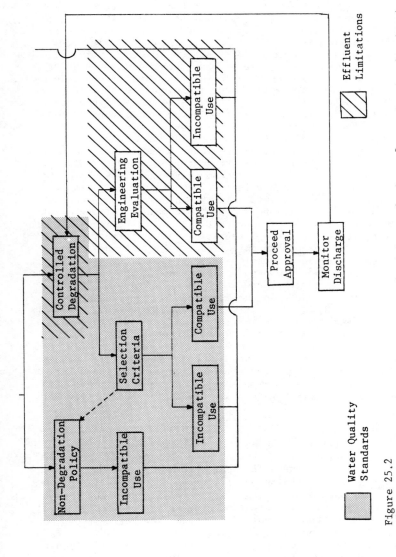

Figure 25.2
Facility planning assessment process emphasizing wetland use for wastewater treatment and discharge.

of such activities as trapping, hunting, and fishing, and the subject wetland's contribution to the aesthetic value of the community and other nonconsumptive values, such as use by nongame species.

Prior to any definitive ecological evaluation of a wetland, a detailed inventory of the plants and animals present should be conducted. Such an inventory can range from simple presence studies to more detailed studies relating to percent plant cover and frequency of occurrence, density, and population of both the plant and animal communities. When evaluating wetland communities for their use as wastewater treatment or discharge recipients, it is recommended that at least an inventory of presence with a relative plant-cover study be conducted. Because of the costs involved in doing plant-cover studies, it is recommended that nonforested wetlands be surveyed using a line intercept method within each plant community. Such a method of data collection is inexpensive and will provide sufficient information for determination of percent plant cover and frequency for later wastewater impact studies.

Determination of quality ratings for plant and animal communities within a wetland are often based on the individual opinions of the biologist reviewing the site. The degree of subjectiveness can be reduced, however, by using plant community assessments, such as the one suggested by Swink and Wilhelm (1979) that uses a rating index, or a natural area classification, such as the one proposed by Tans (1974) in which the wetland quality is a measure of the diversity of plant and animal species present, the structure and integrity of the plant community, and the extent of human disturbance. Also to be considered is the commonness of the particular wetland community within the area or in the presettlement vegetation; for example, a measure of the amount of cover type as compared to the total and the diversity of plant communities and other natural features occurring within the subject wetland. On the basis of these criteria, wetlands can be ranked and classified. For example, natural areas, including wetlands in many Wisconsin counties, have been ranked and classified as follows (Germain et al. 1977; Tans and Dawson 1980):

1. State Scientific Areas (SAs) that include those natural areas, geological sites, archaeological sites, or combination of areas that are of at least statewide significance and that have been designated as such by the Wisconsin Scientific Areas Preservation Council;

2. Natural Areas of Statewide or Greater Significance (NA-1) that include those natural areas slightly modified by human activity or that have sufficiently recovered such that they contain a nearly intact native plant and animal community believed to be representative of the presettlement vegetation;

3. Natural Areas of Countywide or Regional Significance (NA-2) that include those natural areas that have been slightly modified by human activities or have insufficiently recovered from past disturbances. Criteria considered include the degree to which the natural area quality is less than ecologically ideal and evidence of past or present disturbance from such activities as logging, grazing, manipulation of water levels, or pollution. Also considered is the commonness of the wetland type in a region (only the best of which might qualify for SA designation) and the size of the wetland area (it may be too small for consideration as an SA or NA-1); and

4. <u>Natural</u> <u>History</u> <u>Areas</u> (NA-3) that include those natural areas
   that have been modified by human activities. However, such NA-3s
   do retain a moderate amount of natural cover such that exclusion
   from a natural area inventory would be an oversight. Two or more
   of the above-mentioned identifying natural-area criteria may be
   substandard in natural history areas. Natural history areas
   should reflect patterns of former vegetation.

   While the above rating system is habitat based, other methodologies have been or need to be identified, evaluated, and utilized
in wetland evaluations. It should be noted that no single evaluation
methodology will cover all necessary function and value assessments.
However, a habitat-based evaluation methodology is supported here
because the analytical tools of this function are better developed
than for other functional elements (Lonard et al. 1981) and have
properties common to all wetland systems. The need for further
development of evaluation tools is evident. That all evaluation
tools be responsive to local, regional, and statewide wetland variations is essential to sound analysis (Lonard et al. 1981). It is
also apparent that a description or an understanding of the wetland
is not sufficient to make judgments about wastewater discharges.
Hirsch (1976) points out the role of baseline studies in environmental assessments. Benforado (1981) states the necessity for
evaluating the ecological effects (alterations) through identification of impact pathways (alteration mechanisms). How impact
pathways (see Brennan, this volume) can be controlled or eliminated
to achieve protection criteria for a given wetland will determine to
a large extent whether a wetland discharge is approved.
   The natural area criteria that we have identified are obviously
on the upper end of the range of natural values. These wetland types
should be protected through appropriate standards. However, additional criteria should be developed for both higher- and lower-value
wetlands to allow for analysis of various wetland/wastewater discharge alternatives. For instance, areas with existing management
activities, such as waterfowl production and endangered species
recovery, should be identified. A technical determination of their
compatibility with wastewater treatment and discharges should be
made. If it is determined that the receipt of any wastewater discharge would result in unacceptable changes in the composition and
structure of the plant community, manipulation of water levels or
flows, water quality, or the contribution of any substances directly
or indirectly deleterious to the target species being managed
(especially on refuges or similarly valuable areas), then the wetland
site being considered for wastewater discharge should not be used.
Similar considerations should be applied to such existing regulatory
functions as shellfish management, fish management, and designated
critical habitat areas for state and federal endangered and
threatened species.

NONDEGRADATION AND CONTROLLED DEGRADATION POLICIES

   Technically sound management requires that the resource manager
both understand and be able to convey to others what is to be accomplished. This usually requires goal identification and a management
process utilizing evaluation criteria for determining goal attainment.

The challenge facing this workshop and, more important, the practitioners of wastewater, water quality, and wetland management, is the developmemt of criteria that direct planning efforts for wastewater disposal and/or treatment in wetlands. The issue of wetland discharge is obscure in the Section 201 Facility Planning process because criteria for determining acceptability have been inadequately addressed or addressed by generic secondary effluent limits that do not consider the values, functions, or protection goals of an individual wetland. Formalized wetland criteria need to be adopted that will effectively guide the facility planning process. Criteria adopted into administrative rules will become water quality standards that represent the water quality objectives of an individual state and will provide the legal framework for water quality management planning and regulation (U. S. EPA 1976). The state of Wisconsin has recognized this concept and has developed administrative rules that recognize wetlands in their water quality standards (Wisconsin Department of Natural Resources 1979).

Criteria and standards are not necessarily synonymous. Water quality criteria are numerical concentrations or narrative descriptions of the levels of pollutants that have been determined by scientists to provide optimum protection for aquatic life and other uses. They do not reflect attainability and socioeconomic preference. Water quality standards are governmental regulations comprised of designated uses, narrative and numerical criteria to protect those uses, as well as an antidegradation policy to maintain high-quality values (U. S. EPA 1981). Where high-value wetlands occur and where significant resource impacts could occur, an antidegradation policy can be used to protect the high-value uses either through complete effluent prohibition or through effluent treatment requirements for targeted problem parameters (U. S. EPA 1976). High-value wetlands, such as SAs, NA-1s, and NA-2s, should be considered for inclusion in a nondegradation category. Hines (1977) has provided excellent insight into the nondegradation issue that should be of benefit regarding wetland use for wastewater discharges. Other wetlands of national importance could be classified in the nondegradation category, such as Outstanding National Resource Waters (U. S. EPA 1981).

Wetlands offer treatment or assimilative capabilities that argue for controlled degradation. Those wetlands that support limited species diversity, are regionally abundant, lack major beneficial uses, or have benefits that could be enhanced would be likely candidates for further treatment or discharge assessment. All of these terms should be clearly understood and defined in advance of specific proposal evaluations in order to avoid confusion and semantic differences. If a wetland is determined to be potentially acceptable as a receiving water for wastewater discharge, an engineering analysis should be undertaken to determine if additional mitigative efforts are necessary to meet controlled degradation criteria. This analysis may include a water discharge rate assessment to achieve discharge compatibility with maintenance of wetland flora dependent on certain water depths, phosphorus removal requirements to maintain plant diversity, emergent plant/open water ratios, or prevention of nuisance plant species from becoming more abundant. Additional assessment could also be required to adjust the pH of the effluent to mimic the receiving water, such as in a northern black spruce acid bog. Likewise, there may be a criterion that would limit the seasons of discharge to protect sensitive species during critical periods or to maintain plant phenology. Such criteria would require assessment of a gross water budget and seasonal storage requirements.

These specific examples are not intended to be viewed as absolute requirements but, rather, as a glimpse at the wide range of wetland values that need consideration in any formalized evaluation process. How individual wetlands will be protected or abused depends on our ability to transform our understanding of wetland systems and society's values into an evaluation procedure that gives sound guidance on what to look at and what must be done to assure reasonable protection of resource values. Much of this integration of scientific data, evaluation techniques, and substantive criteria must come from the monitoring of existing discharges and additional research if an information feedback loop is to validate assumptions and refine criteria for approving new or recurring discharges. Most likely, the specific criteria will be developed in the context of state or regional needs and only supplemented by very broad generic requirements.

Without the continual injection of this new information into the evaluation procedure, we will not be providing the necessary feedback to refine our technical understanding and improve subsequent decisions. The economic costs of this information may seem high and, indeed, they probably are. However, the ecological costs to society may be much higher in the long run if we do not incur these costs during the initial planning and implementation stages of wastewater discharges.

CONCLUDING RECOMMENDATIONS

Despite the recognition that there is an established institutional framework to evaluate wastewater discharges into wetlands, little in the way of formalized criteria to either evaluate wetlands as wastewater receiving waters or provide reasonable protection of important wetland values have been developed. Such criteria need to be developed in order to provide guidance to those responsible for planning and reviewing wastewater projects. The criteria must remain flexible to accept the widely varying functions and values of individual wetlands on a national, state, and substate (regional) level. They must also be succinct enough to direct planning efforts in a timely manner with sufficient requirements to afford reasonable resource protection.

Clark and Clark (1979) recognized the need for a "National Wetlands Research Assessment." We support the concept generally and submit that a specific assessment task be devoted to the issue of wastewater and wetlands. This assessment would be most beneficial since it could consolidate and refine value assessment methodologies, produce workable guidelines and criteria for planning, be available for standards review, and maintain a strong technical coordination and direction function. We believe the wetlands wastewater issue has been trapped in a "deflection of goals" dilemma as described by Bardach (1977). As such, the Clean Water Act objective to restore and maintain the chemical, physical, and biological integrity of the nation's waters is not being met in an optimum manner.

Finally, we recommend, as an initial task, that planning criteria, water quality standards, and National Pollution Discharge Elimination System programs on both the state and federal level be surveyed to acknowledge what is being done and what may need to be done to refine the process.

REFERENCES

Bardach, E., 1977, The implementation game: What happens after a bill becomes law, MIT Press, Cambridge, Mass.

Bastian, R. K., and S. C. Reed, 1979, Aquaculture systems for waste-water treatment: Seminar proceedings and engineering assessment, EPA 430/9-80-006, U. S. EPA, Office of Water Program Operations, Municipal Construction Division, Washington, D. C.

Benforado, J., 1981, Ecological considerations in wetland treatment of municipal wastewater, in Selected proceedings of the Midwest conference on wetland values and management, B. Richardson, ed., Minnesota Water Planning Board, St. Paul, Minn., pp. 307-323.

Chan, E., J. A. Bursztynsky, N. Hantzsche, and Y. J. Litwin, 1978, The use of wetlands for wastewater pollution control, Municipal Environmental Research Laboratory, Office of Research and Development, U. S. EPA, Cincinnati, Ohio.

Clark, J. R., and J. E. Clark, eds., 1979, Scientists report: The national symposium on wetlands, Lake Buena Vista, Florida, National Wetlands Technical Council, Washington, D. C.

Council on Environmental Quality and U. S. Fish and Wildlife Service, 1980, Biological evaluation of environmental impacts: The pro-ceedings of a symposium, FWS/OBS-80/26, Office of Biological Services, U. S. Fish and Wildlife Service, Washington, D. C.

Cowardin, L. M., V. Carter, F. C. Golet, and E. T. LaRoe, 1979, Classification of wetlands and deepwater habitats of the United States, FWS/OBS-79/31, U. S. GPO, Washington, D. C.

DeWitt, C. B., and E. Soloway, eds., 1978, Wetlands ecology, values, and impacts: Proceedings of the Waubesa conference on wetlands held in Madison, Wisconsin, June 2-5, 1977, University of Wisconsin, Madison, Wis.

Germain, C. E., W. E. Tans, and R. H. Read, 1977, Wisconsin scien-tific areas 1977: Preserving native diversity, tech. bull. no. 102, Wisconsin Department of Natural Resources.

Good, R. E., D. F. Whigham, and R. L. Simpson, 1978, Freshwater wetlands: Ecological processes and management potential, Academic Press, New York, 378p.

Greeson, P. E., J. R. Clark, and J. E. Clark, eds., 1979, Wetland functions and values: The state of our understanding, Proceedings of the national symposium on wetlands, Lake Buena Vista, Florida, tech. pub. 79-2, American Water Resources Association, Minneapolis, Minn., 674p.

Hines, N. W., 1977, A decade of nondegradation policy in Congress and the courts: The erratic pursuit of clean air and clean water, Iowa Law Review 62:643.

Hirsch, A., 1976, The baseline study as a tool in environmental impact analysis, in Biological evaluation of environmental impacts, FWS/OBS-80/26, Council on Environmental Quality and U. S. Fish and Wildlife Service, Washington, D. C.

Horwitz, E. L., 1978, Our nation's wetlands: An interagency task
    force report, U. S. GPO, Washington, D. C.
Lonard, R. I., E. J. Clairain, R. T. Huffman, J. W. Hardy, L. D.
    Braun, P. E. Ballard, and J. W. Watts, 1981, Analysis of method-
    ologies used for the assessment of wetlands values, U. S. Water
    Resources Council, Washington, D. C.
Richardson, B., ed., 1981, Selected proceedings of the Midwest
    conference on wetland values and management, Minnesota Water
    Planning Board, St. Paul.
Sloey, W. E., F. L. Spangler, and C. W. Fetter, Jr., 1977, Management
    of freshwater wetlands for nutrient assimilation, in Freshwater
    wetlands: Ecological processes and management potential, R. E.
    Good, D. F. Whigham, and R. L. Simpson, eds., Academic Press, New
    York, pp. 321-340.
Swink, F., and G. Wilhelm, 1979, Plants of the Chicago region,
    2d ed., rev., Morton Arboretum, Lisle, Ill.
Sutherland, J. C., and R. H. Kadlec, 1979, Wetland utilization for
    management of community wastewater, abstracts of a conference held
    July 10-12, 1979, Higgins Lake, Michigan.
Tans, W. E., 1974, Priority ranking of biotic natural areas, Michigan
    Botanist 13:31-39.
Tans, W. E., and R. Dawson, 1980, Natural area inventory: Wisconsin
    Great Lakes coast, Office of Coastal Management, Wisconsin Dept.
    of Administration and Scientific Areas Section, Wisconsin Dept. of
    Natural Resources.
Tilton, D. L., R. H. Kadlec, and C. J. Richardson, eds., 1976,
    Proceedings of the national symposium on freshwater wetlands and
    sewage effluent disposal, University of Michigan, Ann Arbor.
U. S. Environmental Protection Agency, 1976, Water quality standards,
    in Guidelines for state and areawide water quality management
    program development, chap. 5, Washington, D. C.
U. S. Environmental Protection Agency, 1981, Water quality standards,
    40 CFR 35.1550, U. S. GPO, Washington, D. C.
Weller, M. W., 1981, Freshwater marshes: Ecology and wildlife
    management, University of Minnesota Press, Minneapolis, Minn.
Wisconsin Department of Natural Resources, 1979, Wisconsin adminis-
    trative code: Rules of the department of natural resources,
    Environmental protection: NR 102 and 104, Madison.

DISCUSSION

Odum: You have addressed the objective of creating new wetlands or
taking degraded ones and providing some nutrients to let them create
a new system. Put a fence around it and call it a treatment plant if
you want to. There is, however, another objective, unstated but
implied--that of protecting natural wetlands from adverse impact. It
would be a big mistake for a regulatory agency to confuse your stated
objective with the implied one--that is, protecting the public domain
versus helping create new wetlands or restoring existing wetlands to
meet a special need.

Reed: Your comment is correct. When following the flow chart
(Fig. 25.1) that we presented, we come to a branch where we must
decide between an artificial or natural wetland. If we select an
artificial wetland, we will then undertake a groundwater or effluent

impact evaluation as we would for any other sewage treatment discharge. At that point, the facility planning processes are employed.
The processes that Tim and I are concerned with are the ecological
evaluation of the wetland; that is to say, whether one should decide
to discharge to a natural wetland system rather than to an artificial
system.

Odum: I can see obstructionism getting in the way here. We have two
objectives and they both ought to be accomplished. Unfortunately, I
can see one blocking the other.

26

Management Potential for Nutrient Removal in Forested Wetlands

Mark M. Brinson

Because of the diversity of wetland types, it is difficult to arrive at useful generalizations on the effectiveness of these ecosystems for assimilating nutrients in wastewater. However, workshops, reviews, and symposia during the past decade (Greeson, Clark, and Clark 1978; Good, Whigham, and Simpson, eds. 1978; Johnson and McCormick 1979; Brinson et al. 1981; Wharton et al. 1982; Clark and Benforado 1981) and several ecosystem-level studies on wetlands (Boyt, Bayley, and Zoltek 1977; Odum and Ewel 1977; Richardson et al. 1976; Schlesinger 1978; Kuenzler et al. 1980) have provided us with enough information to allow us to evaluate the capacity of some of these ecosystems to retain and process nutrients. For example, the great majority of wetlands occur in depositional environments that tend to accumulate materials (sediments) from adjacent ecosystems. Although geologic events or human alterations may quickly reverse this trend of accretion, wetlands are depositional environments in contrast to uplands. A wastewater treatment function is an attempt to capitalize on the capacity for material accumulation beyond natural levels. Another attribute of wetlands is that they tend to be more complex structurally than streams, the latter serving as the traditional discharge point for sewage effluent. Discharge lines often bypass wetlands that could provide a filtering function if an effective distribution system were designed.

This paper briefly examines how variations among wetland types may affect their suitability for wastewater treatment. Finally, an experimental approach for assessing sustained nutrient loading is evaluated.

ECOSYSTEM CHARACTERISTICS

Hydroperiod, sediments, and vegetation are the principal features that distinguish one wetland type from another. All should be regarded as having an influence on the capacity of a wetland to remove nitrogen (N) and phosphorus (P) from wastewater.

HYDROLOGY

For a wetland to be effective in nutrient assimilation, waste-water must remain in contact with the wetland long enough to allow removal by microbial transformations, adsorption by surfaces, uptake by plants, or deposition of particulate matter. Knowledge of the hydroperiod (depth, duration, and seasonality of flooding) is essential for insight into the amount of water retention that a wetland may provide. Some examples of hypothetical hydroperiods are given in Figure 26.1. These examples illustrate the variety of conditions that can be encountered in different wetland types. The amount of inflow and outflow can vary greatly for a given hydroperiod. For example, a floodplain with a relatively steep downstream slope will not normally flood deeply, and the velocity of water flow will be high and thus result in only a short retention time. Shallow flooding in a topographically level bog forest may result in little lateral flow and have a long retention time. Thus, water level changes are insufficient information for determining the retention time of water in a wetland.

However, some generalizations may be made based on hydroperiod alone. For example, the tidal swamp (Figure 26.1) will seldom retain water for more than a few days, while the shrub bog may be relatively "closed," or isolated from exchange with adjacent aquatic systems (Daniel 1981). With the exception of tidal swamps, wetlands may undergo a degree of dry-down when inflows are low and evapotrans-piration is high. During dry-down, the retention and contact time of wastewater discharged into wetlands would be maximal.

Wastewater discharge into wetlands may significantly alter the hydroperiod and hydrology depending on the ratio of natural inflows to loading rate. The implications of altered hydroperiod must be understood since an altered hydroperiod can affect the integrity of the vegetation (differential tolerances of species to flooding) and the extent to which aerobic or anaerobic processes predominate in the sediment.

SEDIMENT

Sediment characteristics in wetlands are dependent on the geomorphology and hydrology of the basin in which the wetland exists. Wetlands with low water "turnover," or flow (stillwater wetlands), often have long hydroperiods, insufficient energy to import inorganic sediments, and low decomposition rates. These factors combine to result in deposits with high organic carbon content. Wetlands with more rapid flowthrough and a source of sediments from eroding uplands have predominately inorganic soils. Most wetlands acquire sediments of intermediate composition that depend upon the ratio of biogenic and fluvial deposition rates (Fig. 26.2).

Sediment composition can have a strong influence on the cycling of wastewater nutrients. For example, a high cation exchange capac-ity is necessary to provide adsorption sites for NH_4^+, labile organic matter is necessary for the denitrification of NO_3^-; and iron oxides allow adsorption and precipitation of PO_4^{-3}. Both the physical-chemical environment and the microbial component undergo significant

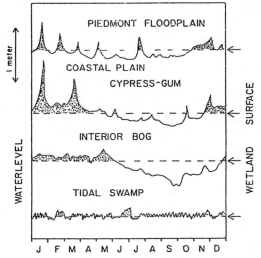

Figure 26.1
Hypothetical hydroperiods that illustrate the variety of
hydrologic conditions in four types of forested wetland.

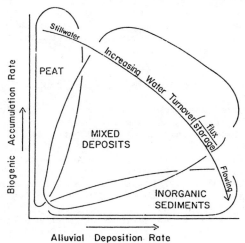

Figure 26.2
Factors influencing the composition of wetland
sediments. Most sediments are intermediate
between peat and inorganic composition, a function
of biogenic accumulation and alluvial deposition as
determined by water flow. (After S. Brown, M.M.
Brinson, and A.E. Lugo, 1979, Structure and function
of riparian wetlands, in Strategies for protection
and management of floodplain wetlands and other
riparian ecosystems, R.R. Johnson and J.F. McCormick,
tech. coords., gen. tech. rep. WO-12, USDA Forest
Service, Washington, D.C., pp. 17-31.)

seasonal shifts due to changes in temperature and moisture status.

VEGETATION

To illustrate the limitations of timber harvesting for nutrient removal from wetlands, we can compare the P accumulation in stem wood of a productive forested wetland with the P production per capita from a municipal effluent. Brown (1981) reported that stem wood production in Florida cypress forests receiving effluent increased by two- to threefold over rates prior to application. Using her wood production value for the most productive of her cypress forests (1060 g $m^{-2}yr^{-1}$) and her highest stem wood P concentration (0.11 mg P g^{-1}) gives 1.2 kg P $ha^{-1}yr^{-1}$ as "available" for harvest through timber removal. The sewage treatment plant of Greenville, North Carolina (population 40,000), has an average discharge of 20.8 x 10^6 1 day^{-1} and a mean concentration of filterable reactive P of 3.6 mg 1^{-1}, or an average annual production of 0.68 kg P per person. Thus, a hectare of highly productive forest could serve only about two people if harvesting was the sole managment approach for P removal.

These results imply that harvest of timber is not an effective management strategy for exporting nutrients added by wastewater. Too many other factors (e.g., fire, growing season, hydroperiod, sediment type), in addition to nutrient inputs, control the structure and growth of forested wetlands. In the loading experiments described below, much of the P is accumulated in sediments rather than being stored by vegetation, similar to the trends reported by Brown (1981). Calculations for N would lead to similar conclusions regarding the effectiveness of timber harvest.

EXPERIMENTAL APPROACH TO EVALUATING NUTRIENT ASSIMILATION

The foregoing discussion of hydroperiod, sediments, and vegetation of forested wetlands focused on some of the considerations for nutrient removal from wastewater. This section describes and evaluates an experimental approach for evaluating the capacity of an alluvial forest in the coastal plain of North Carolina to assimilate nitrate, ammonium, and phosphate in sewage effluent (Brinson, Bradshaw, and Kane 1981). Previous studies on the site demonstrated that 90% of the nitrate added to the surface water was removed in 10 days (through denitrification), that approximately 50% of added ammonium was removed in the same time, and that the transfer of phosphate from surface water to the sediments was by nonbiological processes (Brinson, Bradshaw, and Holmes 1983). However, these preliminary experiments, conducted seasonally, did not assess the effects of sustained nutrient loading that would simulate use of the swamp forest for continuous nutrient removal.

The experiment consisted of adding concentrated solutions of nitrate (NO_3 treatment), ammonium (NH_4 treatment) and phosphate (PO_4 treatment) singly and in combination (PNN treatment) to 1.46 m^2 enclosures on the swamp forest floor at a rate of 1 g $m^{-2}week^{-1}$ over 46 months. Also, secondarily treated sewage effluent was added at weekly intervals to yet another enclosure so that the average weekly loading rate was 0.92 g N m^{-2} for ammonium, 0.08 g N m^{-2} for nitrate, and 0.19 g P m^{-2} for filterable reactive P (78% of total P). This loading rate added about 5 cm week^{-1} of sewage effluent to the forest floor. Treatments were compared to control areas with and without enclosures. Water samples were collected approximately 1 hour after nutrient addition and 7 days later for surface water (when present), weekly for subsurface water, and monthly for sediment analysis (upper 5 cm). Other details of the experiment and analyses can be found in Brinson, Bradshaw, and Kane (1981).

One major objective of the experiment was to assess the response of the forest floor to ammonium loading. Consequently, the reason for choosing the 1 g NH_4-N m^{-2}week^{-1} ammonium treatment rate was to closely simulate the ammonium loading rate (0.92 g N m^{-2}week^{-1}) in the sewage effluent treatment. Ammonium concentrations for various treatments over the study period are shown in Figure 26.3. The difference between the day 0 and day 7 plots indicates the amount of reduction of concentration in the surface water. Also, the sewage treatment had concentrations after 7 days similar to the NH_4 treatment. Higher concentrations were maintained in the PNN treatment (phosphate, nitrate, and ammonium added together in equal amounts) because the chamber was located on a topographic low that had greater water depth. We believe that the greater water depth maintained more anaerobic conditions that in turn suppressed nitrification. However, in all treatments and for most sampling dates, there were substantial accumulations of ammonium on day 7 compared with concentrations in controls.

The nitrate treatment showed an almost unlimited capacity for reducing concentrations to very low levels after 7 days, both when nitrate was added in concentrated form (NO_3 and PNN treatments) and in sewage effluent (Fig. 26.4). The sewage loading rate of nitrate averaged only 8% of that added in "pure" form in the NO_3 and PNN treatments. Thus, the capacity of the system for nitrate removal was not exceeded in the sewage experiments, probably owing to the presence of anaerobic conditions near the soil surface and an ample supply of organic carbon (C) in the sediment for driving denitrification.

The results for exchangeable ammonium (Fig. 26.5) over the treatment period suggest that a summer dry-down period is important for depleting this pool of ammonium through nitrification and subsequent denitrification. Controls also showed substantial fluctuation in exchangeable ammonium, indicating that ammonification, nitrification, and denitrification are all very active microbial processes during seasonal fluctuations of water level under natural conditions.

The results of the PO_4 treatments show that the capacity of the sediments for filterable reactive P removal from surface water was very rapidly exceeded at the loading rates used (Fig. 26.6). However, the results identify a flaw in the experimental design since the loading rate was unrealistically high in relation to that used in the sewage treatment. A better design would have been to add phosphate either at levels present in the sewage treatment (N:P atomic ratio of 2.3:1 rather than the ratio of 0.45:1 that was used) or to use the N:P ratio of the sediment (4:1) as a guide. Regardless, the extractable P fraction in the sediment showed an upward trend even in the sewage treatment when compared to the controls (Fig. 26.7). Since P, unlike N, has no atmospheric sink, P loading results in relatively irreversible accumulation in sediments.

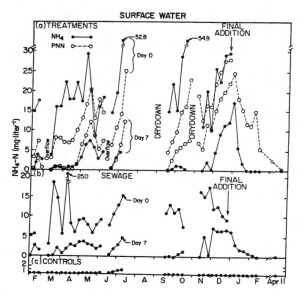

Figure 26.3
Ammonium concentrations of surface water (<u>a</u>) in the NH₄ and PNN
treatments, (<u>b</u>) in the sewage treatment, and (<u>c</u>) in controls not
receiving ammonium loading. In <u>a</u> and <u>b</u>, upper lines are concentrations
< 1 h after addition (day 0) and lower lines are concentrations after
7 days. Control concentrations are the means of the PO₄ treatment,
NO₃ treatment, chamber control, and open area.

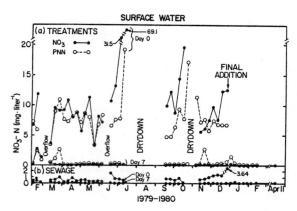

Figure 26.4
Nitrate concentrations of surface water in the (<u>a</u>) NO₃ and PNN
treatments and (<u>b</u>) sewage treatment. Surface water concentrations
of controls not receiving nitrate are not graphed because mean values
exceeded 0.1 mg NO₃-N liter^{-1} only once after the first month.

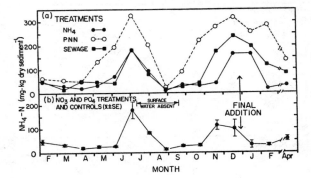

Figure 26.5
Exchangeable ammonium concentrations of the surface sediment for
(<u>a</u>) NH_4, PNN, and sewage treatments, and (<u>b</u>) NO_3 and PO_4 treatments
and controls. (After M.M. Brinson, H.D. Bradshaw, and E.S. Kane,
in press, Nutrient assimilative capacity of an alluvial floodplain
swamp, <u>J</u>. <u>Appl</u>. <u>Ecol</u>. 21:2.)

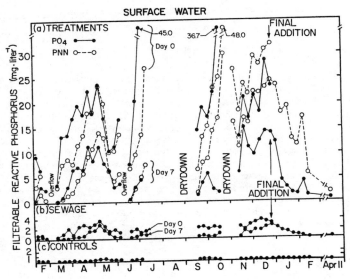

Figure 26.6
Filterable reactive phosphorus (FRP) concentrations of surface water
in (<u>a</u>) PO_4 and PNN treatments, (<u>b</u>) sewage treatment, and (<u>c</u>) controls
not receiving phosphate loading. In <u>a</u> and <u>b</u> upper lines are
concentrations < 1 h after addition (day 0) and lower lines are
concentrations 7 days later. Control concentrations are the mean of
the NH_4 treatment, NO_3 treatment, chamber control, and unenclosed
area.

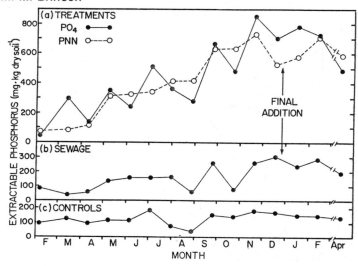

Figure 26.7
Extractable phosphorus concentrations of the surface sediment for
(a) PO₄ and PNN treatments, (b) sewage treatment, and (c) controls.
(After M.M. Brinson, H.D. Bradshaw, and E.S. Kane, in press, Nutrient
assimilative capacity of an alluvial floodplain swamp, J. Appl. Ecol.
21:2)

ASSESSMENT OF THE EXPERIMENTAL DESIGN

The advantages and disadvantages of the experimental approach
described can be compared with the level of effort and quality of
information derived from using other "scales" of experiments. The
weekly additions of concentrated forms of nutrients represent a level
of effort intermediate between laboratory experiments with soil
columns and pilot scale attempts using wastewater effluent in a
wetland ecosystem. In comparison with laboratory experiments, our
intermediate scale experiments have the advantage of studying nutrient behavior during natural water-level and temperature fluctuations
and other seasonal influences that are present in the ecosystem in
question. It would be difficult to duplicate these conditions in the
laboratory and impossible to simulate influences of vegetation.
However, carefully designed laboratory experiments would have the
advantage of being able to control individual variables (temperature,
soil moisture, redox status, and so on). This approach has been the
one used to gain an understanding of N (Patrick and Tusneem 1972;
Reddy and Patrick 1975; Phillips, Reddy, and Patrick, Jr. 1978;
Graetz et al. 1980) and P (Patrick and Mahapatra 1968) dynamics in
wetland soils.

A pilot scale approach using wastewater and existing wetland
conditions would have the advantage of realism over the intermediate
experimental approach described above. The use of enclosures in our
system altered dispersion of nutrients and hydrologic conditions, but
the degree of these effects is either small (open and enclosed
controls were not different) or cannot be completely evaluated.

However, the flexibility of experimentally isolating the behavior of individual nutrients is sacrificed in the pilot scale approach. It also may represent a higher level of resource commitment, depending on the proximity of a suitable wetland area to the wastewater source.

REFERENCES

Boyt, F. L., S. E. Bayley, and J. Zoltek, Jr., 1977, Removal of nutrients from treated municipal wastewater by wetland vegetation, Water Pollut. Control Fed. J. 48:789-799.

Brinson, M. M., H. D. Bradshaw, and R. N. Holmes, 1983, Significance of floodplain sediments in nutrient exchange between a stream and its floodplain, in The dynamics of lotic ecosystems, T. D. Fontaine, III and S. M. Bartell, eds., Ann Arbor Science, Ann Arbor, Mich., pp 199-221.

Brinson, M. M., H. D. Bradshaw, and E. S. Kane, 1981, Nitrogen cycling and assimilative capacity of nitrogen and phosphorus by riverine wetland forests, rep. no. 167, Water Resources Research Institute, University of North Carolina, Raleigh, 90p.

Brinson, M. M., H. D. Bradshaw, and E. S. Kane, in press, Nutrient assimilative capacity of an alluvial floodplain swamp, J. Appl. Ecol. 21:2.

Brinson, M. M., B. L. Swift, R. C. Plantico, and J. S. Barclay, 1981, Riparian ecosystems: Their ecology and status, FWS/OBS-81/17, U. S. Fish and Wildlife Service, Washington, D. C., 153p.

Brown, S., 1981, A comparison of the structure, primary productivity, and transpiration of cypress ecosystems in Florida, Ecol. Mongr. 51:403-427.

Brown, S., M. M. Brinson, and A. E. Lugo, 1979, Structure and function of riparian wetlands, in Strategies for protection and management of floodplain wetlands and other riparian ecosystems, R. R. Johnson and J. F. McCormick, tech. coords., gen. tech. rep. WO-12, USDA Forest Service, Washington, D. C., pp. 17-31.

Clark, J. E., and J. Benforado, eds., 1981, Wetlands of bottomland hardwoods, Vol. II. Developments in agricultural and managed forest ecology, Elsevier Science, New York, 402p.

Daniel, C., III, 1981, Hydrology, geology and soils of pocosins: A comparison of natural and altered systems, in Pocosin wetlands: An integrated analysis of coastal plain freshwater bogs in North Carolina, C. J. Richardson, ed., Hutchinson Ross, Stroudsburg, Pa., pp. 69-108.

Good, R. E., D. F. Whigham, and R. L. Simpson, eds., 1978, Freshwater wetlands: Ecological processes and management potential, Academic Press, New York, 378p.

Graetz, D. A., P. A. Krottje, N. L. Erickson, J. G. A. Fiskell, and D. F. Rothwell, 1980, Denitrification in wetlands as a means of water quality improvement, publ. no. 48, Florida Water Research Center, University of Florida, Gainesville.

Greeson, P. E., J. R. Clark, and J. E. Clark, 1978, Wetland functions and values: The state of our understanding, American Water Research Assoc., Minneapolis, Minn., 674p.

Johnson, R. R. and J. F. McCormick, tech. coords., 1979, Strategies for protection and management of floodplain wetlands and other riparian ecosystems, gen. tech. rep. WO-12, USDA Forest Service, Washington, D. C., 410p.

Kuenzler, E. J., P. J. Mulholland, L. A. Yarbro, and L. A. Smock, 1980, Distribution and budgets of carbon, phosphorus, iron and manganese in a floodplain swamp ecosystem, rep. no. 157, Water Resources Research Institute, University of North Carolina, Raleigh, 234p.

Odum, H. T., and K. C. Ewel, eds., 1977, Cypress wetlands for water management, recycling and conservation, 4th annual report to NSF (RANN) and Rockefeller Foundation, Center for Wetlands, University of Florida, Gainesville, 945p.

Patrick, W. H., Jr., and I. C. Mahapatra, 1968, Transformation and availability to rice of nitrogen and phosphorus in waterlogged soils, Adv. in Agron. 20:323-359.

Patrick, W. H., Jr., and M. E. Tusneem, 1972, Nitrogen loss from flooded soil, Ecology 53:735-737.

Phillips, R. E., K. R. Reddy, and W. H. Patrick, Jr., 1978, The role of nitrate diffusion in determining the order and rate of denitrification in flooded soil: II. Theoretical analysis and interpretation, Soil Sci. Soc. Am. J. 42:272-278.

Reddy, K. R., and W. H. Patrick, Jr., 1975, Effect of alternate aerobic and anaerobic conditions on redox potential, organic matter decomposition, and nitrogen loss in a flooded soil, Soil Biol. Biochem. 7:87-94.

Richardson, C. J., W. A. Wentz, J. P. M. Chamie, J. A. Kadlec, and D. L. Tilton, 1976, Plant growth, nutrient accumulation, and decomposition in a central Michigan peatland used for effluent treatment, in Freshwater wetlands and sewage effluent disposal, D. L. Tilton, R. H. Kadlec, and C. J. Richardson, eds., Proceedings of a symposium held May 10-11, 1976, School of Natural Resources and College of Engineering, University of Michigan, Ann Arbor, 343p.

Schlesinger, W. H., 1978, Community structure, dynamics and nutrient cycling in the Okefenokee cypress swamp forest, Ecol. Monogr. 48:43-65.

Wharton, C. H., W. M. Kitchens, E. C. Pendleton and T. W. Sipe, 1982, The ecology of bottomland hardwood swamps of the Southeast: A community profile, FWS/OBS-81/37, U. S. Fish and Wildlife Service, Washington, D. C., 133p.

DISCUSSION

Miller: It is interesting that you have such a different system from our cattail marshes in some respects, yet our results tend to indicate the very same things: (1) that the phosphorus ends up in the sediment, and (2) you have excellent rates of denitrification; but it is the rate of the nitrification that determines how much ammonia you have left in the effluent. In our system, nitrification is linked to the redox potential of the marsh sediments because when the system gets very anaerobic and redoxes are very negative, nitrification seems to shut down and the ammonia goes right through. That's apparently what's happened in your system, too, because when you dry out the forested wetlands, everything is reoxidized.

Brinson: Exactly, so in any management scheme, you would have to dry out the sediments at some point if you were going to attempt ammonia removal.

Miller: That is exactly what we're doing in one of our cells right now.

Brinson: Another point I would like to make is that this experimental approach is somewhat in between a purely bench-model, soils-type analysis (soil columns) and either a pilot scale or full-fledged scale of application of sewage effluent. So, you could probably take this approach and apply it to different wetlands without a whole lot of commitment or effort. It gives you quite a bit of information. However, there are some limitations in extrapolating results to a larger scale.

Bayley: Do you think you get any burial of that phosphorus? In other words, is it going to reach a steady state for input/output?

Brinson: I think the phosphorus is going to be remobilized. We saw a lot of phosphorus in the surface water and accumulation in sediments, and I believe it will eventually leak out of the system at the rate at which you put it in.

Bayley: So, you do not get any phosphorus burial with sedimentation?

Brinson: No, the rate of sedimentation is not enough to be significant. I'm sure it would be in some systems [see Day and Kemp, this volume].

Bayley: That's why you get those low N:P ratios.

Brinson: Right.

Larson: Mark, this is perhaps not directed at you. Before we leave the section involving hydrology, I wondered if someone, maybe Harry [Hemond], would like to comment on the applicability of Darcy's Law to organic soils? I'm not a hydrologist, so I don't know the answer, but I hear enough questions raised about whether Darcy's Law prevails to ask for some comments.

Hemond: I'll address that. I don't think there's much question about whether it works. There are problems, and Bob [Kadlec] alluded to some of the issues that have to be addressed. One problem is that the conductivity is a function of a number of things. In particular, it can vary as the peat desaturates; that is true of any soil. It can also vary as the peat compresses, so compressibility is yet another complicating factor. It seems very possible that it can be affected by the state of the microbial community. It most certainly is influenced by root growth. So the answer is yes. It works, but we have a time-varying, hydraulic conductivity that we have to understand in order to make a useful model. In some situations, K may vary so much that we have to put special effort into pinning down just what it is. That would be my answer.

Kadlec: Yes, my point earlier, but restated. I question the utility of a constant in an equation that is time-varying, distance-varying, direction-varying, and depends on previous history of water saturation of the medium because once-dried peat is different from never-

dried peat. It depends on too many factors. It was supposed to be a
constant in the original sewage of Paris, you know, but it is not in
wetlands.

Hemond: It depends, obviously, on how much it varies. If it varies
by orders of magnitude, you have real problems; if it varies by a
factor of two, you may still have a very useful tool. I don't think
enough is known about it to . . . .

Odum: I'd like to raise the question of large-scale structure. An
example of that is taken from personal experience on the British
moors, research site of an IBP study [British moors and montane
grasslands, O. W. Heal and D. F. Perkins, eds., Springer-Verlag, New
York, 1978]. As I was tromping in the heather, along with the red
grouse flying up, all of a sudden, I dropped into a hole right up to
my neck. The high moor and its peat build up over a broad surface,
but it is eroded in one spot at a time by water converging as it
drains. Water circulation and peat deposition are not at all represented by selecting a typical place because the water converges
locally and intermittently. Whether there is net deposition is not
discernible on a small scale. You have to look at things on more
than one scale.

Hemond: Yes, that's an excellent point.

Ewel: I think that applies especially to forested wetlands where you
get channels from root decay.

Hemond: Absolutely, also enlarged crab burrows and other things.
You've got to look at all those different scales to find an absolute
equation.

Bedford: The phenomenon that H. T. [Odum] mentioned--of channels
through peat--was also true of the Columbia site. There were areas
where water moved from the cooling lake to distances of 500 m away
within a day through an underground channel in the peat, and there
were other areas where it took months for water from the same source
to move a comparable distance.

Hemond: What all this, probably, is going to mean is that the people
versed with statistics are going to have to get into the act at some
point. I mean, there are meaningful things you can describe proba-
bilistically. There are those worm channels where a lot of the flow
may go through and, conversely, on some scales, there are blocks of
peat where certain important nutrient transformations occur because
of transport limitations, insufficient oxygen availability, or what
have you. These phenomena undoubtedly will have to be described
statistically when we're looking at the full ecosystem scale.

27

Wetland-Wastewater Economics

Jeffrey C. Sutherland

In the mid-1970s, there was a great need in Michigan for economical, postsecondary wastewater treatment, including phosphorus (P) removal. Wetland application for removal of P from this community's stabilization pond effluent seemed an affordable alternative. By early 1976, four years of research under the direction of Dr. Robert Kadlec of the University of Michigan at the Porter Ranch peatland near Houghton Lake, Michigan indicated excellent renovation of wastewater and nitrogen (N) species by wetland application (Kadlec, Tilton, and Kadlec 1976).

If it had not been for National Science Foundation (NSF) support, a full-scale wetland treatment project at Houghton Lake might never have been developed. Nothing like it had been done before in Michigan. The U. S. EPA had never funded design and construction of a wetland treatment project through the Municipal Construction Grants program, and state and federal review agencies had understandable concerns for the integrity of this pristine natural wetland. The factors that brought success to the project were excellent P-removal potential, prospects for net positive environmental responses, an informed local populace, and great projected savings in wastewater treatment costs. The construction cost for the wetland at Houghton Lake was projected in 1976 to be $600,000. The construction of the upland irrigation alternative was projected at $1.1 million, or 83% higher. The construction of the wetland wastewater facilities, in fact, cost approximately $400,000 (1978 dollars).

By early 1976, the NSF project officer, Dr. Edward Bryan, thought it time to develop the general economic feasibility picture. Williams and Works, Inc., was selected to begin such a study, limited to Michigan (Sutherland 1977). At that time, we had more questions than information. One of the questions was: What kind of pretreatment assumption should be made? It seemed reasonable to think of secondary treatment as the minimum pretreatment requirement for natural wetlands. Natural wetlands, in general, were perceived to be highly sensitive to cultural use. Also, there was great need in Michigan for economical, postsecondary treatment, including P removal. The use of either wetlands or upland irrigation for P removal would be achieved primarily by small rural and resort communities in

Michigan. There are only a few large communities located near the
extensive tracts of undeveloped upland or wetlands needed, but there
are many small communities with wastewater flows of less than 0.5
million gallons per day (MGD) that are so located. And, a great
number of these were already committed to secondary treatment by
stabilization ponds.

We found that wetlands of significant size located within ten
miles of secondary treatment facilities were all bordered by streams
and rivers. The size of wetlands we had in mind would be such that
all of a community's wastewater could be applied at <1 in/wk during a
six-month season. Stabilization ponds would store the wastewater
during the six-month off-season. We selected ten communities to
evaluate in detail. The ten had wastewater flows in the range of
about 40,000 gallons per day (GPD) to 350,000 GPD. All were located
near wetlands large enough to apply wastewater at the rate of <1
in/wk. We assumed that state-owned wetlands would be free, that
privately owned wetlands would cost $550/acre (updated to 1982), and
that a gated irrigation pipe would be used for applying wastewater to
the wetland based on the developing design for Houghton Lake. We
also assumed that chlorination and dechlorination facilities would be
needed and, of course, the transmission forceline from ponds to
wetland would be like any other conventional wastewater forceline.
Very few small Michigan communities are served with biomechanical
secondary treatment systems; however, the economic projections should
apply to wetland treatment of secondary effluent in general.

CAPITAL COSTS

All of the costs and equations have been updated to March 1982
(Fig. 27.1 through Fig. 27.3 and Tables 27.1 and 27.2 are based on
the 1977 data). In Figure 27.1, wetland capital costs, exclusive of
scientific and engineering-related costs, are shown plotted against
distance in miles that the wastewater must be pumped to reach the
wetland irrigation pipe. Forceline is the overwhelming cost at
roughly $132,000/mi. The correlation between costs and distance is
quite good, with a coefficient of variation of 0.90. Although five
of the ten communities would need to purchase privately owned wetlands at $550/acre, neither the land costs nor the amount of waste-
water to be treated seem to affect capital costs very much. As an
example, the Houghton Lake system, at a cost of $560,000, falls close
to the line, even though the piping system can handle flows up to 2.5
MGD.

For comparison, spray irrigation system costs were also modeled
for each of the ten communities. For spray irrigation, the estimate
of cost for upland tracts is $1500/acre, including enough total area
to provide 800 ft of owned isolation distance around the wetted spray
perimeter. We assumed, for the spray irrigation system, a fairly
conservative hydraulic loading of 2 in/wk. We also picked the
nearest apparently suitable land for spray irrigation to hold forceline costs down. The average pond-to-spray site distance for the ten
communities was a little less than 1.5 mi. Land costs and mechanical
irrigation facilities were the most prominent costs involved in spray
irrigation. Center-pivot sprinklers would cost around $3000/acre,
and fixed irrigation laterals would cost around $6000/acre. We
picked the lowest-priced combination of the two, related to the shape

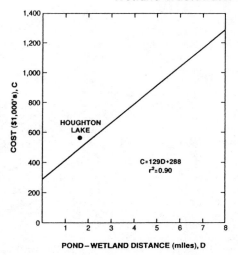

Figure 27.1
Wetland costs versus distance from
ponds.

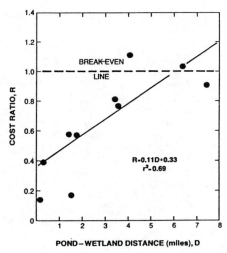

Figure 27.2
Wetland: Spray irrigation cost
ratio versus wetland distance.

of the land tract and the topographic relief.

Figure 27.2 is a plot of the ratio of the capital costs: wetland divided by spray irrigation versus pond-to-wetland distance in miles. The data points are all shown. The correlation is not especially strong, the coefficient of variation being 0.7, but there is a clear trend. Figure 27.2 indicates that the wetland system should be less expensive than the spray irrigation system if a suitable wetland can be found within 6 mi of the secondary effluent point. This conclusion is strictly true only if the spray irrigation land cost is $1500/acre. If the land can be obtained at no cost (the state or other owner being willing to give over uplands at no cost to the community) the break-even distance is around 3.5 mi. Therefore, within the 3.5 mi to 6 mi distance, depending on upland costs, the wetland alternative should be less expensive.

Figure 27.3 compares the electrical energy required to operate wetlands and upland spray irrigation. The electrical energy used in spray irrigation mainly provides sprinkler discharge pressures of around 40 psi and higher. For the wetland system, though, most of the energy is consumed in overcoming friction in the forceline. The relationship shown in Figure 27.3 is the ratio of energy consumed in the wetland operation to that in spray irrigation versus the force-line distance from secondary to wetland on the horizontal axis. The relationship is quite strong. One data point is missing and was not included in developing the least-squares line: The calculated energy was so low that it did not seem valid to include the community as representative of typical conditions. At distances from secondary effluent point to wetland of less than 4 mi to 5 mi, the energy cost of operating the wetland system is likely to be lower than that to operate the spray system. This situation would be the case, no matter how near available land might be for an upland irrigation system.

OPERATION AND MAINTENANCE COSTS

VILLAGE OF VERMONTVILLE

In Table 27.1, operation and maintenance (O&M) costs are given for the wetland system at Vermontville, Michigan (Williams & Works 1979). Vermontville has around 900 residents, and the municipal wastewater is entirely domestic, averaging around 75,000 GPD. Very soon after the village's wastewater stabilization ponds and surface irrigation fields were constructed in 1972, the fields were taken over by cattails, duckweed, and other adventitious volunteers. In other respects, this system is a conventional flood-irrigation opera-tion, relying on soil filtration and soil adsorption to accomplish P removal and other postsecondary treatment. The irrigation season is five and a half months or so, usually June through mid-November. Vermontville's fields are 11 acres to 12 acres in area. If these fields were planted in upland grasses, they would be uneconomical to harvest because of their small size, but cutting and removal of grasses would probably be necessary on some scale to control nuisance animals. In contrast, the adventitious wetland soils seem to stay open without maintenance, in spite of near constant submergence and wetness.

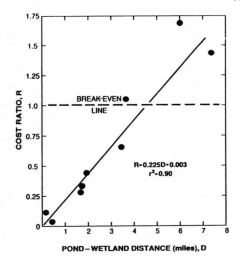

Figure 27.3
Wetland: Spray irrigation power
cost ratio versus wetland distance.

The annual operation and maintenance costs for Vermontville are tabulated in Table 27.1 by the categories of labor, electrical energy, equipment use, repair/replacement, new equipment, and environmental monitoring. These labor costs include overhead and administration. The annual labor involved with the wetland irrigation program is $1700. Most of the labor, around $1200, involves opening and closing the irrigation valves manually on a five-day weekly schedule. About 30% of those costs are accounted for in mowing long, grassed slopes facing the wetland fields. There is a difference of about 22 ft of elevation, on average, between the ponds and the wetland fields. Pumping costs are around $340 annually. Use of a pickup truck, a tractor with a sickle-bar cutter, and fuel totals around $960/yr. There are virtually no costs for repair, replacement, and new equipment. The most significant cost category is labor at 37% of the total wetland O&M cost of $4560. Environmental monitoring is approximately 34% of the total.

HOUGHTON LAKE

O&M costs for the Houghton Lake wetland operation are shown in Table 27.2. Labor at Houghton Lake, including overhead and administration, is $4320, or 2.5 times that at Vermontville. Electrical energy costs last year were $2200 (this estimate may be high) or six times those at Vermontville. The most significant factor in the Houghton Lake electrical budget is friction losses in the forceline. Equipment, repair, and replacement are relatively minor, costing $1500 last year. Environmental monitoring costs were $10,000 last year, $5000 for wildlife studies and $5000 for vegetation-related studies. Those studies are necessary as part of the proof-of-concept documentation for the wetland program.

Table 27.1
Village of Vermontville 1981 Operation and Maintenance
Annual Costs
(in Dollars)

Category	Cost	% of Total
Labor (incl. overhead, adm.)	$1700	37.2
Electrical energy	338	7.4
Equipment use (incl. fuel)	960	21.0
Repair/replacement	30	0.7
New equipment	0	0.0
Environmental monitoring	1532	33.6
Total	$4560	99.9

Table 27.2
Houghton Lake Sewer Authority 1981 Operation and
Maintenance Annual Costs
(in Dollars)

Category	Cost	% of Total
Labor (incl. overhead, adm.)	$4320	24.0
Electrical energy	2200	12.2
Equipment use (incl. fuel)	600	3.3
Repair/replacement	150	0.8
New equipment	750	4.2
Environmental monitoring	10,000	55.5
Total	$18,020	100.0

GENERAL

For both Vermontville and Houghton Lake, the annual wetland-related O&M cost is about $160/MG treated. Apart from environmental monitoring costs, the O&M is $110/MG for Vermontville and $67/MG for Houghton Lake. The operation of the wetland treatment system cost each Vermontville customer, or family, about $15 last year. Each customer in Houghton Lake paid less than $6 to operate the wetland treatment system last year, or around $0.50 per month.

ACKNOWLEDGMENTS

Mr. Brett Yardley of the Houghton Lake Sewer Authority and Mr. Tony Wawiernia and Ms. Natalie Gaedert of the Village of Vermontville provided O&M information for the wetland component of wastewater treatment in their respective communities.

This paper is an edited version of an article by the author (Sutherland 1981), which is presented here with the permission of the State of Minnesota Water Planning Board, the copyright holder. Construction costs have been reestimated to spring 1982 and O&M data have been updated by one year, based on actual costs.

REFERENCES

Kadlec, R. H., D. L. Tilton, and J. A. Kadlec, 1976, Feasibility of utilization of wetland ecosystems for nutrient removal from secondary municipal wastewater treatment plant effluent, Semi-annual report no. 5 to the National Science Foundation, 364p.

Sutherland, J. C., 1977, Investigation of the feasibility of tertiary treatment of municipal wastewater stabilization pond effluent using river wetlands in Michigan, ENV 76-20812, National Science Foundation, NTIS PB 275-283.

Sutherland, J. C., 1981, Economic implications of using wetlands for wastewater treatment, in Selected proceedings of the midwest conference on wetland values and management, B. Richardson, ed., Minnesota Water Planning Board, 660p.

Williams and Works, 1979, Reuse of municipal wastewater by volunteer freshwater wetlands, ENV-20273, National Science Foundation, NTIS PB-299 262/6 wp.

DISCUSSION

Kaczynski: Were your land costs factored in here for your annual costs?

Sutherland: They were not.

Kaczynski: That's straight O&M?

Sutherland: Yes. There's no debt service involved with the O&M costs, as I presented it.

Miller: I was interested in your forceline costs. I'm not an engineer, but I've picked this up in conversations with engineers. Our forceline costs in Canada seem to be much higher than that. You were talking of costs of a million dollars for 7 mi. That seems incredibly cheap.

Sutherland: Well, actually, that would be about right--about $140,000/mi.

R. Kadlec: Jeff, you may be interested to know that recently we've done a little digging to find out the costs at other treatment sites. Your curves are very good for another half a dozen sites that you didn't show up there--based on actual data--for both O&M and capital. In no case are the predictions in your study lower than those actually incurred. Also, the study by Boyle Engineering is right on the money, too. It has land costs as a variable cost rather than fixed as in your study. It also works very well for those half dozen sites.

Male: The monitoring costs are a fair portion. Is it just effluent monitoring?

Sutherland: Yes. Just effluent monitoring.

Shuldiner: I have a question about the effluent monitoring. I wondered if those costs are the result of a set of procedures that are prescribed by the sponsors of the project or whether those are monitoring requirements that a prudent manager would assume ordinarily. Could they be reduced by careful sampling once you had experience?

Sutherland: I think this is true. Monitoring of both Houghton Lake and Vermontville is probably not excessive. At Vermontville, the only monitoring that is done is on a weekly or biweekly basis at the extreme end of the wetland system where there is a surface overflow. There isn't really wetland effluent at that point; it's wastewater that has circulated through the ground and is emerging in the final wetland field. Still, the Michigan Department of Natural Resources thinks that ought to be monitored, and so it is. Nevertheless, it's kind of unrelated to wastewater. Still, those monitoring costs are there. At Houghton Lake, I think approximately $5000 of the $10,000 budget is for plant-related documentation and $5000 for animal-related documentation. Is that correct, Bob?

Bastian: How much monitoring would they be paying for if they had conventional plants there--a pond and a discharge? What kind of costs would the same size plant in Michigan be paying out, assuming that typical monitoring would be required for the effluent?

Sutherland: Well, I should think at Vermontville, the cost would be the same because there's only one point source of effluent. At Houghton Lake, I'm not sure. Houghton Lake does have an NPDES permit now and, frankly, I don't know how that's affected monitoring costs. Probably not greatly because there was a good deal of monitoring prior to that.

R. Kadlec: There are two different types of "monitoring" on these
projects. One relates to those things that are required by regulatory agencies pertaining to water quality. Those items were not in
the $10,000 figure, for example, that Jeff just showed. Type two are
additional requirements placed on the project by regulatory agencies
because it is a wetland. That is the $10,000 figure that Jeff
alluded to at Houghton Lake. So those two types ought to be distinguished: the water quality monitoring and the wetland monitoring and
research.

Jackson: Once we're past this research stage in using wetlands, what
kind of monitoring costs do you expect as being an acceptable routine
overhead for this kind of operation?

Sutherland: Well, in any natural wetland system being used for
wastewater treatment, it may take a number of years to be satisfied
that you have things under control and more or less at a steady
state. And if it turns out to be an acceptable method, by then you
will probably have reduced monitoring to a more or less standard set
of parameters of the type required by an NPDES permit. You'd be
measuring suspended solids, BOD [biochemical oxygen demand], one or
another of dissolved nitrogen species, total phosphorus, perhaps
orthophosphorus, dissolved oxygen; you might have to monitor chloride,
one or another metal, depending upon circumstances, and coliform
bacteria.

Jackson: Do you think it reasonable that the same standards and
requirements that are applied to a conventional plant be applied
across the board to a wetland system? Is this an arbitrary administrative decision that has no logical sense?

Sutherland: I don't believe so. I don't quite understand the
question. I think that the level of monitoring of natural wetland
systems is not excessive, considering how little we know about them
and how they're going to evolve. So I would say no. I can't think
of a reason for thinking that they're excessive.

Jackson: Well, I think one of the reasons for thinking it excessive
is that we're dealing with a system that has a great deal more
complexity, as Tom Odum indicated a little earlier. We are not
dealing with a once-through tank in the ground somewhere. We're
dealing with a very complex system, and as long as that system is
functioning, it's going to do an increasingly complex job of removing
or changing these various components of the aquatic system. I can
see two reasons for monitoring: one, if things are going wrong you
want to determine the cause of that. But, if you're monitoring an
ongoing system that is functioning, you're going to be measuring a
lot of things that have no meaning--you're going to get, essentially,
background values. The second reason for monitoring is to assure
that you've got nothing but background. All right, but I wonder
about the cost effectiveness of doing that.

Sutherland: At the point where you're monitoring on whatever
frequency, and the data show constant conditions, I guess you would
then attempt to relax the frequency of measurement--for example,
instead of monitoring on a monthly basis, maybe you could do it

quarterly or semiannually.

S. Reed: If you have long, narrow channels, shallow in depth of flow and subject to, say, a ten-day retention time, you can achieve an awful lot just by plain gravity sedimentation in the hydraulic system. Would you like to speculate on what function the plants actually provide?

Bastian: Obviously, they shade the surface and prevent algal growth.

Wile: I think they also provide an environment for bacteria that are probably the main sources of the nutrient-removal effects and a lot of physical filtration in the litter.

Reed: So, it's a substrate for attached growth, then. Does that still prevail in the wintertime?

Wile: Well, the litter is still there, so you're still getting physical filtration. Certainly, we have bacterial action--both sulphur-cycle bacteria and nitrogen-cycle bacteria--still available, but on a much lower level.

28

Use of Wetlands for Wastewater Treatment and Effluent Disposal: Institutional Constraints

Frank Rusincovitch

The Environmental Protection Agency's (EPA) Office of Federal Activities (Policy and Procedures Branch and 404 Branch) and the Office of Water Programs, in conjunction with the U. S. Fish and Wildlife Service, are considering the use of wetlands to treat municipal wastewater. The impetus for such use is that many small communities have found that "high-technology," conventional municipal sewage treatment systems are not well suited to their limited needs.

The primary reason that such conventional municipal treatment systems are often not appropriate for small communities is that they use sophisticated biological and chemical processes to remove biochemical oxygen demand (BOD), suspended solids (SS), and nutrients. Such systems are relatively expensive to operate, especially on a scale suited to small communities and, for the most part, require at least one full-time, highly trained operator to work properly. The advantage for larger communites is that higher-technology treatment plants can remove pollutants from municipal waste more rapidly and require much less land area than the less technical, less expensive, more "natural" systems, such as stabilization ponds.

ADVANTAGE OF WETLAND USE

Over the years, EPA has encouraged smaller communities with available land resources to build stabilization ponds. Such treatment systems are designed to hold sewage long enough for natural biological action to break down the organic substances in the waste. The ponds, therefore, must be capable of holding the waste for many days' detention time before the treated effluent is discharged into receiving waters.

More recently, EPA has encouraged greater use of land treatment practices for treating and recycling sewage. Slow-rate land treatment that involves application of partially treated wastewater to a vegetated land surface is similar to conventional crop irrigation practices and is now a relatively popular wastewater treatment technique. Plant and soil interactions in irrigated cropland, old fields, or forests can reclaim wastewater while producing a marketable crop. Gaining popularity is rapid infiltration of wastewater

427

through basins with highly permeable soils. Alternating wet and dry
cycles allow such systems to effectively convert ammonia nitrogen to
nitrate and free N; these systems also achieve sizable reductions of
phosphorus (P), organic matter, SS, and pathogens. In <u>overland flow</u>
land treatment, wastewater is applied at the upper reaches of grass-
covered slopes and allowed to flow over a vegetated but impervious
land surface to runoff collection ditches. The wastewater is
renovated as it flows in a thin film down the length of the slope.
Although overland flow has only recently been applied to municipal
wastewater in the United States, this technology has been effectively
used in treating food processing-wastes for over 25 years. Unfortunately, like ponds, land treatment systems require relatively large
land areas. In addition, not all areas are amenable to the establishment and operation of land treatment systems.

EPA's Office of Water Programs believes that natural and
artificial wetlands can also be utilized effectively to help treat
sewage wastes and, in some cases, handle more wastewater per acre
than some land treatment practices. Initial findings from pilot
projects involving wetland treatment systems indicate that the
federal government should begin encouraging communities to consider
the use of certain wetlands for wastewater treatment. However,
certain institutional barriers may impede the immediate use of
wetlands for effluent treatment and disposal.

Using or creating a wetland to help treat municipal sewage
presents a shift from the philosophy that wetlands must be totally
protected from all human development activity. Instead, it presents
the idea that if used wisely, a wetland can complement both development and the natural environment. Further, it recognizes that water
pollution control is one of the natural functions of existing
wetlands that in many areas have served as an important means of
handling sediment and nutrient loadings from both point and nonpoint
pollution sources.

IMPEDIMENTS TO WETLAND USE

Communities considering the use of wetlands as part of a wastewater treatment system will have the option of either building an
artificial wetland or discharging into an existing natural wetland.
However, federal laws, regulations, and executive orders and numerous
comparable state and local regulations and ordinances exist that
regulate activities in wetlands or potential wetland areas. Most of
these codes were designed to discourage activity that could damage
natural wetlands. They center around the theory that wetlands
provide a rich, unique wildlife habitat that requires special protection from what has historically been indiscriminate abuse by man.

FEDERAL LAWS

Since multiple uses for a wetland were not considered when most
existing laws, regulations, and executive orders were drafted, these
codes could--because of their lack of appropriate language--constrain
a community's consideration of a wetland as part of its sewage
treatment system. A summary of the various federal laws and the
constraints they impose on the use of wetlands for effluent treatment

follows.

Section 404 of the Clean Water Act

Traditionally, dredge and fill permitting under Section 404 of the Clean Water Act is the regulatory action most closely associated with human actions in wetlands. Although Section 404 is intended to prevent the destruction or modification of wetlands generally, its application relates only to dredging or filling. Therefore, Section 404 permits would have very minor applicability--if any--on use of wetlands for municipal wastewater treatment. No Section 404 permit would be needed either to create a wetland or to discharge effluents to a wetland. However, a 404 permit might be required if the treatment system required activity in a natural wetland (earth-moving equipment, and the like). A Section 404 permit might also be needed for the construction of outfall devices in an existing natural wetland because this construction might be interpreted as a "filling" action.

National Pollutant Discharge Elimination System (NPDES)

NPDES permitting under Section 402 of the Clean Water Act and permit monitoring of a wetland municipal treatment system may present several problems, specifically in using natural wetlands for wastewater treatment. The Clean Water Act requires that most municipal treatment facilities provide at least secondary and, in some cases, higher levels of wastewater treatment. If an artificial wetland is designed to provide all or part of this secondary treatment or some treatment beyond secondary, a single discharge pipe would likely be built from the artificial wetland to the receiving waters. Permit conditions could easily be written for the discharge from such a pipe where compliance monitoring would take place.

If, however, a natural wetland were used, the wetland would be an integral part of a lake or river system and there might be no precise location where the wetland ends and the river or lake begins. Therefore, it might be impossible to establish a discrete outfall for monitoring purposes. In such a situation, it would be impossible to write an NPDES permit because of the difficulty in monitoring permit conditions.

A solution to the above problem would be to allow discharge to the wetland, assuming that a natural wetland provides a predictable but not readily measurable amount of treatment and that this treatment is adequate to maintain the water quality of the ultimate receiving stream. This approach would require a permit for the discharge to the wetland and compliance monitoring at this point. Allowing this type of permitting and monitoring would require modification of the Clean Water Act, in a manner similar to either Section 301 (h), which allows secondary treatment waivers for ocean discharges, or Section 318, which allows discharge, under controlled conditions, of certain pollutants associated with approved aquaculture projects.

The advantage of providing specific provisions in the Clean Water Act for treatment in existing wetlands is twofold: It allows communities the flexibility to plan for treatment using existing wetlands together with artificial wetlands, and it provides communities with treatment or pollution control options that might not

otherwise have existed by using existing wetlands. For example, a
community may be located near a receiving stream that is experiencing
pollution problems from both storm-water runoff and municipal wastewater. Diverting the storm water into a conventional wastewater
treatment plant would dilute the influent and could make it impossible to achieve the 85% removal of BOD and SS currently required to
meet secondary treatment standards. An alternative treatment
scheme--discharging the storm water to a wetland--could solve this
problem. The permit could even be for the discharge of all wastewater to the wetland, thus allowing the community to use it for
treatment of both storm water and partially treated municipal waste.

Modification of the Clean Water Act to allow special permit
considerations to municipalities with wetlands treatment systems
offers yet another option that could provide a margin of safety,
especially for systems that would not be using artificial wetlands to
achieve the secondary treatment level. This option could require
some relatively energy-efficient, low-cost, but less effective
conventional biological treatment system, such as a trickling filter
or aeration pond, before discharge to the wetland. Such a trickling
filter/wetland or aeration pond/wetland system could be an economical
and equally effective substitute for the sensitive and more costly to
operate suspended-growth systems, such as activated sludge. Also,
the wetland would be subject to less potential degradation because
the discharge from the trickling filter or aeration pond would be of
higher quality than primary-treated sewage effluent, thereby
relieving some of the uncertainty mentioned earlier over permitting
and monitoring if a community chose to use a natural wetland for
discharge.

Wetlands Executive Order 11990

Executive Order 11990 requires that actions by federal agencies
minimize degradation of wetlands and preserve and enhance their
natural and beneficial values. Some regional EPA and state officials
believe that this executive order not only prohibits the use of a
natural wetland as part of a municipal treatment system but also
restricts the discharge to wetlands of fully treated effluent, except
under special situations. Further, EPA's wetlands treatment
initiatives are only now developing procedures to allow discharges of
completely treated effluent to natural wetlands.

It remains to be seen whether Executive Order 11990 is, in fact,
an impediment to use of wetlands for municipal wastewater treatment.
The reason is that it is not at all clear that wastewater discharges
to a wetland will "degrade" it. It could be argued that in many
cases the addition of nutrients will enhance rather than degrade.
Much depends upon which wetland values are focused on.

Indeed, one might predict future occurrence of an opposite
problem. If the wetland treatment system were to enhance those
values most sought by the local constituency (e.g., increased game
sought by hunters), difficulty could be encountered when the
community, state, or federal agency attempted to abandon the system
in preference for a better treatment option. Therefore, even if a
community outgrows a wetland system and must build a new treatment
plant, the community may be forced to continue to discharge some
portion of its wastewater to the wetland, not because it needs the
additional treatment, but because someone is benifiting from the

enhanced wetland and is forcing its survival by citing the requirements of Executive Order 11990.

Floodplain Management Executive Order 11988

The initiatives outlined in Executive Order 11988 present one of
the more difficult problems with using wetlands to treat municipal
wastewater. Executive Order 11988 directs federal agencies to avoid
direct or indirect support of floodplain development. It further
directs federal agencies to provide leadership and take actions to:
(1) reduce the risk of losses due to flooding, (2) minimize the
impact of floods on human safety and health, and (3) preserve the
beneficial values of floodplains.

Generally, the reference floodplain mapped by the federal and
state agencies responsible define the floodplain as the area
inundated by a 100-year flood. This land area is that which, based
on historic information, has a 1% probability of flooding in any
year. Many natural wetlands are within 100-year floodplains.
Similarly, it would be safe to assume that a significant amount of
the land that communities might consider suitable for an artificial
wetland would also be in a 100-year floodplain.

A conflict could arise over whether EPA was directly supporting
floodplain development by encouraging wetland treatment systems. If
improper floodplain development resulted from use of a poorly
thought-out wetland system, this effect could discourage wetland use
in surrounding areas. Further, communities often receive funding for
water and sewage projects from several federal and state sources.
Therefore, even if both EPA and a community are convinced that a
wetland treatment system is ideally suited to the needs of a locality,
the system may not be installed because another agency may consider
the wetland system an improper use of a floodplain area and refuse
the additional funding necessary to build the system.

Finally, wetland treatment systems in floodplains do not easily
lend themselves to conventional methods of floodproofing. Such
treatment systems could become damaged or destroyed and unavailable
to effectively treat wastewater during or after a floodplain
disaster. Thus, in certain cases, flood disaster problems for
communities with wetland systems would be compounded because the
community might be forced to discharge untreated sewage for some time
following the disaster, whereas communities with conventional flood-
proofed treatment systems would not.

STATE AND LOCAL REGULATIONS, ORDINANCES AND GUIDELINES

In addition to these federally initiated constraints on wetland
treatment systems, comparable state and local regulations, ordinances,
and guidelines may further discourage a community. For example, a
state may have a federally approved coastal zone management plan that
was drafted without considering the possibility that a wetland could
be used for treatment of municipal wastewater. With good intentions
for protecting coastal wetlands, the state plan may contain language
that makes wetland treatment systems impossible in the coastal areas
of a state. Similarly, there may be zoning regulations and health
codes that would either eliminate consideration of a wetland system
or severely restrict some of the secondary uses or benefits that

initially made the wetland treatment alternatives attractive to a community. For example, a locality may want a wetland system to upgrade an area (i.e., to improve local hunting and fishing or tourism), but state health regulations may restrict access to a wetland used to treat municipal wastewater. Or a community may be attracted to a wetland treatment scheme because of the potentially low operation and maintenance costs that this type of "advanced" treatment system requires for the amount of nutrients the system can remove. However, state regulations may require a specially certified operator for advanced wastewater treatment systems. Such an operator would be overqualified for the work necessary to operate the wetland systems and, in addition, the higher salary for the operator could further dampen a community's enthusiasm for the project. Also, of course, local zoning and conservation bylaws may preclude use of wetlands for any purpose at all.

CONCLUSION

In conclusion, while there does not appear to be any unequivocal constraint in any existing laws, regulations, or guidelines that would prevent the sensible use of wetlands in a municipal wastewater treatment system, the idea is new, and it requires a reassessment of whether or not wetlands serve their greatest value when left completely alone. Wetlands used for wastewater treatment promote their multiple use, hopefully, to the benefit of both the residents of the community and the wetland. However, there are those who believe that people and their wastes do not belong in wetlands because the wetland should be left completely natural; or that the only "real" way to treat sewage is in concrete tanks with mechanical mixers and pumps where the chemistry and biology of waste treatment can be carefully controlled and monitored to provide the best-possible level of treatment. Proponents of these arguments have the potential to exert strong pressures on communities to pursue the more-established, conventional wastewater treatment routes, simply because they will seem easier or safer in the absence of complete knowledge about all wetland impacts.

No doubt, over time, the use of wetlands in the treatment and disposal of municipal wastewater by small communities with appro-priate land resources and soils will become an established procedure. However, the amount of time that the establishment of wetlands treat-ment takes will depend to a great degree on how quickly EPA can satisfy a community's concerns for what they perceive as the institu-tional, engineering, and biological constraints to wetland use for municipal waste treatment.

SESSION VIII

Synthesis and Conclusions

29

Responses of Wetlands and Neighboring Ecosystems to Wastewater

Katherine Carter Ewel

The use of wetlands for treatment of wastewater is clearly an attractive alternative to construction and maintenance of advanced wastewater treatment plants. Several types of wetlands have been assayed for their ability to provide an acceptable level of treatment, and patterns are emerging to indicate which kinds of wetlands might be most suitable in different geographic regions. However, full-scale implementation of the use of both natural and artificial wetlands is proceeding more rapidly than is initiation of the additional careful, in-depth research projects needed to verify and round out the information that has already been collected. Consequently, it is necessary to examine as thoroughly as possible the research that has already been conducted to determine patterns of response and to identify important areas that must be addressed more thoroughly before use of such systems becomes widespread.

This workshop has facilitated such an examination. Most of the research that has been presented identifies specific short- and long-term responses of communities and nutrient cycles to wastewater additions, and some general patterns of response may be discerned. Less attention has been paid to changes in ecosystem structure and function that have impacts beyond the boundaries of the receiving ecosystem. Nevertheless, the work that is being done to identify the importance and magnitude of these indirect effects is contributing significantly to our understanding of how regional systems function.

ECOSYSTEM RESPONSES TO WASTEWATER ADDITION

The responses of species composition and productivity in plant communities to wastewater addition show no conspicuous pattern across the spectrum of ecosystems that has been considered in this workshop. However, this may be due to differential abilities of communities that are entrained to respond to particular combinations of hydroperiod and nutrient availability. Nutrient and water additions should certainly elicit very different responses and, therefore, even quite disparate community responses may be the result of a common functional relationship.

		Diversity	Productivity	
			Gross	Net
Hydroperiod	Short	Decrease	Increase	No Change or Increase
	Long	No Change or Increase	Increase	Increase

Figure 29.1
Effect of wastewater on diversity and productivity of a
wetland community.

A proposed generalized pattern of response is outlined in Figure
29.1. Anoxia in the rhizosphere is probably the most stressful
aspect of a wetland plant environment, and diversity usually
decreases with longer anoxic periods. Therefore, in an ecosystem
with a short hydroperiod, a given percent increase in hydroperiod
should produce a greater percent change in species composition than
would the same percent increase in an ecosystem with a longer
hydroperiod because in the latter, plants are already adapted to
lengthy inundation and a long period of anoxia.

On the other hand, nutrient addition has a much more direct
effect on gross primary productivity than on species composition.
Increased gross primary productivity should be accompanied by a net
increase in nutrient uptake, but increased respiration in response to
water additions may negate both. If added nutrients are not taken up
by the existing community, the wetland is susceptible to invasion by
species that can utilize that resource.

Consequently, changes in species composition in wastewater-
enriched ecosystems with a long hydroperiod should largely result
from the inability of existing species to fully utilize added
nutrients. Diversity and productivity are therefore more likely to
increase because the existing community is augmented rather than
replaced. Changes in an ecosystem with a shorter hydroperiod would
be due to the inability of the existing species both to withstand the
increased hydroperiod and to take up nutrients, therefore leaving
room for more water-tolerant species to come in and utilize those
nutrients. Diversity may decrease in these ecosystems; productivity
increases should be less pronounced because of the probability of
increased respiration.

Intolerance of the existing plant community to a longer hydroperiod is probably the major factor involved in immediate changes
that occur in wastewater-enriched wetlands. Changes due to inability
to utilize added nutrients are probably manifested more slowly,
especially in communities with longer hydroperiods.

Several speakers have stressed the physiological and structural
flexibility of wetland plants, including relatively high photosynthesis rates, ability to reallocate carbon (C), and flexibility in

stand architecture. These attributes tend to decrease the magnitude of change in species composition when wastewater is added and are major factors in making wetlands attractive and useful for wastewater disposal.

The long-term projects that were discussed suggest that we can expect steady state in both nutrient-removal rate and vegetation response to be obtained in an area that is impacted (but not over-loaded) by wastewater. The question of permanence of nutrient-removal pathways has not been effectively addressed, particularly with respect to the role of relatively irreversible precipitation mechanisms. Evidence from projects involving upland wastewater disposal suggests that nutrient removal can be more than a simple cation exchange process. Consequently, detailed studies of other nutrient-removal pathways in wetlands are needed, possibly using such procedures as simulated long-term loading of columns of different kinds of soils.

THE ROLE OF WETLANDS IN A LANDSCAPE

Several studies have demonstrated that the role a wetland plays in the landscape must be taken into account in any management scheme. In particular, changes in the movement of water and of animals into and out of wetlands can be crucial to determining the overall accept-ability of wetland discharge.

The value of a wetland depends on the size and nature of the terrestrial system that it drains, as well as the fate of the water that leaves it. Upland palustrine wetlands, which are most likely to recharge groundwater, apparently have low transpiration rates that give them value as water conservation agents in a regional context. Although it is doubtful that wastewater disposal in such wetlands would decrease this value because of the likelihood of net increase in recharge as well as in transpiration, it is clear that the values of undisturbed ecosystems are not always obvious.

The effect of wastewater enrichment on animal populations in and around a wetland can be profound. Increases in herbivory because of the invasion of vertebrates and/or invertebrates from outside the wetland, as well as population increases within, have been associated with increased nutrient concentrations (especially nitrogen [N]) in plant tissues. Increased decomposition rates in wastewater-enriched wetlands are associated with a variety of changes in litter quality and suggest that major changes may be occurring in the detritus-based food web as well. The importance of large, surface-dwelling benthic invertebrates and the relative balance of fungi and bacteria may change considerably when redox potentials and natural water-level fluctuations are altered. Since wetland fauna are seldom confined to wetlands, it is obvious that wastewater enrichment will affect food webs in both upland and aquatic ecosystems.

CONCLUSIONS

   The phenomena and hypotheses described above require greater
attention in order to increase our ability to manipulate and/or
design wetland-based wastewater treatment systems. Our high public
health standards and our shrinking energy supplies indicate that
increasing use of natural ecosystems will be necessary to absorb the
wastes of our growing population. However, our need to understand
and maintain the basic functions of our natural ecosystems dictates
that advances in this direction be made only with the greatest care
and planning.

30

Wetlands, Wastewater, and Wildlife

John A. Kadlec

Wetlands are noted for the diversity and abundance of the
wildlife that inhabit them. Indeed, it was their importance to
wildlife, notably ducks and geese, that led to the first (and, for a
long time, only) efforts to preserve wetlands and marshes. More
recently, other values of wetlands have received increased attention.
Of particular concern to us here is the potential use of wetlands for
wastewater treatment. Wildlife interests view this possibility with
mixed emotions: Having battled so long for wetlands preservation,
they welcome any help they can get in "selling" the need for wetlands, but they are concerned that this use not destroy the very
values they seek to preserve.

Data on the effects on wildlife of wastewater inputs to wetlands
are distressingly scarce. As a consequence, the approach adopted
herein is to examine the documented and predicted effects of wastewater on wetlands and then to interpret those effects in terms of
their potential impact on wildlife. For convenience, I will cate-
gorize wastewater effects on wetland ecosystems in five major
categories: (1) primary production, (2) plant community structure,
(3) water regime, (4) pathogens, and (5) toxic chemicals. These
categories are not mutually exclusive and there are important interactions among categories, but each category serves to focus attention
on a particular set of concerns.

Before commencing a consideration of each of the major effects,
a comment about the ubiquity of change in wetlands is in order.
Natural, pristine wetlands are dynamic and undergo change at various
rates, yet many are essentially permanent features of the landscape.
For example, prairie potholes seem to change in a cyclic fashion with
patterns of drought, with a time scale on the order of 5 to 10 years
(van der Valk and Davis 1978), yet they have persisted over a vast
area of the northern prairies since the last glaciation, approximately 10,000 years ago. Against that background of natural change,
our knowledge of wetland ecosystems strongly indicates that the
discharge of wastewater into a wetland ecosystem will produce changes,
perhaps dramatic, perhaps subtle, in that system. It might very well
change one kind of wetland into a different kind. Both kinds may
well be "natural" in the sense that equivalents are found elsewhere
as essentially undisturbed sytems. A part of this change quite

probably will be a change in wildlife. Some species will increase,
others decrease. In and of itself, such change is neither bad nor
good but becomes so only in the context of human values and the
relative commonness or rarity of these ecosystems in the landscape,
usually at the regional, state, or national level.

PRIMARY PRODUCTION

     Because most wastewaters contain plant nutrients, an observed
effect (e.g., Valiela et al., this volume) is an increase in plant
growth. This effect may result in change in the plant species, and
so forth, but here I want to concentrate on the impact of the
increased production of organic matter per se. Wetlands are usually
considered primarily detritus-based ecosystems--that is, most of the
plant material is not eaten green by herbivores but rather becomes an
energy source for animals only after death and "conditioning" by
microorganisms (Cummins 1974; Hodson, this volume). Invertebrates
play a key role in this food web. For example, waterfowl biologists
now recognize the importance of aquatic invertebrates, such as
immature insects and scuds (amphipods), in the nutrition of breeding
female ducks. And these invertebrates depend directly or indirectly
on detritus (Anderson and Sedell 1979). Although the species of
plants in the wetland affect the quality of its detritus, far more
important is the quantity of detritus in the system. Consequently,
we might expect an increase in primary production to increase the
quantity of material in the detrital food web and thus ultimately
increase the wildlife species, such as birds and fish, that depend on
it. And, indeed, there are some empirical observations, largely
anecdotal, that suggest that this effect, in fact, does occur (R.
Kadlec, personal communication; I. Wile, discussion, this conference).
As a cautionary note, there is some basis in ecosystem theory to
predict a change in ecosystem structure (e.g., number of links in
food web) rather than an increase in a particular species of consumer.
For example, fertilizing lakes has sometimes led to decreases in
desirable fish, although fertilization is a common method of
increasing the yield of fish.
     The notable exception to the importance of detritus food webs is
the direct grazing of muskrats, and sometimes ducks and geese. All
these herbivores are known to substantially impact primary production
of their preferred food plants--for example, cattail by muskrats
(Weller 1981). As was observed at Houghton Lake (R. Kadlec, personal
communication), an increase in cattail production can lead to
increased muskrat herbivory and subsequent cattail reduction.
Studies by Errington, Siglin, and Clark (1963) suggest this sequence
may be followed by decreasing muskrat abundance. Cattail recovery
and a repetition of the cycle depend on water depths that influence
both muskrat survival in winter and cattail regeneration; thus, the
interaction is very complex and the reliability of predictions is
poor.

PLANT COMMUNITY STRUCTURE

Wetland birds seem to respond to the physical architecture of marsh vegetation more than to the particular species of plants (Weller 1981,62), partly because they usually do not rely directly on the plants for food. Rather, vegetation is important for other functions, such as nest sites and places to escape predators. Weller and Spatcher (1965) are usually credited with the idea that a mixture of 50% open water and 50% emergent plants, such as cattail, is near optimum for many marsh birds. Kaminski and Prince (1981) have confirmed the high value of the 50:50 mixture for breeding waterfowl.

Other aspects of the species composition of the wetland vegetation and the distribution in space of those species also may affect wildlife use (Wiens 1976). A recurring thought at this conference was that wastewater additions to wetlands lead to monocultures of cattail. To the extent that cattail persists in dense, even stands, a reduction in abundance and diversity of wildlife will occur. Even birds, such as red-winged blackbirds that prefer cattail marshes, are more abundant when the cattail is in patches and provides "edges" that seem to enhance insect food supplies for the birds (Novy 1973; Orians 1961).

In ecological terms, monocultures tend to occur when ecosystems are dominated by a single influence, be it herbivory, soil salts, or nutrient additions. And, monocultures usually have less diverse and often less abundant fauna, even though primary production may be maximized.

WATER REGIME

Wetlands, by definition, depend on water; moreover, the kind of wetland is strongly, if not entirely, determined by the water regime. Different plants differ in their ability to tolerate flooding in terms of such factors as depth, duration, and frequency (e.g., Brinson, this volume). Because wastewater additions involve the addition of substantial quantities of water, often in a uniform flow compared with natural events such as rains or floods, a substantial impact on the wetland may be anticipated.

To predict the effects of increased and constant water supply, we can look to two sources: (1) experience in wetlands with managed water levels and (2) recent theory about natural vegetation cycles (van der Valk 1981). Both of these sources suggest that prolonged constant water levels lead to a decrease in emergents such as cattail and a conversion of marshes to shallow open ponds, although with perhaps substantial quantities of submerged plants, such as pondweeds (Potamogeton spp.).

Large open ponds are also monocultures and subject to the same constraints on wildlife abundance and diversity as monocultures of cattail.

Note that we have now predicted diametrically opposite effects of wastewater additions, albeit both monocultures: dense cattail and open water ponds. The data to date suggest that the first effect is the most likely outcome. Longer experience may show us when to expect open ponds. I would guess that the outcome will depend on water depth. However, because good nutrient removal from wastewater requires shallow water, we may never see open ponds develop.

PATHOGENS

Friend (this volume) has discussed the potential impact of wastewater additions on pathogens of importance to wildlife. In relatively few cases is it likely that the wastewater will contain the pathogens directly, particularly if the source of the wastewater is domestic human waste and not waste from livestock, especially poultry, sources. And, even then, there must be a large wildlife population available to contact the wastewater in order for a serious problem to develop. At present, we cannot adequately assess the potential hazard of direct pathogen transport--the various factors involved have not been adequately assessed. Potentially, effluent chlorination or other methods of disinfection might be used to reduce this hazard, but there are problems with this approach as well. We might note that lagoon sewage treatment systems are common in the West and, because water is scarce, often attract large numbers of wildlife, particularly waterfowl. Because this sewage treatment system involves raw sewage, one might consider the hazard much higher and, indeed, disease outbreaks do occur. Obviously, more research is required to assess this aspect of wastewater in wetlands.

Friend (this volume) did point out that the impacts of wastewater in altering the environment physically, chemically, and biologically may be more important in creating disease problems than direct pathogen transport. We simply do not know enough yet to be able to make predictions, however.

TOXIC CHEMICALS

The kind of materials included in toxic chemicals probably varies widely from site to site but often includes such things as pesticides and heavy metals. Wildlife species, especially birds and fish, have been a major source of concern--and,indeed, are used as indicators--because of the impacts of these materials. Obviously, a wastewater with unusually high concentrations of such materials should not be considered for discharge or treatment in natural wetlands. Conversely, there is some evidence (e.g., Giblin, this volume) that wetlands may be particularly valuable as treatment sites for some chemicals because of their high adsorptive capacity or unique microbiological properties. Again, we do not have a good knowledge base for assessing these problems. Two observations are encouraging: (1) natural geothermal wetlands in the West have very high concentrations of some chemicals yet seem not to be problems (Kaczynski, this volume); and (2) Risebrough (1978) has suggested that heavy metals are not a particularly serious hazard for wildlife.

CONCLUSION

This consideration of the potential impacts of wastewater on wildlife in wetlands has deliberately tended to emphasize the negative. Clearly, the potential for both positive and negative impacts on species or groups of species exists. Change is most likely, but that change may be good or bad depending on the circumstances and the human value judgments of that time and place. Most worrisome to me, because of our lack of knowledge, are the potential problems with disease and toxic chemicals. Our experience to date suggests that nutrient additions, at least in the relatively short term, are not a problem for wildlife, except that the wildlife community will change as the plant community changes.

REFERENCES

Anderson, N. H., and J. R. Sedell, 1979, Detritus processing by macroinvertebrates in stream ecosystems, Ann. Rev. Entomol. 24:351-377.

Cummins, K. W., 1974, Structure and function of stream ecosystems, Bioscience 24:631-641.

Errington, P. L., R. J. Siglin, and R. C. Clark, 1963, The decline of a muskrat population, J. Wildl. Manag. 27(1):1-8.

Kaminski, R. M., and H. H. Prince, 1981, Dabbling duck and aquatic macroinvertebrate responses to manipulated wetland habitat, J. Wildl. Manag. 45(1):1-15.

Novy, M. E., 1973, Habitat selection by the male red-winged blackbird, M. S. thesis, University of Michigan, Ann Arbor.

Orians, G. H., 1961, The ecology of the blackbird (Agelaius) social systems, Ecol. Monogr. 31:285-312.

Risebrough, R. W., 1978, Pesticides and other toxicants, in Wildlife and America, H. P. Brokaw, ed., U. S. Council on Environmental Quality, Supt. of Documents, Washington, D. C., pp. 218-236.

van der Valk, A. G., 1981, Succession in wetlands: A Gleasonian approach, Ecology 62:688-696.

van der Valk, A. G., and C. B. Davis, 1978, The role of seed banks in the vegetation dynamics of prairie glacial marshes, Ecology 59(2):322-335.

Weller, M. W., 1981, Freshwater marshes, University of Minnesota Press, Minneapolis, 146p.

Weller, M. W., and C. E. Spatcher, 1965, Role of habitat in the distribution and abundance of marsh birds, spec. rep. no. 43, Iowa Agricultural and Home Economics Experiment Station, Iowa State University, Ames, 31p.

Wiens, J. A., 1976, Population responses to patchy environments, Ann. Rev. Ecol. Syst. 7:81-120.

31

Wetlands for Wastewater Treatment:
An Engineering Perspective

Sherwood C. Reed and Robert K. Bastian

The capability of wetlands to renovate wastewater, as well as to
tolerate effluent discharges, has been demonstrated in a variety of
natural wetland types across the country, as well as in artificial,
or constructed, wetlands. This capability has been well documented
elsewhere in a number of state-of-the-art conference and workshop
proceedings (Stowell et al., this volume, Hantzsche, this volume) and
special project reports, such as: a 1978 review by Duffer and Moyer,
the proceedings of a multiagency-sponsored seminar held at the
University of California-Davis in September of 1979 (Bastian and Reed
1980), and a literature review of the vegetative and hydraulic
processes responsible for wetlands removal of selected pollutants
from storm water runoff and treated municipal wastewater by Chan et
al. (1981).
    Under the right conditions, wetland systems can achieve high
removal efficiencies for biochemical oxygen demand (BOD), suspended
solids (SS), trace organics, and heavy metals. They have considerable potential as a low-cost, low-energy technique for upgrading
wastewater effluents (especially for smaller communities located in
areas where natural wetlands abound or ample opportunity for the
construction of new wetlands exists). The concept has been shown to
be viable and qualifies under current EPA definitions as an innovative technology. Projects that incorporate such technologies are
eligible for more federal construction grants assistance than those
projects that rely solely on conventional technologies. However, the
specific factors responsible for such high treatment levels are not
clearly understood at present. Optimum cost-effective criteria
applicable throughout the United States are not now available for
routine design of wastewater treatment systems involving wetlands as
a viable part of the treatment facility.
    The use of constructed wetlands shows great promise for more
general application because potentially they are more reliable and
involve less risk of adverse environmental impact because of better
process control. The potential for a more widespread and routine use
of wetland systems (particularly the constructed type) seems high--if
and when reliable, cost-effective engineering criteria are made
available. While considerable information about these sytems is
already available, the scientific uncertainties over the long-term

impacts (beneficial as well as detrimental) of applying municipal
wastewater to managed wetland systems can only be clarified through
more experience with operational systems. A closely coordinated,
multidisciplinary approach involving the cooperative efforts of
experienced engineers, ecologists, biologists, hydrologists, and the
like will be required to address adequately the full range of
ecological and related issues that must be addressed when designing
wetland treatment sytems.

The engineering design of a wastewater treatment system is based
on a predictable or an assumed set of conditions. The successful
performance of that system is then based upon the maintenance of
those conditions throughout the useful life of the system. For most
municipal wastewater treatment facilities, a useful life of 20 years
is calculated. When the conditions cannot be guaranteed, the engineer incorporates a safety factor and overdesigns the system to
compensate for the uncertainty. Most of these aspects are more or
less under control during the design and operation of a constructed,
or artificial, wetland, so the safety factors for this concept can be
relatively low. These systems can then reliably treat a relatively
large volume of wastewater in a confined space, and the design
criteria used can be extrapolated for use in the design of similar
systems in the same climatic zone. In natural wetlands, the hydrologic regime is difficult to predict and often impossible to maintain. So, a very large safety factor must be incorporated to insure
reliable treatment under all conditions. As a result, the wastewater
loading would have to be relatively low and the required wetland area
relatively large. Since the hydrologic regime of the natural setting
is also usually subjected to other local influences, the extrapolation of design criteria for one natural wetland system to another
setting is usually difficult.

ECOLOGICAL CONSIDERATIONS

Ecological considerations are a major concern where natural
wetlands are involved in wastewater treatment or disposal of treated
effluents, or where wastewater is involved in the enhancement or
restoration of wetlands. The focus of this concern is generally upon
the maintenance of the existing species, community structure and
function, and the wetland's overall values. Caution is therefore
necessary when considering the application of wastewater to any
natural wetland, especially if total preservation of the natural
setting is a goal. The addition of freshwater wastes to a saline
marsh or alkaline effluent to an acid bog could, for example, significantly impact the existing conditions of these natural wetlands.
The additional volume of water applied to any type of wetland generally has a greater initial impact than the effects of nutrients or
other materials in typical municipal wastewaters. Of course, the
potential effects of added nutrients and contaminants on productivity
and the food chain cannot be overlooked. A species change is likely
to result, at least near the point(s) of wastewater application.
Since these responses are not necessarily predictable, pilot testing
or small-scale demonstrations are generally desirable.

Ecological aspects are less critical for constructed, or artificial, wetland systems than for natural wetland treatment systems.
These systems depend upon a balanced biological community to achieve

the desired level of treatment, but if the system is constructed to optimize wastewater treatment, a much lower diversity than that found in a natural wetland can be expected. Before the design of such a constructed wetland system proceeds, these conditions must be made acceptable to all concerned, so that serious conflicts can be avoided later on.

WASTEWATER REGULATIONS

From a regulation standpoint, the use or discharge of treated effluents and providing wastewater treatment are not always the same thing. Natural wetlands are usually considered as a part of the adjacent water body and, therefore, under the Clean Water Act, discharges of municipal effluents to a natural wetland must be treated to at least the level of secondary treatment. This restriction limits the potential use of natural wetlands for municipal wastewater treatment. Except for variations allowed for pond systems and the 301(h) provision that allows for waivers for marine discharges, secondary treatment is currently defined as meeting an arithmetic effluent mean value of 30 mg/l of BOD and SS in a period of 30 consecutive days, (40 Code of Federal Regulations [CFR] 133.102 a1 and b1), 45 mg/l in a period of 7 consecutive days (40 CFR 133.102 a2 and b2), and 85% removal, except treatment works that receive flows from combined sewers (sewers designed to transport both storm water and sanitary sewage), which may be defined on a case-by-case basis (40 CFR 133.102 a3, b3, and 103 a2). However, with the passage of PL 97-117 in December 1981 to amend the Clean Water Act, Section 304(d) of the Act was amended to allow for special consideration to be given to the discharge requirements from biological treatment facilities such as oxidation ponds, lagoons, ditches, and trickling filters, making their discharges equivalent to secondary treatment. Efforts are now underway within the EPA to revise the definition of secondary treatment requirements accordingly.

As noted by Rusincovitch (this volume), federal laws, regulations, and executive orders, as well as numerous state and local regulations and ordinances, exist that were designed to give special protection to wetlands by discouraging any activity that could damage natural wetlands values. Since multiple uses of wetlands were not considered when most of the existing laws and regulations were drafted and adopted, serious restrictions may be imposed on the potential use of natural wetlands as an integral part of wastewater treatment projects. However, the pending changes in the definition of secondary treatment and the ongoing investigations into the potential use of natural wetlands in wastewater treatment may change the current situation.

WETLAND-WASTEWATER COMPATIBILITIES

The term wetland is a broad term encompassing diverse natural ecosystems--swamps, bogs, marshes, and other wet areas--as well as artificially constructed systems, such as shallow ponds that have been taken over by emergent vegetation and referred to as volunteer wetlands. Of course, "natural" and "artificial" wetlands are not sharply divided. Many so-called natural wetlands are man-made, resulting from some manipulation of the environment that alters hydrology, and artificial wetlands can have characteristics that are similar to those of natural systems. Wastewater is a similarly broad term that can range from dilute mixtures of nutrients, organic matter, and SS, to a complex, hazardous mixture of industrial wastes. The kind of wastes and the level of pretreatment they receive prior to discharge into the treatment system in part determine the nature of the wastewater. The matching of wetlands and wastewater management is further complicated by the purpose of the operation, the level of management sophistication, and other site-specific factors.

The functional role of a wetland in a particular wastewater management system needs to be clarified early in project planning. This role might involve increased removal of organic matter and solids, conversion or removal of various nitrogen (N) and/or phosphorus (P) forms, or as a location for the disposal of the effluent into surface waters or groundwater. This role could also involve the recycling of treated effluent as a source of water and nutrients to stimulate the production of wetland vegetation which, in turn, would establish improved wildlife habitat, production, and so forth. In this scenario the main task becomes the careful matching of the capabilities of candidate natural wetlands or artificial systems to tolerate or process the wastewater quality and quantity involved while remaining within the constraints dictated by public and regulatory agency acceptance. Some of the possibilities appear not to be worth further consideration, while others do appear worth developing. Discharging a complex industrial waste into a valuable natural wetland is probably not worth pursuing, while developing engineering technology for using artificial wetlands to treat municipal wastewater is a promising area of current research.

Rather than a single concept for the involvement of wetlands in wastewater treatment, there appears to be a continuum of possibilities for the tolerable (if not beneficial), as well as functional, combining of wastewater and wetlands. We have observed four major categories of wetland-wastewater combinations that have been successful on the basis of ecological, engineering, economic, political, and social criteria:

Category A: Use of natural wetlands for the disposal of treated effluents;

Category B: Use of wastewater for wetland enhancement, restoration, or creation;

Category C: Use of natural wetlands for wastewater renovation; and

Category D: Use of a constructed wetland for relatively high rate wastewater treatment.

Although the emphasis shifts from optimization of a wetland's overall values to optimization of the costs of wastewater treatment, all four categories either directly or indirectly do provide for some treatment of the wastewater.

NATURAL WETLANDS FOR WASTEWATER DISPOSAL

The use of natural wetlands for the discharge or disposal of treated effluents (Category A) reflects the legal status of wetlands as "surface waters of the United States." In some areas, sewage

treatment plants have been designed and permitted to discharge their
treated effluents into wetlands as "receiving waters" in lieu of
building long pipelines and expensive diffuser systems to achieve
appropriate dilution in surface streams, lakes, embayments or estuaries. While considerable wastewater "treatment" may actually take
place in the wetland, for regulatory purposes these dischargers are
required to fully treat the wastewater prior to disposal in the
wetland.

WETLAND RESTORATION

In some situations, the use of wastewater provides an opportunity to create new wetland areas, enhance existing degraded wetlands, or restore wetlands previously lost to development (Category
B). It can even provide for the possible recovery of wetlands in
arid areas where the historical streamflow may have been diverted for
other purposes.
The wetland created by the Mountain View Sanitary District in
Martinez, California, is an excellent example of Category B. This
project involved the restoration of a previously diked and drained
wetland area that also served as a discharge site for treated
effluent. The wetland restoration and enhancement project was
designed to provide maximum wildlife habitat while avoiding nuisance
situations. At the same time, it served as a cost-effective alternative to the construction of an expensive, deep-water outfall into
Suisun Bay.

NATURAL WETLANDS FOR WASTEWATER TREATMENT

The peatland system in Houghton Lake, Michigan, the bog system
in Drummond, Wisconsin, and the cypress dome and strand systems in
Florida are examples of using natural wetland systems to treat wastewater (Category C). The intent of these systems is to achieve wastewater treatment in a way that minimizes ecological disturbance.
Maintaining this balance is critical to the performance of the system
and, therefore, its acceptance by regulatory agencies and neighbors.
Typically, the only engineering or construction activity undertaken
is the installation of the wastewater application system. The renovation capability of these systems has been established, and general
concept-level criteria are transferable from one situation to another.
As a result, it can, for example, be claimed with relative confidence
that peatlands and bogs in the north central states can treat wastewater, but the safety factors involved have not been completely
defined.
Application of wastewater to a natural system is likely to cause
some changes, unless the volume added is insignificant compared to
the wetland's base flow. The most likely impact is species change
near the point(s) of application. The acceptability of any such
change is an issue for public and institutional decisions. More
natural wetland types in a variety of settings should be tested for
their response to wastewater additions. Long-term studies should
continue at the existing natural wetland wastewater treatment sites
to document long-term effects and to determine limits of these
systems to renovate wastewater to the desired level. Additional

studies would also help verify and optimize design criteria. A prime research need is to relate final water quality from particular natural wetlands receiving wastewater to the residence time and hydroperiod.

ARTIFICIAL WETLANDS FOR WASTEWATER TREATMENT

Constructed, or artificial, wetlands systems (Category D) probably have the greatest potential for application in the relatively near future since, in theory, they can be constructed anywhere, they provide an opportunity to optimize control over the treatment process, and they do not interfere with natural wetlands values. Such systems, which generally involve maximizing application rates while minimizing acreage requirements, have recently been built and studied in Ontario, Canada, as well as in New York, Pennsylvania, Maryland, Michigan, Iowa, Idaho, and California. Volunteer wetlands that resulted by accident (e.g., infiltration basins that did not perk) or by design (e.g., shallow ponds that developed emergent vegetation) are also functionally in this same category of man-made wetland systems. Such systems exist in Michigan, South Carolina, and elsewhere.

The principal goal of such constructed systems, in contrast to natural wetlands, centers on the cost-effective treatment of wastewater. A shallow excavation, an impermeable liner or clay seal, and the replacement of some soil may be all that is required for their construction. They may require a balanced aquatic ecosystem as a component of the system, but maintenance of a wetland ecosystem with special social, wildlife habitat, or other values is generally of secondary importance to wastewater treatment. Besides, such ecosystems did not exist prior to construction, and therefore, there are no original wetland values to "preserve." Wetland ecosystem values created by their construction are a bonus. They are pure lagniappe.

The operation of these systems is low in cost and energy requirements. The differences between an artificial wetland system designed to treat wastewater efficiently (Category D) and the creation of a wetland using treated effluent (Category B) will vary depending upon the degree of wastewater pretreatment, wastewater loading rates, public access to the wetland, uses allowed, and other factors. Some studies have indicated a seasonality in the performance capabilities of some constructed systems. However, recent work in Canada with long, narrow channels suggests the possibility of consistent performance in winter as well as summer months, although the treatment mechanisms in the winter have not been clearly identified (Wile, this volume).

CONCLUSIONS

Considerable insight into the process of matching needs and capabilities while living within the constraints of public acceptance and regulatory approval can be gained from the engineering community's increasing experience in addressing the potential use of land treatment technologies. Similar advantages and disadvantages mentioned by many at this workshop have been cited for land treatment projects. The potential applicability to wetlands treatment systems

of the extensive data, information, and insights that have been
developed by research on various agricultural and land treatment
practices regarding anaerobic/aerobic (saturated/unsaturated) soil
conditions and their relationship to N conversion, P and heavy-metals
adsorption, primary productivity, organic-matter decay rates, invertebrate population responses, and other areas should also be carefully considered. In addition, the growing body of information
concerning public health issues associated with land treatment, as
well as conventional wastewater treatment technologies, should be
applicable to wetland systems to a considerable extent.

Further research is needed to help define and optimize engineering design criteria for constructed wetlands. Additional experimental systems in a variety of geographical settings are recommended.
Research is also needed to better define the long-term performance
capabilities and operational problems, the role of emergent vegetation, the need to harvest vegetation and the disposal or beneficial
use of the harvested material, and the impact and control of potential vectors (e.g., mosquitoes) and other nuisance animal populations
(e.g., birds and muskrats) on facility operations. When these design
criteria are known, constructed wetlands can be installed as a component in a new treatment system, easily incorporated as a replacement
or upgrading component in existing systems, or added as a low-cost
method for achieving treatment of increased flows. Artificial
systems appear to have potential for achieving secondary or advanced-
secondary levels of treatment and for nitrification. They may be
somewhat less effective for tertiary removal of nutrients, but
further definitive research is needed in this area.

Finally, while detailed design criteria that would be applicable
to a wide range of potential wetland wastewater treatment projects in
various geographic locations are lacking and probably not attainable,
some good information and even first-order modeling of certain
wetland-wastewater interactions are already available. However, in
most circumstances, site-specific pilot testing and small-scale
demonstration projects are still recommended when attempting to
develop wetland treatment systems involving either natural or
artificial wetlands.

REFERENCES

Bastian, R. K., and S. C. Reed, project officers, 1979, Aquaculture
systems for wastewater treatment: Seminar proceedings and engi-
neering assessment, EPA 430/9-80-006, U. S. EPA, Office of Water
Program Operations, 485p.
Chan, E., T. A. Bursztynsky, N. Hantzsche, and Y. J. Litwin, 1982,
The use of wetlands for water pollution control, EPA 600/S2-82-
086, U. S. EPA, Washington, D.C.
Duffer, W. R., and J. E. Moyer, 1978, Municipal wastewater aqua-
culture, EPA 600/2-78-110, U. S. EPA, Robert S. Kerr Environmental
Research Laboratory, Ada, Okla., 46p.

Open Discussion

Berger: The discussion is now open. I urge those of you who feel that you failed to learn something you came to learn to speak up, not by way of complaint, but by way of guidance for the future.

Lively-DeBold: I'd like to make a comment about something that hasn't been brought up. I work at EPA, and we have a lot of executive orders and other things that we have to address for wetlands, and one is to mitigate adverse impacts to wetlands. Now, I can see that this applies not only to wastewater treatment where an effluent would go into a wetland and adversely impact it, but also, what happens at the end of the life of a project when you stop putting effluent into a natural wetland? How do we mitigate those impacts? Do we continue to pump water in? Proper mitigation, such as returning it to the state it was in before wastewater effluent addition, may not be easy. In other words, a whole range of moral and regulatory issues, even legal ones, are raised by the use of natural and artificial wetlands for wastewater treatment. That's one aspect that hasn't been addressed here, and I mention this as another whole area that people should do some thinking about because we can get ourselves in a bind. In fact, we have a number of projects now where wetlands have been used for disposal of wastewater--not for treatment--and a permit grantee is required to supplement the wetland and maintain it. The question is, how many years does the grantee have to pay for this maintenance? Do responsibilities for a project end exactly at the end of the project's life, or do you have to mitigate for an eternity?

Ewel: I'd like to respond to that. I think one of the advantages of using wetlands in almost all the cases we've looked at--except maybe Ivy Wile's cases where they're treating for very large numbers of people with a highly engineered system--is that they are low-intensity systems that can pretty much run themselves. We are depending on the self-organizing capability of that system to take care of itself. I think that one of the joys and beauties of such a system is the fact that it can also heal itself. It's a wetland to begin with; it has its natural inflows of water; it has a seed bank; it's invaded by seeds; it exports and imports wildlife. If you cut

451

off the wastewater input at the end of the project, say 20 years, it will revert to its "normal" flow of water; and although there may be some undesirable changes--for example, species composition--they are part of a natural process. The wetland is going to take advantage of what energy it has available to it, and I simply don't see the need for ingenious engineering that might require management beyond those 20 years. I think we need to look toward setting up these systems with as little engineering as possible, and I think that is part of the benefit.

Kaczynski: Theoretically, what you've said is great, but from a practical viewpoint, you have to make sure that you are protecting the downstream rights of whoever owns abutting property. Providing that kind of protection guarantee is very hard to achieve, and it is even harder to convince local regulatory agencies that you can do that.

Ewel: When you cut off wastewater flow, do you think there is a "threat?"

Kaczynski: No. I wasn't talking from the point of view of cutting it off. I was talking primarily about putting it on.

Ewel: Okay, but I was addressing the mitigation aspect. You're right. If you "use" a wetland, then you have to warrant downstream protection. But Bobbie was worrying about what happens after waste-water application.

Kaczynski: Any wetland that you are enhancing is not going to encroach on your neighbor's property. But, there has been a prac-tical problem in getting the agencies at the state, local, and county levels to agree with this and to respond to it in a timely fashion. For instance, when you've designed a program, you've got to get it on-line by a certain date, and you can't get a simple response from a county or local agency to an innocuous query: Your problem is insti-tutional, not real. These agencies have been extremely reluctant to go along with this technique. You're almost put in the position of advising your client that you can't do this, simply because you can't guarantee the timetable. Never mind problems of maintenance <u>after</u> application; we can't even get approval to put worthy programs into operation. So we need some help on the other end, from the top down through the system.

Bedford: I'd like to reemphasize some of the points that I made following H. T. Odum's comments a while back. He pointed out that after all, we really aren't wanting to use the valuable, unique resources that we have in wetlands but, rather, are attempting to identify those wetlands that the addition of wastewater may upgrade or improve. I think there are problems in evaluating which wetlands fall in the category of acceptable. For instance, is it acceptable to choose wetlands for use because they are already degraded or don't serve some of the other functions with which we're concerned? Along those lines, I think there are some general criteria that have come out in several of the discussions here that I want to talk about. Building on Kathy Ewel's comments about the importance of the role of the wetland system in the landscape, there are a couple of things

that follow from consideration of the configuration of the landscape.
First, in terms of evaluating the system, you really need (1) to
examine the commonness of wetlands in general within the region--for
example, the comment about Iowa having lost nearly 90% of its wetlands so those remaining, whatever their nature, are highly valuable,
which contrasts with Florida where cypress domes are abundant--and
(2) the commonness of the particular wetland type vis-à-vis other
wetland types--for example, cattail marshes are much more abundant
than good sedge meadows, fens, and so on. Those kinds of things need
to be considered. We can almost inadvertently change the overall
landscape diversity in an area. Wisconsin, Minnesota, and Michigan,
for example, all exhibit a wide range of types, and we should be
careful to maintain such diversity.

A second consideration is that in order to maintain the value of
the overall landscape you need to think about the proportion of
different types of wetlands and their species composition in an area.
This pertains to Bob Bastian's comments about keeping things within a
range of natural variability. Bud Heinzelman has done some really
interesting work in Minnesota comparing the change in a landscape
over a number of years in terms of diminishing the landscape
diversity--the relative number of types of communities that you have
there. We may keep them all within the range of natural variability
but end up with a much higher proportion of one type than of other
types. This issue of species composition relates, also, to wildlife
considerations. Frank Golet found that, in terms of criteria for
wildlife, one of the most important characteristics is richness of
classes--that is, the number of different kinds of wetland community
types that are present in the overall system.

Along the same lines, thinking still in a landscape sense, one
of the things that we need to consider is the number of discharges
that might be permitted within any particular watershed--in any
particular river system. In essence, I've heard people saying,
"Don't even consider using river/stream wetlands." I think many of
the existing discharges are, indeed, on small streams and along
rivers. We need to think about the number of discharges to a given
system along with any other stresses on the system. This statement
should not be interpreted as a denial of the importance of the relationships within any given wetland. I merely wish to say that our
concern for the wetland as a wetland may reduce our awareness of
threats to the entire watershed.

Finally, there are political considerations. One issue that
many people here are aware of--and I am aware because of my previous
position in Wisconsin--is the issue of a generic EIS [Environmental
Impact Statement]. It's been very clearly stated that wetlands are
highly variable. There are some general principles, certainly, but
we need to look at each situation on an individual basis. If the
purpose of developing a generic EIS by EPA is to point out principles, fine; but one of the things that we've been seeing in the
present regulatory climate is a tendency to give general permits for
whole classes of activities. I would advise caution on that. Also,
keep in mind the upcoming revisions to the Clean Water Act. One of
the big issues will be pretreatment requirements prior to discharge
to publicly owned wastewater treatment works. We've talked a lot
about nitrogen and phosphorus and much less about public health and
metal availability. If pretreatment requirements for industrial
discharges to publicly owned wastewater treatment facilities are

relaxed, discharges from publicly owned treatment works to wetlands
should be examined more cautiously for toxics and heavy metals, not
just nitrogen and phosphorus.

Guntenspergen: I'd like to take off on one thing that Barbara
[Bedford] said, in terms of diversity of wetland types--species
diversity in a wetland. One of the things frequently mentioned in
this conference has been that it might be good to apply wastewater
effluent to degraded wetlands to enhance the quality of the wildlife
refuge. I've also heard in this conference that when you apply
wastewater to a lot of wetland types, the result is invasion by
cattails. In Wisconsin, wildlife managers are trying to get rid of
cattails to enhance the quality of the wildlife refuge. So, I would
say that if you are trying to enhance the quality of a wetland, and
you do get the invasion of cattails or some analogous species, have
you enhanced the quality of what was there before or not? I guess
what I'm trying to say is that you don't necessarily enhance the
quality of a degraded wetland by applying wastewater. It depends on
what your management objectives for enhancement are.

Kappel: I'm glad to see we're all in the same ballpark as regards
what we're talking about, but I think we have to remember the spec-
tators in the crowd. We can talk about all the technical and other
aspects of wetlands--how to use them, which ones to put aside, and
everything else--but we still have to sell this to the public, the
politicians, and everybody else. Our experience at Drummond Bog and
experiences of others tell us that it is not enough to conceive great
ideas, conduct good experiments, and accumulate information. We have
to bring this information to the people, either the people who make
the decisions, as well as to the National Science Foundation, or just
generally to the people. I think there has to be more dissemination
to the public to gain their understanding. We can talk at length
among ourselves, but few others are going to understand what we're
trying to say.

Finn: I have a question for Milt Friend and any of the wildlife
people. I think you said that in order to prevent wildlife diseases--
at least some of them--a stable hydroperiod regime is needed, but at
the same time, variability in hydroperiod is needed for high waterfowl productivity. Is there a way to look at the trade-off between
those two so that you can have the variability--that is, high waterfowl productivity--and still not get epizootic outbreaks?

Friend: The comment was made relative to a specific disease, avian
botulism. As I've indicated, avian botulism requires a certain set
of circumstances for its development. You must have the host and
agent together and at a time when the water temperature is within a
certain range before you can have a problem. In Wisconsin--for
example, in the Grand River Marsh, where drawdowns are used to
increase vegetative regeneration in a manner desirable for the production of food plants for waterfowl--we are timing the drawdown
relative to the appearance of birds and weather conditions. I think
that the answer to your concern is that most of the problems can be
managed around.

I detect from various comments that some of you still continue
to place disease in a rather simplistic perspective. During the last

two days, we've heard descriptions of some rather elaborate and sophisticated research concerning nutrient transport and some other things but very little about sophisticated health research. By comparison, it's still the Stone Age when we talk about concepts relative to disease. We cannot make valid generalizations about what is good or bad without going to the same degree of effort as has been expended for other research.

Niering: Here is one interesting observation that I've made in the Northeast, and I'll just toss it out for people to think about. We've been talking about piping wastewater from a sewage treatment plant and distributing it on a wetland, but I think that in New Jersey, Connecticut, Massachusetts, and other highly populated northeastern states, wetlands are being used for wastewater treatment in an unnoticed and unplanned manner. That is, many houses with inground septic systems are located very near to wetlands, especially red maple swamps. I think this would be a very interesting research area.

Terrell: I want to make a couple of comments about the agricultural side. We have heard a few references to agriculture in one or two of the papers in the last two days, but I don't think we have heard enough when we remember that agriculture is the largest water user in the United States; we should start to think of its significance for our wetlands. We also should be thinking about where our wetlands occur. They are mostly in rural areas and are not urban phenomena for the most part. That's where our agriculture is. What I would suggest as a research direction for the next few years is an assessment of the wetland treatment potential for agricultural runoff and non-point source pollution. Ideally, perhaps, we should investigate how to utilize wetlands as a source of inexpensive treatment if the situation is appropriate, but at least we should be able to recommend to farmers whether or not wetlands should be utilized in their farming scheme instead of filling and turning them into many acres of grain or something like that. So I hope we see a little bit more on the agricultural research side over the next year or two.

Jackson: We've all been using one word. That word is enhancement, and I would bet that if we went around the circle, we would probably have as many definitions of enhancement as we have people here. When we talk about stopping erosion and being satisfied with creating a kudzu-filled stream bank by spraying effluent, maintaining a climax community, maintaining a marsh in its pristine condition, taking a degraded environment and not returning it to its pristine condition but to some biologically functioning condition, or biomass production, we are using different criteria for enhancement. I sense that when we try going to the political scene--to the local town councils, to the state governments--and try to sell this kind of program as enhancing wetlands, we are going to have to be very careful to define exactly what we mean by enhancement because I suspect our definitions will be very different from theirs. Bob Bastian alluded to some of these considerations in his comments. I think we need to define for ourselves whether enhancement means something that stays within "normal bounds"--for example, biomass enhancement--or whether we put some new kinds of constraints on this term that are pragmatic in their form and definition but are very "unnatural," perhaps, in the strictest sense of the word.

Stearns: That was a good comment because I was also disturbed about the use of the word enhancement. In lots of cases, modification might be a better term because to many local people, enhancement means getting out the dragline and, if you are trying to improve fish production, maybe that's something that may need to be done.

   May I take a moment to point up a few, somewhat unrelated ideas? One thought: Milt Friend, in his discussion of improving wildlife productivity versus wildlife disease, suggested that there are problems. Perhaps we shouldn't be shooting for a maximum number of ducks. We ought to be trying to compromise and optimize. We may produce fewer but lose even fewer to mortality. A second thought (and I think one of the most forward-looking, perhaps) relates to agriculture, as well as other things: I refer to Curt Richardson's comment that "ten years from now we may need to consider nitrogen and phosphorus as a resource rather than waste." Perhaps in our planning for the use of wetlands, we ought to keep in mind the possibility that future problems may be quite different.

   Well, it's been a very fascinating conference, and I guess I'd like to point out one thing inherent in all of our discussions but perhaps one that we need to remind ourselves of every now and then. It is that wetlands are naturally in the process of change. From the time that the glacier, for instance, left the hole in the pitted outwash for the ice to melt in until today when it's filled with cattails, having progressed through five or six other types of vegetation, it's been changing through nutrient addition, through accretion, and so forth. This natural change is something that we ought to be looking at and considering.

Friend: I'd like to comment relative to the issue of the number of ducks for the reason that I think it is relevant to the whole issue being discussed here. For a long time, the value of wildlife management and refuge practices were measured in terms of how many birds one could sustain on a particular area. We have gotten away from that in recent times, in part because of disease problems, but we are still faced with the reality that wildlife management goals are being set at increasingly higher levels to sustain and even increase migratory bird populations in this country. And, that cannot be ignored because it is happening despite the present existence of only 40% of the wetlands that were in existence when the white settlers came to this country and an annual loss of 300,000 acres of wetlands. So, the quality of the remaining habitat really becomes critical because in essence, we're going to have to meet those management goals of sustaining more on less for the foreseeable future. I'm not sure that's possible. Therefore I suspect your comment is valid--that is, that we should aim for quality of what we have rather than numbers.

Stearns: It's the same problem with the human population.

Friend: Exactly, and so, when you start now to talk about placing wastewater, secondarily treated effluent, raw sewage--anything--into those wetland environments, you immediately enter a conflict situation and need to recognize its causes. That's the real world.

Hemond: I'd like to say another thing about this question of the

value of nutrients. We've been speaking as though removal of
nutrients were more or less universally the goal of wastewater
management. The point that nutrients may have value in agriculture,
for example, is something that we may need to consider ten years
down the line, I think. That may be translated into different sorts
of criteria for wetland management. In some situations, we may want
to use a wetland in the hope that we will, perhaps, remove some
sediment and thus, perhaps, create conditions ideal for, let's say,
the hydrolysis of some toxic materials or BOD reduction, but we may
not want to optimize for removal of some of the nutrients. I even
offer the almost theoretical notion that in future years, some
receiving waters may actually have other impacts, particularly
acidification, that could be ameliorated if we can keep nutrient
removal down to a minimum. We may actually mitigate other disturbances by judicious tailoring of the wetland treatment process.

R. Kadlec: I don't think we ought to get too enthusiastic about the
fertilizer value of secondarily treated wastewater. You have available less than a half pound of phosphorus per person per year, for
example. That's one way to look at it. In other words, it is a
fertilizer, but it is far, far inferior to any used agriculturally.
Furthermore, the cost of recovery, which has been studied a number of
times, is great, and this particular resource is available only in
very small quantities.

Hemond: I think, though, that they are in quantities sufficient to
biostimulate receiving waters. In fact, the whole notion of tertiary
treatment is that very fact. If we have waters that are impacted by
acid input and productivity is decreasing due to the dynamics of the
system, which is a complicated story in itself, this degree of bio-
stimulation may be of some value.

Giblin: I'd just like to say that I think all the contaminants and
toxins in sewage are going to be with us, even in rural areas:
Someone is going to dump DDT or something down his drain. I just
hope that in a lot of the ongoing studies that you people are doing,
you will try to monitor—now that the analytical capabilities are
better—some of the metals, pesticides, and other contaminants
because eventually we're going to have to make a decision based upon
lessening these impacts as well.

Kaczynski: One additional thing would have helped me: techniques to
enhance the type of vegetation you want when you're designing a new
wetland. Maybe if you could get a certain type of vegetation established, you could at least retard the invasion of Typha for a period
of time.

Berger: I have a personal remark. There's a presumption, I believe,
that the wetland as a waste treatment plant can meet effluent standards as regards EPA or state requirements at lower overall cost than
a conventional system. I would say that for many types of communities, particularly small communities, this is probably more than just
a presumption—it's a promise. But I don't think the promise can be
realized until we have answers to some very important research questions. I think it is particularly important that these research
questions be set out explicitly. Perhaps this has been done—it

would be surprising indeed, if the research questions of high priority had not been set forth at one of our workshops or another. If it hasn't, it certainly should be done. One step beyond that: I think a state-of-knowledge report should be issued together with the statement of research needs. It would be a most useful outcome of this workshop if some group could be appointed to do this very task. And I would make one further suggestion; that if this is done, the state-of-knowledge report, together with the statement of research needs, be submitted to the <u>Journal of the Water Pollution Control Federation</u>. There are 25,000 engineers and scientists of all kinds associated with the Water Pollution Control Federation. Their interest and support could be most useful to a future research program on the use of wetlands as waste treatment systems. I urge you to think about this seriously.

   We've come now to the end of the terminal session and I turn the meeting back to Paul Godfrey.

<u>Godfrey</u>: Thank you, Bernie. I hope someone picks up on your charge to the group. I know you are all tired; this has been an extremely intensive session for the two days. I want to take just a moment to thank those people who made it possible: you, in general, for making it such a successful session; my fellow moderators for keeping everybody on schedule, cracking the whip when necessary--although it wasn't necessary very often--and helping to make a success of it. There have been a number of people behind the scenes who have been invaluable. In fact, the workshop could not have gotten together or been conducted without their help, and I'd just like to mention some of those people: Ron Reed, who has been our UMass Conference Services expediter and has provided us with the food, provided us with the facilities here, and seen that things have gone well; the people in the Water Resources Research Center, my staff: Marge Ferris, Mary Henry, Sheila Pelczarski, and Judy Kaylor; and some graduate students who have pitched in on a purely voluntary basis to help out with transportation: Pete Saunders, Hisachi Ogawa, and Hank Kaylor. All of these people have been instrumental. I'd like to pick on one person in particular as probably being the key person in organizing this workshop and keeping it running at full steam--that is my assistant, Ed Kaynor. He's done an excellent job! I'd like to commend him and hope you will all do the same. Thank you, have a safe journey home, and perhaps we'll see you again.

INDEX